Forest Ecosystems: Nutrient Uptake and Cycling

Forest Ecosystems: Nutrient Uptake and Cycling

Editor: Austin Campbell

MURPHY & MOORE
www.murphy-moorepublishing.com

www.murphy-moorepublishing.com

MURPHY & MOORE

Cataloging-in-Publication Data

Forest ecosystems : nutrient uptake and cycling / edited by Austin Campbell.
p. cm.
Includes bibliographical references and index.
ISBN 978-1-63987-760-7
1. Forest ecology. 2. Plant nutrients. 3. Forest plants--Nutrition. 4. Trees--Nutrition.
5. Biogeochemical cycles. I. Campbell, Austin.
QH541.5.F6 F67 2023
577.3--dc23

Murphy & Moore Publishing
1 Rockefeller Plaza,
New York City,
NY 10020, USA

ISBN 978-1-63987-760-7

Contents

Preface

Forest ecosystem is a self-sustaining functional unit of nature wherein living organisms interact among themselves and with the surrounding physical environment. It is a type of terrestrial ecosystem. Both biotic and abiotic elements such as soil, trees, insects, animals, birds and humans are a part of the forest ecosystem. There are four major components of a forest ecosystem, namely, productivity, decomposition, energy flow, and nutrient uptake and cycling. Nutrient cycle refers to a system that involves the movement of substances and energy between living organisms and non-living elements of the environment. This happens when plants and animals ingest nutrients present in the soil, which are then released back into the ecosystem after they die and decompose. Nutrient cycling is crucial for meeting the nutrient supply of plants in the forest and increasing productivity of forests. This book unfolds the important aspects of nutrient uptake and cycling in forest ecosystems, which will be crucial to develop a complete understanding of the subject matter. It is a resource guide for experts as well as students.

All of the data presented henceforth, was collaborated in the wake of recent advancements in the field. The aim of this book is to present the diversified developments from across the globe in a comprehensible manner. The opinions expressed in each chapter belong solely to the contributing authors. Their interpretations of the topics are the integral part of this book, which I have carefully compiled for a better understanding of the readers.

At the end, I would like to thank all those who dedicated their time and efforts for the successful completion of this book. I also wish to convey my gratitude towards my friends and family who supported me at every step.

Editor

1

Impact of Global Change on Nutrient Dynamics in Mangrove Forests

Daniel M. Alongi

Tropical Coastal & Mangrove Consultants, 6 Cobb Court, Annandale, QLD 4814, Australia; dmalongi@outlook.com

Abstract: The cycling of essential nutrients is central to mangrove productivity. A mass balance shows that mangroves rely on soil ammonification, nitrification, and dissimilatory reduction to ammonium for available nitrogen. Mangroves are often nutrient limited and show tight coupling between nutrient availability and tree photosynthesis. This relationship and, thus, forest productivity can be disrupted by various disturbances such as deforestation, changes in hydrology due to impoundments, land-use change, increasing frequency and intensity of storms, increasing temperatures, increasing atmospheric CO_2 concentrations, and a rising sea-level. Deforestation and hydrological changes are the most devastating to soil nutrient-plant relations and mangrove productivity. Land-use changes can result in positive and negative impacts on mangroves and can also results in increasing frequency of storms and intensity of storms. Increasing temperatures and atmospheric CO_2 levels have an initially enhanced effect on mangroves and microbial transformation rates of nitrogen and phosphorus. The effects of rising seas are complex and depend on the local rate of sea-level rise, the soil accretion rate, the subsidence or uplift rate, and the tidal position. If mangroves cannot keep pace with a sea-level rise, seaward mangroves will likely drown but landward mangroves will expand and show enhanced growth and more rapid nutrient cycling if space permits.

Keywords: climate change; disturbance; global change; mangrove; nitrogen; nutrients; phosphorus

---

1. Introduction

Mangrove forests and their associated waterways are dominant ecosystems in the subtropics and tropics and line most of the world's coastlines at low latitudes. Mangroves inhabit over 152,000 km^2 of coastal area but occupy <1% of the world's coastal zones [1,2]. These tidal forests are the only woody halophytes that live in salt water and provide a wide variety of ecological and economic goods and services [3]. They act as habitats for terrestrial and marine flora and fauna, as ecosystem engineers shaping and maintaining the intertidal zone, as a renewable resource of wood and fuel, and as key accumulation sites for soils, sediment particles, and associated elements [3].

Mangrove forests are the most productive plants in the sea with average rates of gross and net primary production of 4.6 kg C ha^{-1} year^{-1} and 1.9 kg C ha^{-1} year^{-1}, respectively [4]. These rapid rates of productivity drive high requirements for nutrients and trace elements to sustain growth, production, and physiological maintenance. These high requirements translate into rapid uptake of soil nutrients that are transformed by Archaea, bacteria, fungi, protozoa, and microalgae. There are tight links between carbon fixation and nutrient cycling [5].

Mangrove forests, living at the dynamic interface between land and sea, are subject to various disturbances that vary in time and space and occupy a harsh environment, which is daily subjected to tidal changes in temperature, water and salt exposure, and varying degrees of anoxia [3,4]. Mangroves are, thus, robust and highly adaptable to changing conditions [6]. A variety of features of mangroves contribute to their resilience to disturbance. These features are: (1) a large reservoir of below-ground nutrients that serve to replenish nutrient losses, (2) rapid rates of nutrient flux and microbial decomposition that facilitate rapid biotic turnover, (3) complex and highly efficient rates of water-use and nutrient-use efficiency that allow internal reuse of resources to augment recovery, (4) simple architecture that leads to rapid reconstruction and rehabilitation post-disturbance, (5) redundancy of keystone species, which can lead to restoration and recovery of key structures and function, and (6) negative and positive feedback that provides stability to help dampen changes in the ecological steady-state [3,6]. This is not to say that mangroves always recover from disturbances. Deforestation and impoundments are devastating for mangrove forests.

Mangroves are currently being strongly affected by various human-induced disturbances including increased sedimentation and nutrient inputs, impoundments, deforestation, increased frequency and intensity of storms, increasing temperatures, increasing atmospheric CO_2, and a sea-level rise. Several reviews have focused on the impact of climate change on mangrove forests [6–8], but none have examined in detail the impact of both climate and various other disturbances on these tidal ecosystems. The object of this review is to critically assess the actual and potential impacts of global change on nutrient dynamics in mangrove ecosystems. The focus is on nutrient dynamics because of the close links between nutrients and mangroves and its centrality to mangrove ecosystem dynamics.

2. Mangroves and Nutrients

Mangroves present an apparent paradox in that they are highly productive and rich in carbon while nutrient-poor. Many studies of mangrove nutrition have focused on how mangroves can sustain high rates of productivity in the face of nutrient limitation [9]. Evidence to date suggests that high productivity in mangroves is achieved when nutrients limit growth through highly efficient nutrient cycling and nutrient retention mechanisms [10].

Owing to their estuarine and marine existence, mangroves are generally not limited by the relatively large quantities of sulfur, boron, potassium, magnesium, and sodium in seawater but are frequently limited by nitrogen and phosphorus. Iron and copper have been found to be limiting for mangroves in mesocosm studies [11,12]. The critical need for nitrogen and phosphorus has been repeatedly demonstrated in field and laboratory studies [9]. A key role in mangrove nutrition is the interactive effects between different nutrients and environmental factors such as salinity, soil type and texture, and frequency of tidal inundation. The effects of added nitrogen and phosphorus vary among species. Fertilization studies demonstrate either nitrogen or phosphorus limitation or both depending on species composition, the extent of terrigenous input, soil fertility and texture, the soil redox status, salinity, and the frequency of tidal inundation [13]. Plant-soil relations can be greatly disturbed by altering hydrology and the construction of impoundments. Flooding and draining of tidal waters are of prime importance on mangrove growth and nutrient use since tidal changes relate to an exchange of nutrients and other materials and to rates of sediment and nutrient input.

High photosynthetic rates by mangroves is supported by a high requirement for nutrients, which implies high nutrient-use efficiencies and high rates of leaf resorption. Nutrient-use efficiency for nitrogen in mangroves is, at the upper end of the range, for other tropical forests [4]. However, nutrient-use efficiency for phosphorus in mangroves is well within the range of values of other tropical forests [4]. Mangroves must be efficient to survive when living in oligotrophic, anoxic soils and waters and being continually inundated by tides. In such habitats, it is clearly advantageous to have mechanisms to conserve limiting nutrients. Efficient utilization of nutrients as an apparent strategy relates well to the generally low concentrations of nutrients in mangrove leaves and other tree parts and the generally high efficiency with which nitrogen and phosphorus are resorbed from leaves [9,10].

Mangroves have other mechanisms to retain nitrogen and phosphorus including a large reservoir of dead roots and tight immobilization of available solutes [4]. Phosphorus is tightly immobilized to inorganic phases such as soil aluminum, iron, calcium, and sulfides and dissolved nitrogen in soil is mainly in the form of ammonium rather than nitrite and nitrate [14,15]. Residence times of the available nitrogen and phosphorus pools are short in keeping with the generally rapid rates of nutrient recycling [16].

There are species-specific differences in nutrient-use efficiency especially in arid-zone forests. In arid Western Australia, Alongi et al. [17] measured mean residence times for both nitrogen and phosphorus of 2–4 years in *Rhizophora stylosa* Griff. forests, whereas the residence times were shorter (<2 years) in *Avicennia marina* (Forsk.) Vierh. stands. These differences may reflect the fact that *A. marina* trees are more efficient at utilizing nitrogen and phosphorus solutes since *A. marina* soils are less anoxic than soils of *R. stylosa*. Other reasons may explain species differences in nutrient use such as differences in the way species allocate nutrients, differences in the proportion of energy and nutrients vested in chemical defenses, differences in leaf life spans, and differences in soil physico-chemistry and biogeochemistry [9].

Most mangrove trees are evergreen with sclerophyllous leaves and high root/shoot ratios. Sclerophylly is a trait related to low soil nutrient availability especially low phosphorus as well as to low water availability and to high-salinity environments since sclereophyllous leaves can lose much of their water content and can exhibit extremely low leaf water potentials [10]. Sclerophylly has also been linked to leaf longevity and to ecosystem nutrient retention. The capacity to sustain low growth rates and reduced nutrient requirements as mangroves do in arid or high-intertidal environments are an adaptation to low-nutrient habitats [18]. Mangroves have root/shoot ratios that are often an order of magnitude greater than those of tropical terrestrial forests. Such high root/shoot ratios are in part an adaptation to salty environments but can vary between mangrove species over time and with forest structure. This results in non-linear relationships between root/shoot ratios and soil conditions. The high root biomass in mangroves is generally conducive to nutrient capture and uptake from low-nutrient soils [9,10].

The efficiency of soil microbial metabolic processes is also an effective nutrient conservation strategy. A large proportion of root respiration goes towards the uptake and assimilation of nitrogen especially ammonium. Ammonium is preferentially taken up when compared with nitrite and nitrate as the uptake of ammonium requires the least energy investment for mangroves [9,10]. As pointed out by Reef et al. [9], the large root biomass in mangroves may overcome the relative immobility of soil ammonium by the large volume of soil covered by the roots.

The nitrogen cycle is understood well enough to construct a preliminary global nitrogen budget (Figure 1). The budget [19] shows that (1) 2687 Gg N $year^{-1}$ is required to sustain mangrove net primary production, (2) nitrogen burial equates to about 25% of total nitrogen input, (3) nitrogen fixation is <5% of total nitrogen input, (4) about 40% of nitrogen input required for total mangrove net primary production is vested in below-ground root production, (5) about 50% of nitrogen required for mangrove net primary production is vested in litter, (6) net tidal export equates to about 55% of nitrogen vested in mangrove net primary production, (7) <10% of mangrove nitrogen is lost via denitrification and N_2O emissions, and (8) nitrogen fluxes are in net balance since the net gain (351 Gg N $year^{-1}$) is very small compared to total inputs and outputs and sources of error [19].

The mangrove nitrogen mass balance estimates, although preliminary, have some implications for the impacts of global change in mangroves. First, it is reasonably clear that most nitrogen to sustain mangrove net primary production is the net result of soil ammonification, nitrification, and/or dissimilatory nitrate reduction to ammonium. Global changes that affect rates of ammonification or dissimilatory nitrate reduction such as massive soil disturbance will likely have significant impacts on mangrove productivity. Second, soil microbial processes such as nitrogen fixation and denitrification and anammox play a comparatively small role in nitrogen dynamics in mangroves so factors that can change rates of such processes will probably have little impact on mangrove nitrogen dynamics. Third, a substantial fraction of total nitrogen input is buried in mangrove soils (Figure 1), which suggests that burial is a significant nitrogen conservation mechanism. Fourth, global changes that affect below-ground root production and plant litter will likely have a substantial impact on net primary productivity. Fifth, global changes that affect tidal exchange such as changes to hydrology and construction of impoundments will very likely have significant impacts on the balance of nitrogen dynamics of mangrove forests. Sixth, at present, nitrogen fluxes in mangrove ecosystems are roughly in overall balance. Global changes may enhance or negatively affect this balance and, thus, the survival of mangrove ecosystems. Some global changes such as temperature increases will likely have little impact on mangroves whereas some changes such as deforestation have a substantial impact.

In contrast, the cycling of phosphorus and other possible limiting nutrients such as iron and copper are poorly known in mangrove soil and waters. The phosphorus cycle is presumably tightly linked to soil geochemical processes and the low available phosphorus concentrations (phosphate, organic phosphorus) suggest low nutrient availability [15]. It is clear that more research is needed regarding the impact of disturbances on the cycling of phosphorus, iron, and copper and their interactions with soil geochemistry.

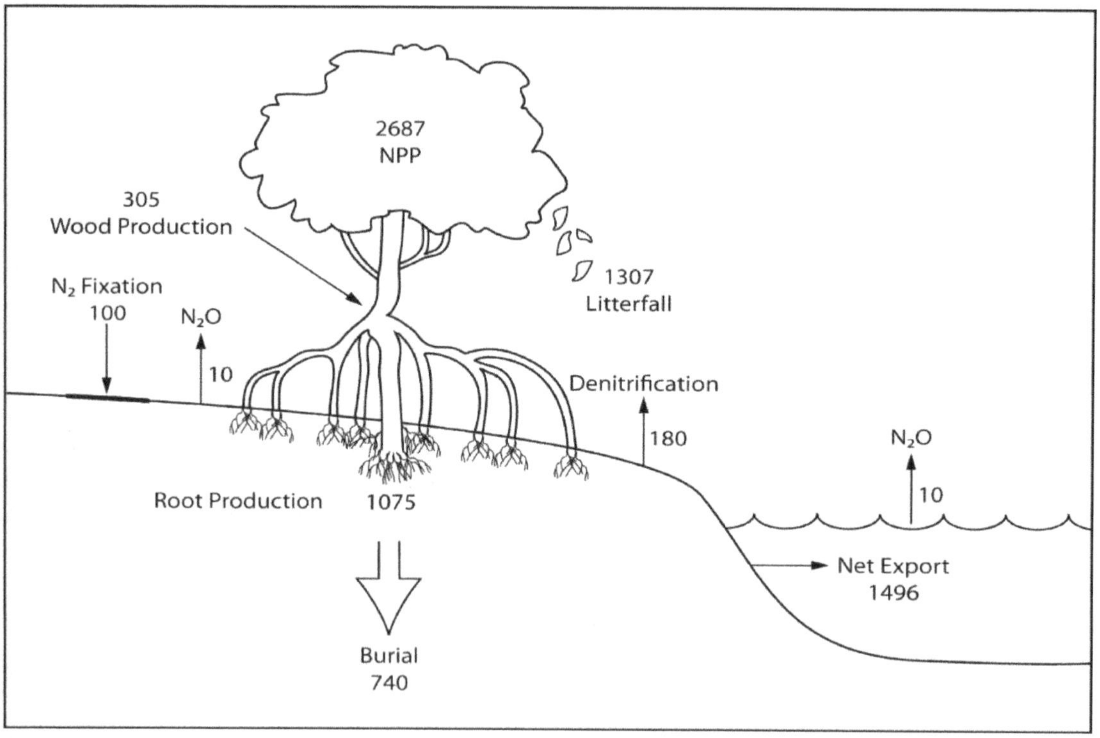

Figure 1. Nitrogen mass balance for the world's mangrove ecosystems. Units are Gg N year^{-1}. NPP = net primary production. The model assumes a global mangrove area of 152,000 km^2 [2]. Data modified from Alongi [19].

3. Construction of Impoundments and Deforestation

Two of the most devastating impacts on mangrove ecosystems are deforestation and the construction of impoundments. Impoundments involve the enclosure of mangroves for the purpose of limiting the movement of water to control problems such as insect infestations, e.g., mosquitoes. Impounding mangroves involves the alteration and disturbance of soil horizons and subsequent change in soil biogeochemistry and geochemistry.

Deforestation is occurring at a rate of about 1% to 2% of total area per year [2]. The loss of trees results in the decline of forests but, in the process of cutting trees, the soil horizons are greatly disturbed, which results in the loss of soil and changes in rates and pathways of soil microbial processes. In deforested mangrove soils, concentrations of porewater ammonium are high due to lack of uptake by tree roots but nitrogen fixation rates are low, which suggests surface soil disturbance [20–23]. Removal of mangroves results in equivalent removal of reactive nitrogen [24] as well as an increase in dissolved inorganic nitrogen concentrations in tidal creeks due to drainage of degraded mangroves [25]. The few data available also indicate that concentrations of soluble reactive phosphate are lower in deforested areas due to changes in soil geochemistry [26].

Nitrogen transformation processes such as denitrification decline in impounded and cleared forests due to surface disturbance of the soil [23,27]. In impounded forests where nitrate was added over a two-year period, the responses of mangroves were insignificant or minor when compared with microbial responses [27]. Denitrification rates increased significantly and nitrous oxide emissions were nearly six times higher in fertilized plots. Nitrogen fixation was lower in fertilized plots. These results suggest that nitrogen-limited impounded mangroves are likely to be long-term sinks for any increases in nitrogen loading [28]. In impoundments where alteration of hydrology results in increased seasonal flooding, the waterlogged soil shows higher soil ammonium, lower nitrate, and lower nitrification and denitrification while plant growth and nitrogen uptake increased especially in dwarf and sparse forest types [23].

More stressful physicochemical conditions result from either incomplete tidal flushing or excessive removal (export) of debris and surface soil due to hydrologic alteration [29]. Local or regional factors such as the salinity regime act together with forest history to control turnover rates of organic matter including soil nitrogen.

4. Land-Use Change

Changes in land-use in coastal watersheds is increasingly resulting in rises in inputs of up-river sediments and associated nutrients to mangrove forests. These changes are most often due to poor land practices and deforestation and subsequent erosion of forest soil horizons. While sediment and nutrient enrichment can result in an increase in mangrove mortality [30], such changes can also result in the enhancement of mangrove growth and photosynthesis. The net positive or net negative result likely depends on the amount of sediment and nutrient loading.

Mangroves receiving higher than normal nutrient and sediment loads tend to accumulate organic matter with high rates of sediment accumulation [31]. In an impacted mangrove forest in Brazil, for example, Sanders et al. [31] found that organic carbon and total nitrogen stable isotopes reflect an increased presence of organic matter originating with either phytoplankton, benthic algae, or other allochthonous source within the more rapidly accumulated sediments. This result suggests that the accumulation of organic matter in impacted mangroves may be enhanced through the addition of autochthonous and allochthonous organic matter.

Land-use change resulting in eutrophication may negatively affect mangrove forests by modifying internal carbon and nutrient cycling. Litter decomposition exerts a strong influence on these processes and is potentially modified by eutrophication. In a long-term experiment with nutrient fertilization, Keuskamp et al. [32] found that decomposers were always nitrogen-limited regardless of the effects on the net production of the trees, which leads to a differential nutrient limitation between decomposers and mangroves where the forests were phosphorus-limited. In these latter stands, phosphorus fertilization caused a change in litter quality, which resulted in higher rates of decomposition.

Eutrophication often results in higher rates of nitrogen transformation in mangroves. Weng et al. [33] found that ammonium and nitrate addition resulted in faster rates of ammonification and denitrification from mangrove soils. Furthermore, nitrogen additions change the composition of root exudates. Maher et al. [34] observed that pristine mangrove waters are a sink of nitrous oxide but noted that measurements in impacted mangroves suggest that mangrove waters are a source of nitrous oxide. Thus, eutrophication appears to change mangrove waters from a sink to a source of nitrous oxide to the atmosphere, which represents a "positive feedback to climate change."

Reduction in freshwater flow also results in significant changes in mangrove nutrient cycles. For instance, in the southern Everglades and Florida Bay which have experienced a nearly 50% reduction in freshwater flow resulting in increased salinity and landward expansion of the mangrove forest [35], soils showed uptake of soluble reactive phosphorus, dissolved organic carbon and nitrite + nitrate, and release of ammonium to the overlying tidal water. Enrichment of phosphorus had no effect on dissolved inorganic nitrogen or dissolved organic carbon flux, which suggests that rapid phosphorus uptake may be geochemically mediated with the potential for the mangroves to sequester phosphorus.

Eutrophication impacts on mangroves can be complex and influenced by such factors as tidal inundation frequency, intertidal position, soil type and texture, salinity, and species composition. In fertilization experiments along latitudinal and tidal gradients, Feller et al. [36] found that patterns of nutrient limitation and primary productivity varied by tidal zone and latitude. Nutrient enrichment had complex effects on nutrient conservation with the addition of nutrients having dramatic effects on growth and nutrient recycling even though the pattern "depended on location, site characteristics, and the nature of nutrient limitation." Rates of nutrient transformation vary in pristine and impacted mangroves depending on factors such as the degree of tidal inundation frequency, tidal position, and soil and nutrient input. Reis et al. [37], for example, found that fringe forests exhibit higher rates of nitrogen transformation than basin forests and this, undoubtedly, will hold true if the mangroves were affected by a land-use change.

Ultimately, the effects of land-use change and the resulting eutrophication will likely lead to changes in rates and pathways of nutrient transformation processes in mangroves. These effects will likely, in turn, affect rates of net primary productivity and survival of specific mangrove species. Clearly, more data are needed regarding the effects of eutrophication on soil-plant relationships in relation to global change.

5. Increased Frequency and Intensity of Storms

Climate change predictions are for the increased frequency and intensity of storms especially in the wet tropics. Arid regions will become more arid [38]. The impacts of such disturbances on mangroves can be gleaned from studies on the impacts of floods and cyclones (hurricanes, typhoons) on mangrove structure and function. Intense storms can have deleterious effects of mangrove forests especially on canopy structure. What effect such disturbances have on nutrient cycling and the relationship between soil nutrients and mangrove productivity is poorly understood.

Intense storms and flooding may relieve nutrient limitation in mangroves [39]. Tropical cyclones usually result in pulses of freshwater and sediments delivered in floodwaters to the coastal zone. In arid regions, these pulses may be particularly important for the maintenance of mangrove productivity. On the arid coast of Western Australia, Lovelock et al. [39] found that growth of trees was enhanced by a cyclone and nutrient limitation was eased by sediment and associated nutrients delivered by the cyclone. In addition, before the cyclone, the efficiency of resorption of phosphorus and nitrogen from senescent leaves was higher than after the cyclone, which suggests an enhancement of nutrient availability. In impounded mangroves subject to increased summer flooding, there was higher soil ammonium, lower soil nitrate, and lower nitrification and denitrification due to intense soil disturbance [23]. Therefore, mangroves disturbed by impoundments and land-use change will likely be severely impacted by increased storm frequency and intensity.

Monsoons play an important role in mangrove seasonality and this is likely to increase in importance with increasing precipitation. For instance, denitrification is lowest during the monsoon season when compared with pre-monsoon and post-monsoon periods [40]. The influence of limiting substrates is most important during the monsoon season and post-monsoon season. Soil disturbance is likely to be the main cause of the lower denitrification rates due to lower salinity and dilution of soil porewater solutes.

Flooding has been found to affect soil extractable phosphate in mangroves, which, in turn, affects mangrove nutrient limitation and availability [41]. In mangroves of the Bragança Peninsula, North Brazil, patterns of soil redox and soil extractable phosphate confirmed that both parameters are greatly influenced by flooding. Under waterlogged, flooded conditions, there was a net release of extractable phosphate due to reduction-induced solubilization of P-Fe/Al and increased phosphorus availability for trees. Variations in redox and extractable phosphate were reflected in patterns of leaf phosphorus concentrations. The forest structure appeared to be influenced by waterlogged conditions, pore water salinity, and phosphorus dynamics. Phosphorus availability plays an important role in controlling mangrove species distributions even though comparatively few data exist on this topic. Thus, disturbances to the nutrient dynamics of mangrove ecosystems by increased frequency and intensity of storms may have a significant effect on mangrove community structure and photosynthesis.

6. Increasing Temperatures

The planet is being subject to increasing temperatures and this trend is predicted to continue in forthcoming decades and into the next century [38]. Mangroves will respond similarly to other organisms following a sigmoid curve in which an initial rapid rise in physiological processes (e.g., growth rate, respiration) slows, plateaus, and then declines when a critical lethal threshold is reached and then exceeded [10]. Presumably, mangroves and their associated food webs respond so, but the critical temperatures at which organisms exceed the functional threshold and die are likely >33 °C [10,42]. Photosynthetic rates for most mangrove species peak at temperatures at or below 30 °C. CO_2 assimilation rates of many species decline sharply or gradually when the temperature increases from 33 °C to 35 °C [42]. Photo-inhibition often depresses photosynthesis since mid-day declines have been observed, which ensures the survival of leaf photochemical machinery.

Increases in temperature are likely to result in faster growth, reproduction, photosynthesis and respiration, changes in community composition, diversity, and a poleward expansion of latitudinal limits [43,44]. Mangroves are currently expanding into higher latitudes in North America, New Zealand, Australia, Southern Africa, and Southern China. This global expansion is most likely the result of increases in mean sea surface temperatures. Expansion of mangroves at the expense of salt marshes is the net result of complex ecological interactions [44]. Mangrove expansion results in more carbon, nitrogen, and phosphorus storage in soils and above-ground and below-ground biomass than in the salt marshes they are displacing, which suggests greater nutrient retention with rising temperatures [43,44].

Temperature increases are likely to result in faster rates of nutrient cycling including rates of soil nitrogen and phosphorus transformation processes. Microbial growth and rates of transformation processes are closely and positively linked to changes in temperature [45]. However, it is unlikely that nutrient-use efficiency is tightly linked to temperature since photosynthetic efficiency likely asymptotes at temperatures below 33 °C [10,18]. Tropical organisms including plants are closer to their upper thermal thresholds than boreal and temperate organisms and are, thus, more vulnerable to increases in temperature.

7. Increasing Atmospheric CO_2

Higher atmospheric CO_2 concentrations can result in increased growth of mangroves even though the responses are species-specific. Species' responses are confounded by variations in salinity, nutrient availability, and water-use efficiency [46–49]. Laboratory data show that mangrove responses in most instances will be complex with some species thriving while others decline or exhibit no or little change.

The coastal location may be an important determinant in the mangrove response given the interactive effects of increased CO_2 with salinity and nutrient availability. Based on the ability of each species to respond to spatial and temporal differences in drivers such as salinity and nutrient availability in relation to increasing CO_2 levels, species patterns within estuaries may change.

Nutrient availability will vary in relation to frequency of tidal inundation, soil type, and texture as well as soil accretion rates and the geochemical and biogeochemical composition of dissolved and particulate nutrients and other compounds such as sulfides and iron oxide complexes [14,15]. The inputs of nutrients ultimately depend on the accumulation of nutrient elements from the land and the sea. Transformation of organic nitrogen to ammonium via ammonification and release of phosphate and dissolved organic phosphorus are similarly dependent on microbial breakdown of extractable phosphorus compounds [15].

Nutrient availability and cycling are unlikely to be directly affected by increases in atmospheric CO_2 but will probably be affected depending on the response of mangrove species. Enhanced growth of mangroves will likely result in increased utilization of nutrients such as ammonium and phosphate, which results in more rapid cycling of these essential elements [10,42]. Experiments by Reef et al. [49] show that the effects of CO_2 on mangrove growth are more greatly enhanced in the presence of additional nutrients, which is high nutrient availability. Conversely, a decline in mangrove growth will probably translate into decreased utilization and a possible buildup of dissolved soil nutrients, which is similar to the buildup of soil ammonium when mangroves are deforested [12]. Most nitrogen and phosphorus will likely remain immobilized.

8. Rising Sea-Level

Sea levels are rising due to thermal expansion of the oceans and the melting of polar and land ice due to increasing temperatures [50]. The rise in the sea level is not uniform due to differences in water density due to varying sea temperatures and the weight of ice sheets as well as water buildup. Mangroves have had to adjust to rises and falls in the sea level since the sea level has waxed and waned over geological time in relation to alternating periods of warming and glaciation [51].

Mangrove surface elevation relative to the sea level is affected by several local and regional short-term and long-term processes. These include subsidence, eustatic changes in the sea level, isostatic changes, changes in the subsurface volume of fluids, accretion of fine sediment particles, litter, algal mats, erosion, subsurface expansion due to root growth, subsidence due to compaction, and drainage of pore water and microbial decomposition [51]. How mangroves may change in relation to the rates of the sea level rise depends greatly on differences in the rates of soil accretion and tectonic uplifting or subsidence [52]. Terrestrial plants will invade from landwards as the mangrove forest recedes, but intertidal mud and sand flats will likely continue to accrete if mangrove forest growth rises at a rate exceeding the local rate of the sea level rise. Conversely, the forest will likely survive if the rate of soil accretion equals the rate of a sea level rise. If the rate of soil accretion is slower than the rate of the sea level rise, the forest will probably drown seawards but invade land newly flushed by tides. Because of local geomorphological features, actual changes may deviate from these scenarios especially in the face of severe storms that may affect short-term rates of the sea level.

The existence of relic pollen and peat deposits suggests that mangroves have undergone dramatic changes due to changes in the sea level. The geological record and present distribution of mangroves indicate that they have survived past changes in the sea level [51]. Peat deposits indicate that mangroves have not kept pace with a sea level rise since the rate of the sea level increased beyond a critical threshold at which mangroves were not able to keep up. Modern evidence indicates that mangrove responses to a sea level rise corresponds roughly to patterns of a surface elevation change. López-Medellin et al. [53] found, for example, that, along the Pacific coast of Mexico, a rise in the sea level due to warm waters of El Niño have drowned mangroves fringing the shore but has resulted in a net increase in mangroves being driven inland. Whether or not mangroves will survive in a given locale depends on local factors such as geomorphology, which makes it difficult to offer simple prognostications on survival or loss. However, there are two regions where survival will be difficult, which includes river deltas and low islands. On low islands in the Pacific, mangroves are being drowned due to a lack of sediment input and due to landward expansion being obstructed because of coastal development. In many deltas of large tropical rivers, subsidence, which is a decline in sediment input due to damming and sea-level rise, storms, and cyclones enhance subsidence. This results in a shift of mangroves landwards but with a net contraction. An analysis of the soil accretion rate versus the rise on a mean sea level shows that roughly half the world's mangroves will not easily survive unless they can invade landwards [7].

Mangroves vary greatly in how they respond to a sea level rise and the effect of tidal flooding and such responses appear to be species-specific [52]. For example, *Avicennia marina* is highly tolerant to prolonged water logging, but responses are highly variable in relation to the water depth of immersion and the length of time as well as salinity, temperature, and other environmental factors [54,55]. There may be other biological responses of mangroves to waterlogging. Mangroves have been shown to exhibit differences in morphology and anatomy in relation to global change. Leaf anatomy, vascular vessel densities, diameter, grouping and length, and fiber wall thickness in a large number of species are affected by waterlogging and salinity. *Avicennia marina* and *Laguncularia racemosa* L. also demonstrate modified bark anatomy in response to prolonged flooding as well as the formation of hypertrophied lenticels, adventitious roots, and increased aerenchyma development in the bark. Undoubtedly, there are physiological adaptations to prolonged waterlogging that have direct relevance to a sea level rise.

If the rate of soil accretion and uplift does not keep pace with the sea level rise, seaward mangroves will likely drown and high intertidal mangroves are predicted to progress landwards assuming that space permits [51]. In this scenario, there are likely to be drastic changes in nutrient dynamics including exchange with coastal seas and land. Drowning of seaward mangroves will result in the formation of peat deposits. Once available nutrients are exhausted and trees are dead. These peat deposits will result in a net burial of large amounts of refractory dissolved and particulate nitrogen and phosphorus and other nutrient elements, but no net exchange with the land and the sea [56]. The survival of inland mangroves will be enhanced once frequency of tidal inundation increases with an enhancement of rates of nutrient cycling with an increase in tidal wetting. Wetting will likely result in an increase in N_2O emissions and an increase in transformation rates of nitrogen such as ammonification, nitrification, and denitrification [23]. In degraded mangroves, Weis et al. [57] found that seaward mangroves still not covered by a sea level rise received their nitrogen and phosphorus supply from nitrogen fixation and phosphorus import from the sea while landward mangroves increasingly covered their nitrogen and phosphorus demand from allochthonous sources and recycling of soil organic matter. Such differences are likely to hold true for mangroves subject to a sea level rise even though there will continue to be large spatial and temporal variations in emissions and rates of dissolved and particulate nutrient flux. For example, inland mangroves that are newly and more frequently wetted by rising tides would be expected to initially export comparatively large amounts of litter that built up on the dry forest floor. Similarly, there will be large litter losses from drowned forests. Gaseous nitrogen fluxes (N_2O and nitrogen fixation) will probably be the greatest during a low tide when soils are exposed in inland mangroves.

Mangroves growing on peat deposits show severe phosphorus limitation [58]. Low phosphorus availability is associated with reduced remobilization of nitrogen and accumulation of potassium in senescent leaves. Remobilization of nitrogen and phosphorus indicate complete resorption before leaf abscission. Peat deposits are, thus, poor substrates for regrowth of mangroves, which means it is unlikely that newly formed peat deposits during a sea level rise will result in any significant growth of marine plants including microalgae or waterlogged mangroves or significant transformation of nutrients via microbial growth. Only a small fraction of nitrogen will be lost from the peat deposits to adjacent tidal waters [59].

In areas where mangroves are slowly drowning such as in the Sundarbans, living forests still act as a sink for atmospheric nitrogen and water column dissolved inorganic nitrogen [60]. The nitrogen stored by plant uptake of NO_x, NH_3 from the atmosphere, nitrogen fixation, and sediment-water exchange are about twice as large as that of recycled nitrogen from litter and could account for 74% of the nitrogen required for net production. Most nitrogen is conserved in living biomass rather than in soil. Sundarbans mangroves generally conserve nitrogen but conserve only 0.2% of the annual river flux of nitrogen. Eventual loss of these mangroves to a sea level rise will likely result in little conservation of nitrogen since most nitrogen in biomass will be lost as floating debris to the adjacent coastal ocean.

9. Predictions

In pristine mangroves, the largest inputs of nitrogen are soil ammonification, nitrification, uptake of dissolved inorganic nitrogen via tides, and dissimilatory nitrate reduction to ammonium while the largest loss of nitrogen is through tidal export. Global change impacts on these processes are predicted to affect nutrient cycling and mangrove productivity. Like tides, the impacts of global change on mangroves wax and wane. Some impacts are predicted to have deleterious consequences on nutrient cycles and, thus, on forest productivity. Other impacts are expected to have positive effects on nutrient cycles and production while some impacts are unlikely to have any clear effects on mangrove ecosystems. Deforestation and changes in hydrology in the construction and maintenance of impoundments have severe consequences for mangrove survival (Table 1). Subsequent formation of peat deposits implies that subsequent mangroves will be dwarf forests due to peat being a poor substrate for growth.

Table 1. Impacts of global change on mangrove nutrient cycling. "+" = stimulatory, "−" = deleterious.

Disturbance	Nitrogen Cycling	Phosphorus Cycling
Deforestation/Construction of Impoundments	−	−
Land-Use Change	+ −	+ −
Increased Frequency/Intensity of Storms	+ −	+ −
Increasing Temperatures	+	+
Increasing Atmospheric CO_2	+	+
Rising Sea Level	+ − 0	+ − 0

Some land-use changes such as additional inputs of sediment and nutrients will probably have positive effects on mangrove nutrient cycles while some changes such as eutrophication will have negative consequences (Table 1). The same holds true for increased frequency and intensity of storms. Increases in rainfall in arid-zone areas are predicted to relieve nutrient limitation and enhance mangrove growth while more severe tropical cyclones will likely have negative effects. Increasing temperatures and atmospheric CO_2 concentrations are predicted to initially have positive consequences for nutrient cycles and mangrove productivity even though more data are needed to properly assess their effects. The effects of a rising sea level are complex likely due to both positive and negative consequences depending on several co-factors such as the rates of subsidence, the soil accretion rate, and the tidal position. If soil accretion rates and uplift keep pace with a rise in the sea level, there may be no clear impacts on nutrient cycling and mangrove forest structure and function. If accretion and uplift do not keep pace, seaward mangroves will drown but high intertidal mangroves will likely progress landwards assuming space is available.

At present, there are still substantial limitations to our knowledge of the impacts of global change on mangroves and their relationships with nutrients. More data are needed on the impacts of various scenarios of a sea level rise on mangrove-nutrient relations as well as the impacts of the increased frequency and intensity of storms. Future research should focus on development of experimental approaches to test the effects of these global impacts on mangrove-nutrient relations, as most data is descriptive.

References

1. Alongi, D.M. Present state and future of the world's mangrove forests. *Environ. Conserv.* **2002**, *29*, 331–349. [CrossRef]
2. Spalding, M.M.; Kainuma, M.; Collins, L. *World Atlas of Mangroves*; Earthscan: London, UK, 2010; pp. 1–302, ISBN 1844076571.
3. Saenger, P. *Mangrove Ecology, Silviculture and Conservation*; Kluwer: Dordrecht, The Netherlands, 2002; pp. 1–360.
4. Alongi, D.M. *The Energetics of Mangrove Forests*; Springer: Dordrecht, The Netherlands, 2009; pp. 1–216, ISBN 978-1-4020-4270-6.

5. Alongi, D.M. Carbon cycling and storage in mangrove forests. *Ann. Rev. Mar. Sci.* **2014**, *6*, 195–219. [CrossRef] [PubMed]
6. Alongi, D.M. Mangrove forests: Resilience, protection from tsunamis, and responses to global climate change. *Estuar. Coast. Shelf Sci.* **2008**, *76*, 1–13. [CrossRef]
7. Alongi, D.M. The impact of climate change on mangrove forests. *Curr. Clim. Chang. Rep.* **2015**, *1*, 30–39. [CrossRef]
8. Gilman, E.L.; Ellison, J.; Duke, N.C.; Field, C. Threats to mangroves from climate change and adaptation options: A review. *Aquat. Bot.* **2008**, *89*, 237–250. [CrossRef]
9. Reef, R.; Feller, I.C.; Lovelock, C.E. Nutrition of mangroves. *Tree Physiol.* **2010**, *30*, 1148–1160. [CrossRef]
10. Ball, M.C. Ecophysiology of mangroves. *Trees* **1988**, *2*, 129–142. [CrossRef]
11. Alongi, D.M. Dissolved iron supply limits early growth of estuarine mangroves. *Ecology* **2010**, *91*, 3229–3241. [CrossRef]
12. Alongi, D.M. Micronutrients and mangroves: Experimental evidence for copper limitation. *Limnol. Oceangr.* **2017**, *62*, 2759–2772. [CrossRef]
13. Lovelock, C.E.; Feller, I.C.; Ball, M.C.; Engelrecht, B.M.J.; Ewe, M.L. Differences in plant function in phosphorus- and nitrogen-limited mangrove ecosystems. *New Phytol.* **2006**, *172*, 514–522. [CrossRef]
14. Alongi, D.M.; Boto, K.G.; Robertson, A.I. Nitrogen and phosphorus cycles. In *Tropical Mangrove Ecosystems*; Robertson, A.I., Alongi, D.M., Eds.; American Geophysical Union: Washington, WA, USA, 1992; pp. 251–292, ISBN 0-87590-255-3.
15. Sherman, R.E.; Fahey, T.J.; Howarth, R.W. Soil-plant interactions in a neotropical mangrove forest: Iron, phosphorus and sulfur dynamics. *Oecologia* **1998**, *115*, 553–563. [CrossRef] [PubMed]
16. Kristensen, E.; Andersen, F.Ø.; Holmboe, N.; Holmer, M.; Thongtham, N. Caron and nitrogen mineralization in sediments of the Bangrong mangrove area, Phuket, Thailand. *Aquat. Microb. Ecol.* **2000**, *22*, 199–213. [CrossRef]
17. Alongi, D.M.; Clough, B.F.; Robertson, A.I. Nutrient-use efficiency in mangrove forests along the arid tropical coast of Western Australia. *Aquat. Bot.* **2005**, *82*, 121–131. [CrossRef]
18. Lovelock, C.E.; Krauss, K.W.; Osland, M.J.; Reef, R.; Ball, M.C. The physiology of mangrove trees with changing climate. In *Tropical Tree Physiology: Adaptations and Responses in a Changing Environment*; Goldstein, G., Santiago, L.S., Eds.; Springer: New York, NY, USA, 2016; pp. 149–179.
19. Alongi, D.M. Cycling and global fluxes of nitrogen in mangroves. *Glob. Environ. Res.* **2013**, *17*, 173–182.
20. Alongi, D.M.; de Carvalho, N.A. The effect of small-scale logging on stand characteristics and soil biogeochemistry in mangrove forests of Timor Leste. *Forest Ecol. Manag.* **2008**, *255*, 1359–1366. [CrossRef]
21. Vovides, A.G.; López-Portillo, J.; Bashan, Y. N_2-fixation along a gradient of long-term disturbance in tropical mangroves bordering the Gulf of Mexico. *Biol. Fertil. Soils* **2011**, *47*, 567–576. [CrossRef]
22. Vovides, A.G.; Bashan, Y.; López-Portillo, J.A.; Guevara, R. Nitrogen fixation in preserved, reforested, naturally regenerated and impaired mangroves as an indicator of functional restoration in mangroves in an arid region of Mexico. *Restor. Ecol.* **2011**, *19*, 236–244. [CrossRef]
23. Verhoeven, J.T.A.; Laanbroek, H.J.; Rains, M.C.; Whigham, D.F. Effects of summer flooding on nitrogen dynamics in impounded mangroves. *J. Environ. Manag.* **2014**, *139*, 217–226. [CrossRef]
24. Jordan, S.J.; Stoffer, J.; Nestlerode, J.A. Wetlands as sinks for reactive nitrogen at continental and global scales: A meta-analysis. *Ecosystems* **2011**, *14*, 144–155. [CrossRef]
25. Mandal, S.; Ray, S.; Ghosh, P.B. Impact of mangrove litterfall on nitrogen dynamics of virgin and reclaimed islands of Sundarbans mangrove ecosystem, India. *Ecol. Model.* **2013**, *252*, 153–166. [CrossRef]
26. Sjöling, S.; Mohammed, S.M.; Lyimo, T.J.; Kyaruzi, J.J. Benthic bacterial diversity and nutrient processes in mangroves: Impact of deforestation. *Estuar. Coast. Shelf Sci.* **2005**, *63*, 397–406. [CrossRef]
27. Whigham, D.F.; Verhoeven, J.T.A.; Samarkin, V.; Megonigal, P.J. Responses of *Avicennia germinans* (black mangrove) and the soil microbial community to nitrogen addition in a hypersaline wetland. *Estuar. Coasts* **2009**, *32*, 926–936. [CrossRef]
28. Feller, I.C.; Whigham, D.F.; McKee, K.L.; Lovelock, C.E. Nitrogen limitation of growth and nutrient dynamics in a disturbed mangrove forest, Indian River Lagoon, Florida. *Oecologia* **2003**, *134*, 405–414. [CrossRef]
29. McKee, K.L.; Faulkner, P.L. Restoration of biogeochemical function in mangrove forests. *Restor. Ecol.* **2000**, *8*, 247–259. [CrossRef]

30. Lovelock, C.E.; Ball, M.C.; Martin, K.C.; Feller, I.C. Nutrient enrichment increases mortality of mangroves. *PLoS ONE* **2009**, *4*, e5600. [CrossRef]
31. Sanders, C.J.; Eyre, B.D.; Santos, I.R.; Machado, W.; Luiz-Silva, W.; Smoak, J.M.; Breithaupt, J.L.; Ketterer, M.E.; Sanders, L.; Marotta, H.; et al. Elevated rates of organic carbon, nitrogen, and phosphorus accumulation in a highly impacted mangrove wetland. *Geophys. Res. Lett.* **2014**, *41*, 2475–2480. [CrossRef]
32. Keuskamp, J.A.; Hefting, M.M.; Dingemans, B.J.J.; Verhoeven, J.T.A.; Feller, I.C. Effects of nutrient enrichment on mangrove leaf litter decomposition. *Sci. Total Environ.* **2015**, *508*, 402–410. [CrossRef] [PubMed]
33. Weng, B.; Xie, X.; Yang, J.; Liu, J.; Lu, H.; Yan, C. Research on the nitrogen cycle in rhizosphere of *Kandelia obovata* under ammonium and nitrate addition. *Mar. Pollut. Bull.* **2013**, *76*, 227–240. [CrossRef]
34. Maher, D.T.; Sippo, J.Z.; Tait, D.R.; Holloway, C.; Santos, I.R. Pristine mangrove creek waters are a sink for nitrous oxide. *Sci. Rep.* **2016**, *6*, 25701. [CrossRef]
35. Liu, K.J.; Li, H.P.; Davis S.E., III. Benthic exchange of C, N, and P along the estuarine ecotone of Lower Taylor Slough, Florida (USA): Effect of seasonal flows and phosphorus availability. *Wetlands* **2014**, *34*, 113–122. [CrossRef]
36. Feller, I.C.; Lovelock, C.E.; Piou, C. Growth and nutrient conservation in *Rhizophora mangle* in response to fertilization along latitudinal and tidal gradients. *Smithson. Contrib. Mar. Sci.* **2009**, *38*, 345–358.
37. Reis, C.R.G.; Nardoto, G.B.; Rochelle, A.L.C.; Vieira, S.A.; Oliveira, R.S. Nitrogen dynamics in subtropical fringe and basin mangrove forests inferred from stable isotopes. *Oecologia* **2017**, *183*, 841–848. [CrossRef] [PubMed]
38. Collins, M.; Knutti, R.; Arblaster, J.; Dufresne, J.-L.; Fichefet, T.; Friedlingstein, P.; Gao, X.; Gutowski, T.; Johns, T.; Krinner, G.; et al. Long-term climate change: Projections, commitments and irreversibility. In *Climate Change 2013: The Physical Basis, Contribution of Working Group I to the Fifth Assessment Report of the Intergovernmental Panel of Climate Change*; Stocker, T.F., Qin, D., Plattner, G.-K., Tignor, M., Allen, S.K., Boschung, J., Nauels, A., Xia, Y., Bex, V., Midgley, P.M., Eds.; Cambridge University Press: Cambridge, UK, 2013; pp. 1029–1136.
39. Lovelock, C.E.; Feller, I.C.; Adame, M.F.; Reef, R.; Penrose, H.M.; Wei, L.; Ball, M.C. Intense storms and the delivery of materials that relieve nutrient limitations in mangroves of an arid zone estuary. *Funct. Plant Biol.* **2011**, *38*, 514–522. [CrossRef]
40. Fernandes, S.O.; Gonsalves, M.-J.; Michotey, V.D.; Bonin, P.; Loka Bharathi, P.A. Denitrification activity is closely linked to the total ambient Fe concentration in mangrove sediments of Goa, India. *Estuar. Coast. Shelf Sci.* **2013**, *131*, 64–74. [CrossRef]
41. Mendoza, U.N.; da Cruz, C.C.; Menezes, M.P.; Lara, R.J. Flooding effects on phosphorus dynamics in an Amazonian mangrove forest, Northern Brazil. *Plant Soil* **2012**, *353*, 107–121. [CrossRef]
42. Ball, M.C.; Sobrado, M.A. Ecophysiology of mangroves: Challenges in linking physiological processes with patterns of forest structure. In *Physiological Plant Ecology*; Scholes, J.C., Press, M., Barker, M.G., Eds.; Blackwell: Oxford, UK, 2002; pp. 331–346, ISBN 0521549299.
43. Cavanaugh, K.C.; Kellner, J.R.; Forde, A.J.; Gruner, D.S.; Parker, J.D.; Rodriguez, W.; Feller, I.C. Poleward expansion of mangroves is a threshold response to decreased frequency of extreme cold events. *Proc. Nat. Acad. Sci. USA* **2014**, *111*, 723–727. [CrossRef]
44. Saintilan, N.; Wilson, N.; Rogers, K.; Rajkaran, A.; Krauss, K.W. Mangrove expansion and salt marsh decline at mangrove poleward limits. *Glob. Chang. Biol.* **2014**, *20*, 147–157. [CrossRef]
45. Alongi, D.M. Bacterial production and microbial biomass in tropical mangrove sediments. *Microb. Ecol.* **1988**, *15*, 59–79. [CrossRef]
46. Farnsworth, E.J.; Ellison, A.M.; Gong, W.K. Elevated CO_2 alters anatomy, physiology, growth, and reproduction of red mangrove (*Rhizophora mangle* L.). *Oecologia* **1996**, *108*, 599–609. [CrossRef]
47. Ball, M.C.; Cochrane, M.J.; Rawson, H.M. Growth and water use of the mangroves *Rhizophora apiculata* and *R. stylosa* in response to salinity and humidity under ambient and elevated concentrations of atmospheric CO_2. *Plant Cell Environ.* **1997**, *20*, 1158–1166. [CrossRef]
48. McKee, K.L.; Rooth, J.E. Where temperate meets tropical: Multifactorial effects of elevated CO_2, nitrogen enrichment, and competition on a mangrove-salt marsh community. *Glob. Chang. Biol.* **2008**, *14*, 971–984. [CrossRef]

49. Reef, R.; Slot, M.; Motro, U.; Motro, Y.; Adame, M.F.; Garcia, M.; Aranda, M.F.; Lovelock, C.E.; Winter, K. The effects of CO_2 and nutrient fertilization on the growth and temperature response of the mangrove *Avicennia germinans*. *Photosynth. Res.* **2016**, *129*, 159–170. [CrossRef] [PubMed]
50. Church, J.A.; Clark, P.U.; Cazenave, A.; Gregory, J.M.; Jevrejeva, S.; Levermann, A.; Merrifield, M.A.; Milne, G.A.; Nerem, R.S.; Nunn, P.D.; et al. Sea level change. In *Climate Change 2013: The Physical Basis, Contribution of Working Group I to the Fifth Assessment Report of the Intergovernmental Panel of Climate Change*; Stocker, T.F., Qin, D., Plattner, G.-K., Tignor, M., Allen, S.K., Boschung, J., Nauels, A., Xia, Y., Bex, V., Midgley, P.M., Eds.; Cambridge University Press: Cambridge, UK, 2013; pp. 1137–1216.
51. Woodroffe, C.D.; Rogers, K.; McKee, K.L.; Lovelock, C.E.; Mendelssohn, I.A.; Saintilan, N. Mangrove sedimentation and responses to relative sea-level rise. *Annu. Rev. Mar. Sci.* **2016**, *8*, 243–266. [CrossRef] [PubMed]
52. Krauss, K.W.; McKee, K.L.; Lovelock, C.E.; Cahoon, D.R.; Saintilan, N.; Reef, R.; Chen, L. How mangrove forests adjust to rising sea level. *New Phytol.* **2013**, *202*, 19–34. [CrossRef] [PubMed]
53. López-Medellin, X.; Ezcurra, E.; González-Abraham, C.; Hak, J.; Santiago, L.S.; Sickman, J.O. Oceanographic anomalies and sea-level rise drive mangroves inland in the Pacific coast of Mexico. *J. Veget. Sci.* **2011**, *22*, 143–151. [CrossRef]
54. Lu, W.; Chen, L.; Wang, W.; Tam, N.F.-Y.; Lin, G. Effects of sea level rise on mangrove *Avicennia* population growth, colonization and establishment: Evidence from a field survey and greenhouse manipulation experiment. *Acta Oecol.* **2013**, *49*, 83–91. [CrossRef]
55. Mangora, M.M.; Mtolera, M.S.P.; Björk, M. Photosynthetic responses to submergence in mangrove seedlings. *Mar. Freshw. Res.* **2014**, *65*, 497–504. [CrossRef]
56. Parkinson, R.W.; DeLaune, R.D.; White, J.R. Holocene sea-level rise and the fate of mangrove forests within the wider Caribbean region. *J. Coast. Res.* **1994**, *10*, 1077–1086.
57. Weis, C.; Weiss, J.; Boy, J.; Iskandar, I.; Mikutta, R.; Guggenberger, G. Soil organic carbon stocks in estuarine and marine mangrove ecosystems are driven by nutrient colimitation of P and N. *Ecol. Evol.* **2016**, *6*, 5043–5056. [CrossRef] [PubMed]
58. Medina, E.; Cuevas, E.; Lugo, A.E. Nutrient relations of dwarf *Rhizophora mangle* L. mangroves on peat in eastern Puerto Rico. *Plant Ecol.* **2010**, *207*, 13–24. [CrossRef]
59. Keukamp, J.A.; Feller, I.C.; Laanbroek, H.J.; Verhoeven, J.T.A.; Hefting, M.M. Short- and long-term effects of nutrient enrichment on microbial exoenzyme activity in mangrove peat. *Soil Biol. Biochem.* **2015**, *81*, 38–47. [CrossRef]
60. Ray, R.; Majumder, N.; Das, S.; Chowdhury, C.; Jana, T.K. Biogeochemical cycle of nitrogen in a tropical mangrove ecosystem, east coast of India. *Mar. Chem.* **2014**, *167*, 33–43. [CrossRef]

2

Resource Utilization by Native and Invasive Earthworms and their Effects on Soil Carbon and Nitrogen Dynamics in Puerto Rican Soils

Ching-Yu Huang [1,2,*], Grizelle González [3] and Paul F. Hendrix [1]

1 Odum School of Ecology, University of Georgia, Athens, GA 30602, USA; hendrixp@uga.edu
2 Department of Biology, University of North Georgia, Dahlonega, GA 30597, USA
3 USDA Forest Service, International Institute of Tropical Forestry, Jardín Botánico Sur, 1201 Ceiba St., Río Piedras 00926, Puerto Rico; ggonzalez@fs.fed.us
* Correspondence: ching-yu.huang@ung.edu

Academic Editors: Scott X. Chang and Xiangyang Sun

Abstract: Resource utilization by earthworms affects soil C and N dynamics and further colonization of invasive earthworms. By applying ^{13}C-labeled *Tabebuia heterophylla* leaves and ^{15}N-labeled *Andropogon glomeratus* grass, we investigated resource utilization by three earthworm species (invasive endogeic *Pontoscolex corethrurus*, native anecic *Estherella* sp., and native endogeic *Onychochaeta borincana*) and their effects on soil C and N dynamics in Puerto Rican soils in a 22-day laboratory experiment. Changes of ^{13}C/C and ^{15}N/N in soils, earthworms, and microbial populations were analyzed to evaluate resource utilization by earthworms and their influences on C and N dynamics. *Estherella* spp. utilized the ^{13}C-labeled litter; however, its utilization on the ^{13}C-labeled litter reduced when cultivated with *P. corethrurus* and *O. borincana.* Both *P. corethrurus* and *O. borincana* utilized the ^{13}C-labeled litter and ^{15}C-labeled grass roots and root exudates. *Pontoscolex corethrurus* facilitated soil respiration by stimulating ^{13}C-labeled microbial activity; however, this effect was suppressed possibly due to the changes in the microbial activities or community when coexisting with *O. borincana.* Increased soil N mineralization by individual *Estherella* spp. and *O. borincana* was reduced in the mixed-species treatments. The rapid population growth of *P. corethrurus* may increase competition pressure on food resources on the local earthworm community. The relevance of resource availability to the population growth of *P. corethrurus* and its significance as an invasive species is a topic in need of future research.

Keywords: carbon and nitrogen mineralization; invasive earthworms; Luquillo mountains; microbial respiration; Puerto Rico; stable isotope; tropics

---

1. Introduction

Invasive earthworms have caused significant effects on local biota and ecosystem processes (such as nutrient dynamics) in the invaded areas, e.g., European Lumbricids in North America [1–3]. Population declines of native earthworms, particularly in remote and non-fragmented forests, have contributed to a result of competitive exclusion by expanding invasive earthworm populations [2,4,5]. Lachnicht et al. [6] observed that invasive *Pontoscolex corethrurus* (Müller, 1856) earthworms, when incubated with native *Estherella* sp., utilized different N resources, possibly avoiding direct competition on food resource. Winsome et al. [7] found that invasive *Aporrectodea trapezoides* (Dugès, 1828) lost its competition advantage when co-existing with native *Argilophilus marmoratus* (Eisen, 1893) in the resource-poor habitat of a Californian grassland. Interactions between native and invasive earthworms varied with resource utilization of earthworm species and resource availability [6,7]. Earthworms are categorized into three ecological groups, epigeic, endogeic, and anecic, based on their preferences on space and food resources [8]. Epigeic earthworms mainly consume leaf litter (and microbial populations colonizing on it) and inhabit the litter layer, while endogeic earthworms occupy mineral soils and use soil organic matter as their main food resources. Anecic earthworms utilize mainly leaf litter but with the ability to build burrows deep in the soil [8]. Earthworms with same feeding strategies are expected to evolve stronger competitive interactions because they share the same food resources [2,9,10]. Hence, resource utilization of earthworms could serve as a determinant for the success of earthworm invasions and its effects on the native earthworm community [7].

Earthworm invasions have significantly altered nutrient dynamics (e.g., carbon (C) and nitrogen (N)) in invaded soils [1,11,12]. A mixed-species of European Lumbricid earthworm assemblage has been documented to lessen organic layers and relocate leaf litter and humus fragments (C) into the deeper mineral soils, as well as to cause an increase of N loss in the soil adjacent to plant roots in the temperate forests of North America [1]. The effects of earthworms on soil C and N dynamics may vary with the feeding strategies of earthworms and composition of earthworm assemblages [13]. For example, epigeic earthworms may have stronger effects on nutrient fluxes between leaf litter layers and microbial populations that colonized on it (detritusphere) from their comminution and digestion of the leaf litter substrate [1,11,12]. Endogeic/anecic earthworms, on the other hand, may play a significant role in regulating nutrient dynamics in mineral soil and plant root zones (rhizosphere) by their consumption of soil organic matter and root exudates (and depositions) and their active burrowing activity [14–16]. In an area inhabited by a mixture of earthworms (either different feeding strategies or native co-existing with invasive worms), whether earthworm effects on soil nutrient dynamics can be explained by a summation of individual earthworm effects or disproportionally dominated by one aggressive earthworm species is a topic of interest, yet still in need of more research.

Stable isotope ^{13}C and ^{15}N techniques, including ^{13}C- and ^{15}N-labeled plant materials and a natural abundance of ^{13}C and ^{15}N isotopes, have recently provided invaluable information for studying earthworm feeding strategies and their effects on soil C and N dynamics [6,17–20]. For example, Hendrix et al. [17] suggested an inter-specific competition for N resources based on their observation of overlapped natural abundance ^{15}N in both *Estherella* sp. and *P. corethrurus* in a lower altitude tabonuco forest, Puerto Rico. Neilson et al. [18] found that a natural abundance of ^{13}C and ^{15}N in earthworms can be used to assess the availability and diversity of food resources in the environment. With the application of ^{13}C- and ^{15}N-enriched plant materials, how earthworms utilize different type of food resources and the corresponding effects on soil C and N dynamics can be evaluated by tracking changes of δ^{13}C and δ^{15}N associated with ^{13}C and ^{15}N-labeled plant materials in soils, earthworms, and the microbial populations. In this study, we applied ^{13}C-labeled *Tabebuia heterophylla* (DC.) Britton leaves and ^{15}N-labeled *Andropogon glomeratus* (Walter) Britton, Sterns, & Poggenb. grass to investigate resource utilization of three earthworm species from Puerto Rico (invasive *Pontoscolex corethrurus*, native *Estherella* spp., and native *Onychochaeta borincana* (Borges, 1994) and their effects on soil C and N dynamics in Puerto Rican soils.

Pontoscolex corethrurus has invaded multiple habitats in Puerto Rico, in contrast to the restricted distribution of the native earthworms in mature forests [21,22]. Competition pressure from invasive *P. corethrurus* to native earthworms has been suggested to be responsible for the absence of native earthworms in most disturbed areas, i.e., pasture and young forests [22–24]. Lachnicht et al. [6] observed that endogeic *P. corethrurus* and anecic *Estherella* sp. showed resource partitioning (in terms of space and food) to avoiding direct competition in a 19-day laboratory experiment. The interactions observed between *P. corethrurus* and *Estherella* sp. have also caused differential influences on soil C and N mineralization [6]. In this study, we investigated feeding strategies of endogeic *P. corethrurus*, anecic *Estherella* sp., and endogeic *Onychochaeta borincana* (single-species earthworm treatments) on ^{13}C-labeled *Tabebuia* leaves and ^{15}N-labeled *Andropogon* grass. Changes in resource utilization of individual earthworm species would be evaluated by comparing earthworm tissue ^{13}C and ^{15}N of single-species earthworm treatments to those of mixed-species earthworm treatments (co-existed with other anecic/endogeic earthworms). Influences of individual earthworm species and inter-specific earthworm interactions on soil C and N dynamics would be assessed by tracking the changes of ^{13}C and ^{15}N in soils, earthworms, and microbial populations in single- and mixed-species earthworm treatments. Anecic *Estherella* spp. was expected to utilize more ^{13}C-labeled *Tabebuia* leaves, as compared with endogeic *O. borincana* and endogeic *P. corethrurus*. Given that *P. corethrurus* is believed to exhibit flexible feeding behaviors and enhance soil mineralization [6,17], we expected that *P. corethrurus* would utilize more leaf litter (detritusphere) than plant roots (rhizosphere) resources, when incubated with endogeic *O. borincana*, to avoid competition with *O. borincana*. Higher population growth would be observed in a *P. corethrurus* population, which would enhance soil C and N mineralization. However, the presence of anecic *Estherella* sp. and endogeic *O. borincana* would weaken enhanced soil mineralization caused by *P. corethrurus*.

2. Materials and Methods

2.1. Experiment Design and Setup

The experiment was conducted at Sabana Field Research Station in Luquillo, Puerto Rico, from November to December 2006. A total of 60 soil mesocosms (Polyvinyl Chloride (PVC) material, 11 cm in diameter and 20 cm in depth) were set up with 15-cm-deep field soils with the bottoms sealed with a 1 mm mesh fiberglass window screen. Experimental treatments included (1) control mesocosms (n = 4, no earthworms; Control) with isotope-labeled *Tabebuia* litter and *Andropogon glomeratus* grass; and (2) seven earthworm treatments (each treatment: n = 4) with isotope-labeled *Tabebuia* litter and *A. glomeratus* grass: single and mixed earthworm treatments (two- and three-species earthworm combination; see below). Four soil mesocosms with no isotope-labeled plant materials and no earthworms (Soil; n = 4), four soil mesocosms with ^{15}N-labeled grass plants (Grass; n = 4), and four soil mesocosms with ^{13}C-labeled leaf litter (Litter; n = 4) were also analyzed as reference data to evaluate the efficiency of ^{13}C- and ^{15}N-labeled methods.

Experimental soil was collected from the forest at the Bisley Experimental Watersheds (BEW) in the Luquillo Mountains (18°18′ N; 65°50′ W). The forest at BEW is mostly dominated by a secondary growth of tabonuco trees, and its soils are clayey and well weathered Ultisols. Detailed description of BEW can be found in Scatena [25]. The collected soils were separated by three depths of 0–5, 5–10, and 10–15 cm to air-dry for 48 h and sieved through a 5 mm mesh size sieve to exclude plant roots, rocks, cocoons, and earthworms. Three depths of air-dry soils were used to set up the 0–5, 5–10, and 10–15 cm depth in the mesocosms. Total soil C and N in 0–5 cm were 3.96 ± 0.05% and 0.37%, respectively. Three *Andropogon glomeratus* seedlings (ca. 8 cm tall), the common grass species in Puerto Rico, were transplanted into each control and earthworm mesocosm a week before the beginning of the experiment.

The *Andropogon* grass leaves were brushed with 2 atom % ^{15}N-urea solution every day to establish ^{15}N-labeled plant roots and root-derived substrates (the rhizosphere) during the experiment [26]. Seedlings of *Tabebuia heterophylla*, one of the common, native woody species (Family: Bignoniaceae) in Puerto Rico, were incubated in a growth chamber with pulse injection of 99 atom% $^{13}CO_2$ to acquire ^{13}C-labeled *Tabebuia* leaves through photosynthesis cycles during June–July 2006. After labeling procedures, *Tabebuia* senescent leaves were collected, air-dried for 48 hours, and then shredded into 1 cm^2 pieces ($\delta^{13}C$ varied from 385‰ to 804‰). A total of 3.7 g of dry ^{13}C-labeled *Tabebuia* litter (calculated based on field litterfall data) was applied to the soil surface of each control and earthworm mesocosm to establish ^{13}C-labeled litter and related microbial populations (detritusphere).

2.2. Earthworm Species and Collection

Three earthworm species from Puerto Rico were chosen for this experiment. Two native species, *Estherella* spp. and *Onychochaeta borincana*, were collected from the BEW forests (18.5°18′51.893″ N, 65.5°44′41.694″ W) and a riparian forest in Almirante Norte (18°41′ N, 65°38′ W; alluvial soil) in Puerto Rico [27], respectively; while *Pontoscolex corethrurus* was collected from the pasture at the Sabana Field Research Station (18°18′ N, 65°50′ W) in the town of Luquillo, Puerto Rico. Anecic *Estherella* spp. has dark pigmentation on the dorsal side and stays in leaf litter and upper soil layers. Endogeic *O. borincana* has pale coloration and stays in the subsoil layer. The invasive earthworm species, *P. corethrurus*, as an endogeic species, is the dominant peregrine earthworm that has colonized most habitats of Puerto Rico [6,17]. Before introducing into the earthworm mesocosms, gut contents of all earthworms were voided for 24 h, and their fresh biomass was recorded as the initial biomass data at the beginning of the experiment. Earthworms were introduced to assigned single or mixed earthworm treatments as followed: single species treatments—*O. borincana* only (O; 4 worms), *Estherella* spp. only (E; 4 worms), and *P. corethrurus* only (P; 4–5 worms); two-species mixed treatments—*Estherella* and *P. corethrurus* (E + P; 3 worms from each species), *Estherella* and *O. borincana* (E + O; 4 *Estherella* worms and 3 *O. borincana* worms), and *O. borincana* and *P. corethrurus* (O + P; 3 worms from each species); and three-species mixed treatments—*Estherella, P. corethrurus*, and *O. borincana* (E + P + O; 2 *Estherella*, 2 *O. borincana*, and 3 *P. corethrurus*). Four soil mesocosms were assigned to the control and each earthworm treatment as experimental replicates. The earthworm species were introduced into the experimental mesocosms following the order of *O. borincana, Estherella* spp., and *P. corethrurus*. Average fresh biomass of earthworms for each earthworm treatment is listed in Table 1. Each mesocosm was watered with 35 mL of water every day to maintain soil moisture during the 22-day experiment. The mesocosms were rotated randomly every week during the experiment.

Table 1. Average fresh biomass of *Estherella* spp. (E), *Onychochaeta borincana* (O), and *Pontoscolex corethrurus* (P) earthworms introduced into different earthworm mesocosm (g per mesocosm).

		Earthworm Treatments			
Variables	Single species (E, O, P)	E + O	E + P	O + P	E + O + P
Estherella spp.					
Fresh weight (before)	5.3 (0.5)	4.2 (0.6)	3.2 (0.4)	n/a	2.2 (0.2)
Fresh weight (after)	4.6 (1.2)	3.9 (1.1)	2.7 (0.6)	n/a	2.3 (0.4)
Onychochaeta borincana					
Fresh weight (before)	4.9 (0.6)	3.6 (0.5)	n/a	2.7 (0.6)	2.2 (0.4)
Fresh weight (after)	2.9 (1.8)	2.4 (0.6)	n/a	2.3 (0.7)	1.6 (0.3)
Pontoscolex corethrurus					
Fresh weight (before)	2.0 (0.3)	n/a	1.5 (0.3)	1.4 (0.1)	1.2 (0.1)
Fresh weight (after)	1.8 (0.4)	n/a	1.5 (0.1)	1.7 (0.2)	1.4 (0.2)

Capital letters (E, O, and P) represent treatments with different earthworm assemblages. Single-species: E = *Estherella* spp.; O = *Onychochaeta borincana*; P = *Pontoscolex corethrurus*. Two-species: E + O = *Estherella* spp. and *O. borincana* assemblage; E + P = *Estherella* spp. and *P. corethrurus* assemblage; O + P = *O. borincana* and *P. corethrurus* assemblage. Three-species: E + O + P = *Estherella* spp., *O. borincana*, and *P. corethrurus* assemblage. Value is shown as mean (S.D.) (n = 4) at the beginning of the experiment (before) and after the 22-day experiment (after). "n/a" indicates the particular earthworm species was not introduced into the corresponding experimental mesocosm.

2.3. Experiment Responding Variables

2.3.1. Soil CO_2 and ^{13}C-CO_2

At Day 21 of the experiment, soil carbon dioxide (CO_2) evolution was collected using the alkali absorption technique [28]. At each sampling, a circular area (5 cm in diameter) in between the center and the edge of the mesocosm was randomly chosen for each mesocosm, and the *Tabebuia* litter within was gently removed to the side. A PVC chamber (10 cm tall and 5 cm in diameter) was inserted

1 cm into the soil surface of each mesocosm with a scintillation vial containing 10 mL of a 1 mol/L NaOH solution placed inside each PVC chamber. The chamber was sealed with plastic wrap and aluminum foil on the top for soil CO_2 absorption. Five NaOH solution vials (control) were kept closed during the 24 h absorption, except to open only at the beginning and the end of absorption to assess sampling contamination. Twenty-four hours later, each alkali solution was removed from the chamber, and 2 mL of 1 mol/L $BaCl_2$ was added to form $BaCO_3$ precipitate. Total CO_2 trapped by alkali solution was determined by titration with 1 mol/L HCl to reach a pH neutral point (phenolphthalein endpoint) [28]. $BaCO_3$ precipitate from each sample was air dried and packed in tin capsules for ^{13}C-CO_2 analysis.

2.3.2. The Remaining Mass of the *Tabebuia* Litter

Soil mesocosms were deconstructed at Day 22 to collect final data of the experiment. *Tabebuia* litter was carefully picked up and oven-dried at 60 °C for 48 h. The litter samples were ground, and a subsample of 0.5 g litter was burned at 550 °C for 4 h to obtain ash-free dry matter (AFDM) data. The data were used to calculate the remaining litter mass at the end of the experiment.

2.3.3. Survivorship, Growth, and the ^{13}C and ^{15}N Composition of Earthworms

The number of earthworms that survived at the end of the experiment was used to determine earthworm survivorship. All earthworms were put into separate containers to void their gut contents for 24 h. Final fresh biomass was recorded after gut-voiding. Earthworms were killed by dipping in boiling water for 3 seconds. One-third of the earthworm body (tail part) was cut and rinsed with deionized water with the gut content removed. Earthworm tissue was then freeze-dried and ground. Two milligrams of earthworm tissue was packed into a tin capsule and analyzed by dry combustion on a Carlo Erba NA1500 CN analyzer (Thermo Scientific, Waltham, MA, USA) for earthworm total C, N, and ^{13}C and ^{15}N.

2.3.4. Soil and Soil Microbes

Soil was separated into three soil depths, 0–5, 5–10, and 10–15 cm. Ten grams of soil from each depth was oven-dried at 105 °C for 48 h to calculate soil moisture. Subsamples of soils were ground and packed into tin capsules (ca. 20 mg) for total soil carbon (C) and nitrogen (N) and isotopic analysis (^{13}C and ^{15}N) by dry combustion on a Carlo Erba NA1500 CN analyzer. Two sets of 20 g 0–5 cm soils were extracted with 60 mL of a 0.5 mol/L potassium sulfate (K_2SO_4) solution (3:1 solution to soil mass ratio) for soil microbial biomass analysis by using the fumigation–extraction method [29,30]. Total microbial biomass C and ^{13}C was analyzed from K_2SO_4-extracted samples using an OI analytical TIC/TOC analyzer (Shimaduz, Kyoto, Japan) coupled with a Thermo-Finnigan Delta Plus Isotope Ratio Mass Spectrometer (IRMS) (Thermo Scientific, Waltham, MA, USA). The persulfate digestion method was adapted to obtain microbial N data [31]. The K_2SO_4-extracted samples and persulfate digestion samples were analyzed with an Alpkem nitrogen autoanalyzer (OI analytical, College Station, TX, USA). Dissolved inorganic N (DIN; NH_4^+-N and NO_3^--N) was calculated from a non-fumigated K_2SO_4 extract. Microbial biomass N (MBN) was calculated from the difference between total persulfate nitrogen from fumigated and non-fumigated samples. Total persulfate nitrogen from fumigated samples was used to determine total dissolved nitrogen (TDN).

Delta ^{15}N data for each portion (DIN, MBN, and TDN) were obtained by running the samples through the isotope diffusion method [32]. The $\delta^{13}C$/ $\delta^{15}N$ value is calculated based on the measure isotope ratios between the samples and the standard:

$$\delta\ ^{13}C\ (‰) = \left((R_{sample} - R_{standard})\ /\ R_{standard}\right) \times 10^3 \quad (1)$$

$$\delta\ ^{15}N\ (‰) = \left((R_{sample} - R_{standard})\ /\ R_{standard}\right) \times 10^3 \quad (2)$$

where $\delta^{13}C$ ($\delta^{15}N$) unit is the parts per thousand and R is the mass ratio of $^{13}C/^{12}C$ ($^{15}N/^{14}N$) in the sample and standard [33].

For DIN (K_2SO_4) extracts, KCl was added along with MgO and Devarda's alloy to increase the ionic strength of the solution. For microbial N and TDN (persulfate digests) samples, 10 M NaOH was added to raise pH (>13) of the solution instead. Pairs of glass filter disks (Whatman GF/D) were prepared by baking in a muffle furnace at 500 °C for 4 h. They were acidified with 35 μL of 2M H_2SO_4 and then wrapped with Teflon tape. The Teflon-filter packages were incubated in the solutions for 6 days. After the incubation, the packages were dried over concentrated H_2SO_4 for at least 48 h, then packed in silver capsules for dry combustion on a Carlo Erba NA1500 CN analyzer and IRMS for total N and ^{15}N data.

2.4. Statistic Analysis

The differences of litter remaining mass (data transformed), soil respiration (C-CO_2, ^{13}C-CO_2, and $\delta^{13}C$), total C/N concentration, atom percentage of $^{13}C/^{15}N$, and $\delta^{13}C/\ \delta^{15}N$ in soil, microbial biomass, and earthworm tissue, dissolved inorganic nitrogen (DIN), DIN-^{15}N, and total dissolved nitrogen (TDN) between control and earthworm treatments were analyzed by a one-way ANOVA procedure (a generalized linear model (GLM) was used if data were not balanced) in SAS statistical software [34]. A GLM was also used to compare the differences of earthworm biomass and survivorship (data transformed) between earthworm treatments. If significantly different, Tukey's HSD method was applied for the comparisons between treatments. Student's t-test and GLM were applied to compare worm ^{13}C and ^{15}N differences between earthworm species in two-species and three-species mixed earthworm treatments, respectively. The significance level was set as $\alpha = 0.05$.

3. Results

3.1. Litter Mass Loss and Soil C and N

The remaining mass of the *Tabebuia* litter (ash-free dry weight), ranging from 21.6% in the control treatment to 45.3% in the E + O earthworm mesocosms, was not significantly different between control and earthworm treatments (data transformed; GLM, $F_{7,31} = 2.1$, $p = 0.08$). At the end of the experiment, soil total C and total N concentrations were not significantly different between the initial soil, the soil samples (with no worms), and earthworm treatments (soil C: $F_{8,27} = 1.0$, $p = 0.43$; soil N: $F_{8,27} = 0.2$, $p = 0.9$; Table 2). Soil carbon in the O + P earthworm mesocosms showed a significantly higher soil ^{13}C percentage (1.0786 ± 0.002%) and soil ^{15}N percentage (0.36915 ± 0.00072%), as compared with those in the initial soil (soil ^{13}C = 1.0753 ± 0.0001% and soil ^{15}N = 0.36815 ± 0.00005%) and the control soil (soil ^{13}C = 1.0752 ± 0.0002% and soil ^{15}N = 0.36810 ± 0.00005%) (both $p < 0.01$; Table 2). Soil C and N from the earthworm treatments showed stronger $\delta^{13}C$ (average = −25.9 ± 0.9‰) and $\delta^{15}N$ (average = 6.5 ± 1.0‰) signatures as compared with the control soil ($\delta^{13}C$= −27.9 ± 0.2‰ and $\delta^{15}N$ = 4.5 ± 0.1‰).

Table 2. Total soil carbon (mg C/ g soil) and nitrogen (μg N/ g soil), atom percentages of ^{13}C and ^{15}N (%) and delta ^{13}C ($\delta^{13}C$; ‰) and delta ^{15}N ($\delta^{15}N$; ‰) from the initial soil samples (no isotope-labeled materials and no worms at Day 0; Initial), control treatment (no isotope-labeled materials and no worms at Day 22; Soil) and earthworm treatments at the end of the 22-day mesocosm experiment with Puerto Rican soils.

	Earthworm Treatment									
Variables	Initial	Soil	E	O	P	E + O	E + P	O + P	E + O + P	Statistics
Soil Carbon										
Total C	39.6 (0.5)	43.5 (2.5)	42.5 (3.7)	43.8 (2.0)	41.0 (1.6)	42.9 (3.8)	42.1 (2.7)	42.0 (1.5)	42.1 (2.2)	$F_{8, 27} = 1.0; p = 0.43$
Atom ^{13}C (%)	1.0753 a (0.0001)	1.0752 a (0.0002)	1.0767 bc (0.0003)	1.0771 abc (0.0009)	1.0767 ac (0.0004)	1.0768 abc (0.0002)	1.0776 bc (0.0005)	1.0786 b (0.0020)	1.0769 abc (0.0007)	$F_{8, 27} = 7.2; p < 0.0001$
$\delta^{13}C$	−27.8 a (0.1)	−27.9 a (0.2)	−25.8 bc (0.3)	−26.1 abc (0.9)	−26.5 ab (0.4)	−26.4 abc (0.2)	−25.6 bc (0.4)	−24.8 c (1.8)	−26.3 abc (0.6)	$F_{8, 27} = 7.2; p < 0.0001$
Soil Nitrogen										
Total N	371.5 (2.3)	367.9 (19.0)	375.0 (19.5)	370.8 (21.4)	364.3 (11.0)	369.1 (19.6)	362.3 (19.7)	367.4 (11.4)	365.6 (8.3)	$F_{8, 27} = 0.2; p = 0.9$
Atom ^{15}N (%)	0.36815 a (0.00005)	0.36810 a (0.00005)	0.36885 ab (0.00042)	0.36858 ab (0.00012)	0.36866 ab (0.00036)	0.36889 ab (0.00025)	0.36885 ab (0.00004)	0.36915 b (0.00072)	0.36882 ab (0.00033)	$F_{8, 27} = 4.3; p = 0.002$
$\delta^{15}N$	4.6 a (0.1)	4.5 a (0.2)	6.5 ab (1.1)	5.8 ab (0.3)	6.0 ab (1.0)	6.6 ab (0.7)	6.5 ab (0.1)	7.3 b (2.0)	6.4 ab (0.9)	$F_{8, 27} = 4.3; p = 0.002$

Capital letters (E, O, and P) represent treatments with different earthworm assemblages. Single-species: E = Estherella spp.; O = Onychochaeta borincana; P = Pontoscolex corethrurus. Two-species: E + O = Estherella spp. and O. borincana assemblage; E + P = Estherella spp. and P. corethrurus assemblage; O + P = O. borincana and P. corethrurus assemblage. Three-species: E + O + P = Estherella spp., O. borincana and P. corethrurus assemblage. Value is shown as mean (S.D.) (n = 4). Statistics shows the statistical results (F ratios and p values) from one-way ANOVA (GLM for unbalanced data). Different letters indicate significant difference among earthworm treatments (Tukey's HSD, $p < 0.05$).

3.2. *Earthworm Populations*

3.2.1. Earthworm Biomass and Survivorship

Average fresh biomass of the surviving earthworms for each earthworm treatment at the end of the 22-day mesocosm experiment is listed in Table 1. The endogeic earthworm, *Onychochaeta borincana*, showed significantly lower survivorship (71.8 ± 25.0%) than the other two earthworm species (epi-endogeic *Pontoscolex corethrurus*: 96.9 ± 8.3%; anecic *Estherella* spp.: 93.8 ± 13.0%) (data-transformed, GLM, $F_{2, 47} = 9.56$, $p = 0.0003$). However, the survivorship of individual earthworm species did not significantly differ between the single or the mixed-earthworm treatments (GLM, *Estherella*: $F_{3, 15} = 1.6$, $p = 0.2$; *O. borincana*: $F_{3, 15} = 0.4$, $p = 0.8$; *P. corethrurus*: $F_{3, 15} = 0.7$, $p = 0.6$), nor did the biomass changes (%) of individual earthworm species (*Estherella*: $F_{3, 15} = 0.6$, $p = 0.7$; *O. borincana*: $F_{3, 15} = 1.1$, $p = 0.4$; *P. corethrurus*: $F_{3, 15} = 2.1$, $p = 0.2$). A total of eight *P. corethrurus* were reproduced during the 22-day mesocosm experiment.

3.2.2. Tissue C/^{13}C and N/^{15}N in Native *Estherella* spp.

Percentage of tissue biomass C of anecic *Estherella* spp. showed no significant difference between its single species treatment (cultivated alone; tissue C = 46.3 ± 0.3%) and the mixed-species treatments (cultivated with *O. borincana* and/or *P. corethrurus*; tissue C (%) = 45.7%–47.2%) ($F_{3, 39} = 2.2$, $p = 0.1$; Table 3). However, *Estherella* spp. when cultivated alone was found to have significantly higher ^{13}C enrichment (as in δ^{13}C and atom percentage of ^{13}C) as compared with the mixed-species treatments (for δ^{13}C: E + P and E + O + P mesocosms; $F_{3, 39} = 2.0$, $p = 0.04$) (for tissue ^{13}C (%): E + P mesocosms; $F_{3, 39} = 2.9$, $p = 0.047$) (Table 3). *Estherella* spp. did not show a significant difference in worm tissue N (%), δ^{15}N, and ^{15}N (%) between its single species and the mixed-species mesocosms (all $p > 0.4$; Table 3).

Table 3. Earthworm tissue total carbon (C) and nitrogen (N) percentages (%), atom percentages of ^{13}C and ^{15}N (%), and delta ^{13}C (δ^{13}C; ‰) and delta ^{15}N (δ^{15}N; ‰) in native earthworms *Estherella* spp. (E) and *Onychochaeta borincana* (O) at each earthworm mesocosm from different earthworm treatments at the end of the 22-day experiment with Puerto Rican soils.

	Earthworm Treatments					
Variables	**Single species (E or O)**	**E + O**	**E + P**	**O + P**	**E + O + P**	**Statistics**
Estherella spp.						
Total C (%)	46.3 (0.8)	45.7 (1.3)	46.2 (1.1)	n/a	47.2 (1.8)	$F_{3, 39} = 2.2$; $p = 0.10$
Atom^{13}C (%)	1.0805 a (0.0039)	1.0785 ab (0.0004)	1.0781 b (0.0006)	n/a	1.0788 ab (0.0007)	$F_{3, 39} = 2.9$; $p = 0.047$
δ^{13}C	−23.0 a (3.5)	−24.8 ab (0.4)	−25.2 b (0.5)	n/a	−24.6 b (0.7)	$F_{3, 39} = 2.9$; $p = 0.040$
Total N (%)	12.4 (0.5)	12.3 (0.8)	12.2 (1.0)	n/a	12.7 (0.4)	$F_{3, 39} = 0.8$; $p = 0.5$
Atom^{15}N (%)	0.3690 (0.0002)	0.3688 (0.0003)	0.3689 (0.0004)	n/a	0.3688 (0.0002)	$F_{3, 39} = 1.0$; $p = 0.4$
δ^{15}N	6.8 (0.6)	6.2 (0.9)	6.6 (1.0)	n/a	6.5 (0.6)	$F_{3, 39} = 1.0$; $p = 0.4$
Onychochaeta borincana						
Total C (%)	46.0 (1.2)	46.6 (1.5)	n/a	46.6 (1.3)	46.5 (1.2)	$F_{3, 25} = 0.4$; $p = 0.8$
Atom^{13}C (%)	1.0823 (0.0046)	1.0812 (0.0016)	n/a	1.0845 (0.0102)	1.0812 (0.0006)	$F_{3, 25} = 0.5$; $p = 0.7$
δ^{13}C	−21.4 (4.2)	−22.4 (1.5)	n/a	−19.3 (9.4)	−22.3 (0.5)	$F_{3, 25} = 0.5$; $p = 0.7$
Total N (%)	11.8 (0.8)	12.5 (0.7)	n/a	12.3 (0.7)	12.4 (0.5)	$F_{3, 25} = 1.8$; $p = 0.2$

Table 3. *Cont.*

Variables	Earthworm Treatments					Statistics
	Single species (E or O)	E + O	E + P	O + P	E + O + P	
Atom ^{15}N (%)	0.3693 (0.0013)	0.3694 (0.0006)	n/a	0.3705 (0.0035)	0.3697 (0.0004)	$F_{3,25} = 0.4$; $p = 0.7$
δ^{15}N	8.69 (3.6)	8.2 (1.6)	n/a	11.0 (9.5)	8.9 (1.0)	$F_{3,25} = 0.4$; $p = 0.7$

Capital letters (E, O, and P) represent treatments with different earthworm assemblages. Single-species: E = *Estherella* spp.; O = *Onychochaeta borincana*; P = *Pontoscolex corethrurus*. Two-species: E + O = *Estherella* spp. and *O. borincana* assemblage; E + P = *Estherella* spp. and *P. corethrurus* assemblage; O + P = *O. borincana* and *P. corethrurus* assemblage. Three-species: E + O + P = *Estherella* spp., *O. borincana* and *P. corethrurus* assemblage. Value is shown as mean (S.D.). Statistics shows the statistical results (*F* ratios and *p* values) from one-way ANOVA (GLM for unbalanced data). Different letters indicate significant difference among earthworm treatments (Tukey's HSD, $p < 0.05$).

3.2.3. Tissue C/^{13}C and N/^{15}N in Native *O. Borincana*.

For endogeic *O. borincana*, there was no significant difference in worm tissue C and N (%), δ^{13}C and δ^{15}N signatures, and tissue ^{13}C and ^{15}N (%) between its own single species and the mixed-species mesocosms (all $p > 0.2$; see Table 3).

3.2.4. Tissue C/^{13}C and N/^{15}N in Invasive *P. corethrurus*

Invasive *P. corethrurus* earthworms did not show significant differences in worm tissue C and N (%), δ^{13}C and δ^{15}N enrichments, and tissue ^{13}C and ^{15}N (%) between its single species and the mixed-species mesocosms (all $p > 0.2$; Table 4). However, the juvenile *P. corethrurus* reproduced during this 22-day mesocosm experiment did show significant lower tissue C (41.0 ± 4.7%) and N percentages (9.2 ± 2.0%), as compared with the adult *P. corethrurus* worms (tissue C (%): $F_{4,52} = 3.9$, $p = 0.007$; tissue N (%): $F_{4,52} = 6.0$, $p < 0.001$) (Table 4). Juvenile *P. corethrurus* worms also showed lower enrichment of δ^{13}C (−24.3 ± 1.4‰) and tissue ^{13}C (1.0791 ± 0.0011%), as compared with the adult *P. corethrurus* worms (δ^{13}C: $F_{4,52} = 4.7$, $p = 0.002$; tissue ^{13}C (%): $F_{4,52} = 4.8$, $p = 0.002$) (Table 4). There was a significantly higher enrichment of ^{15}N (as in δ^{15}N = 7.6 ± 0.9‰ and an atom percentage of ^{15}N = 0.3692 ± 0.0003%), as compared with the adult *P. corethrurus* worms (δ^{15}N: $F_{4,52} = 7.2$, $p < 0.001$; atom percentage ^{15}N: $F_{4,52} = 7.2$, $p < 0.001$) (Table 4).

Table 4. Earthworm tissue total carbon (C) and nitrogen (N) percentages (%), atom percentages of ^{13}C and ^{15}N (%) and delta ^{13}C (δ^{13}C; ‰) and delta ^{15}N (δ^{15}N; ‰) in native earthworms *Estherella* spp. (E) and *Onychochaeta borincana* (O) at each earthworm mesocosm from different earthworm treatments at the end of the 22-day experiment with Puerto Rican soils.

Variables	Earthworm Treatments					Statistics
	Single Species (P)	PJ	E + P	O + P	E + O + P	
Pontoscolex corethrurus						
Total C (%)	47.0 (1.0) a	41.0 (4.7) b	44.7 (6.1) ab	46.3 (1.6) a	46.9 (0.3) a	$F_{4,52} = 3.9$; $p = 0.007$
Atom ^{13}C (%)	1.0809 a (0.0008)	1.0791 b (0.0011)	1.0812 a (0.0009)	1.0813 a (0.0012)	1.0812 a (0.0005)	$F_{4,52} = 4.8$; $p = 0.002$
δ^{13}C	−22.6 a (0.8)	−24.3 b (1.4)	−22.4 a (0.8)	−22.2 a (1.1)	−22.3 a (0.5)	$F_{4,52} = 4.7$; $p = 0.002$
Total N (%)	11.9 (0.5) a	9.2 (2.0) b	11.3 (1.5) a	11.4 (1.1) a	11.9 (0.4) a	$F_{4,52} = 6.0$; $p < 0.001$
Atom ^{15}N (%)	0.3686 a (0.0001)	0.3692 b (0.0003)	0.3686 a (0.0002)	0.3687 a (0.0003)	0.3686 a (0.0001)	$F_{4,52} = 7.2$; $p < 0.001$
δ^{15}N	5.9 (0.4) a	7.6 (0.9) b	5.9 (0.5) a	6.1 (0.9) a	5.9 (0.3) a	$F_{4,52} = 7.2$; $p < 0.001$

Capital letters (E, O, and P) represent treatments with different earthworm assemblages. Single-species: E = *Estherella* spp.; O = *Onychochaeta borincana*; P = *Pontoscolex corethrurus*. Two-species: E + O = *Estherella* spp. and *O. borincana* assemblage; E + P = *Estherella* spp. and *P. corethrurus* assemblage; O + P = *O. borincana* and *P. corethrurus* assemblage. Three-species: E + O + P = *Estherella* spp., *O. borincana* and *P. corethrurus* assemblage. Value is shown as mean (S.D.). Statistics shows the statistical results (*F* ratios and *p* values) from one-way ANOVA (GLM for unbalanced data). Different letters indicate significant difference among earthworm treatments (Tukey's HSD, $p < 0.05$).

3.3. Microbial Biomass Carbon and Soil Respiration

There was no significant difference in microbial biomass carbon (MBC) and MBC-$_{13}$C between the soil (MBC = 741.3 ± 103.0 ug C g^{-1} soil; MBC-^{13}C = 8.0 ± 1.1 ug C g^{-1} soil), control (MBC = 332.1 ± 183.2 ug C g^{-1} soil; MBC-^{13}C = 3.6 ± 1.3 ug C g^{-1} soil), and earthworm treatments (MBC ranged from 340.1 to 532.1 ug C g^{-1} soil; MBC-^{13}C from 3.7 to 7.2 ug C g^{-1} soil) (MBC: $F_{10,\,29} = 1.3$, $p = 0.26$; MBC-^{13}C: $F_{10,\,29} = 1.3$, $p = 0.27$; Table 5). Microbial biomass ^{13}C (%) was significantly higher in the control treatments, as compared with those in the Soil Only or Grass mesocosms ($F_{10,\,29} = 2.8$, $p = 0.015$; Table 5), which suggested the microbes utilized and incorporated the ^{13}C-labeled litter into their biomass. The microbial biomass δ^{13}C enrichment from P (−28.8 ± 3.4‰), O + P (−28.8 ± 3.4‰), and E + O + P (−29.1 ± 2.5‰) treatments were significantly higher than the Soil Only treatment (−36.1 ± 1.4‰) ($F_{10,\,29} = 3.3$, $p = 0.006$; Table 5).

At the end of the experiment (Day 21), soil respiration C-CO_2 and ^{13}C-CO_2 (%) from the control (soil with both ^{13}C- and ^{15}N-labeled materials but no worms) and the earthworm treatments were significantly higher than the Soil Only and Grass treatments (both $p < 0.0001$; Table 5). This suggested that the input of ^{13}C-labeled leaf-litter and earthworms facilitated microbial respiration. However, different earthworm treatments showed differential effects on ^{13}C-CO_2 (%) evolved in the microbial respiration. The *P. corethrurus* earthworm treatment showed higher ^{13}C evolved from the microbial respiration (as in ^{13}C-CO_2 and δ^{13}C; Table 5) as compared with that from the O + P earthworm treatment at the end of the experiment (both $p < 0.0001$; Table 5).

Table 5. Microbial biomass total carbon (MBC, μg C /g soil), carbon-^{13}C (MBC-^{13}C; μg ^{13}C /g soil), atom percentage of ^{13}C (%), and soil delta ^{13}C ($\delta^{13}C$; ‰), soil respiration C-CO_2 (μg C per day), atom percentage of ^{13}C-CO_2 (%), and delta ^{13}C-CO_2 ($\delta^{13}C$; ‰) from the control treatments (Soil Only, soil with ^{15}N-labeled grass (Grass), soil with ^{13}C-labeled leaf litter (Litter), and Control (soil with both grass and leaf litter but no worms)) and earthworm treatments at the end of the 22-day mesocosm experiment with Puerto Rican soils.

					Earthworm Treatments							
Variables	**Soil Only**	**Grass**	**Litter**	**Control**	**E**	**O**	**P**	**E + O**	**E + P**	**O + P**	**E + O + P**	**Statistics**
Microbial biomass												
MBC	741.3 (103.0)	570.2 (167.9)	568.3 (166.3)	332.1 (183.2)	659.8 (119.2)	340.1 (115.3)	528.8 (91.4)	479.4 (119.2)	448.1 (110.3)	483.4 (199.5)	431.7 (224.4)	$F_{10,29} = 1.3; p = 0.26$
MBC-^{13}C	8.0 (1.1)	6.2 (1.3)	6.2 (1.1)	3.6 (1.3)	7.2 (1.3)	3.7 (1.2)	5.7 (1.0)	5.2 (1.3)	4.9 (1.1)	5.3 (2.2)	4.7 (1.1)	$F_{10,29} = 1.3; p = 0.27$
Atom^{13}C (%)	1.078 a (0.002)	1.079 a (0.0004)	1.083 ab (0.004)	1.09 b (0.009)	1.083 ab (0.003)	1.082 ab (0.003)	1.086 ab (0.004)	1.085 ab (0.002)	1.083 ab (0.002)	1.086 ab (0.004)	1.086 ab (0.003)	$F_{10,29} = 2.8; p = 0.015$
$\delta^{13}C$	−36.1 a (1.4)	−34.9 ab (0.4)	−31.1 ab (3.5)	−30.6 ab (1.4)	−31.1 ab (2.3)	−32.0 ab (2.9)	−28.8 b (3.4)	−29.4 ab (1.8)	−31.5 ab (1.6)	−28.8 b (3.4)	−29.1 b (2.5)	$F_{10,29} = 3.3; p = 0.006$
Variables	Soil Only	Grass	Litter	Control	E	O	P	E + O	E + P	O + P	E + O + P	Statistics
Soil respiration at day 21												
C-CO_2	1.73 a (0.79)	3.87 ab (0.91)	5.24 abc (0.92)	9.51 c (4.95)	8.01 bc (2.05)	6.35 abc (1.58)	9.32 bc (1.50)	9.50 c (0.99)	8.90 bc (1.09)	9.99 c (3.17)	7.88 bc (2.23)	$F_{10,32} = 5.2; p < 0.001$
^{13}C- CO_2 (%)	1.085 a (0.002)	1.088 a (0.004)	1.228 b (0.014)	1.223 b (0.012)	1.206 bc (0.020)	1.209 bc (0.004)	1.220 b (0.023)	1.205 bc (0.007)	1.215 bc (0.018)	1.183 c (0.008)	1.195 bc (0.010)	$F_{10,32} = 53.6; p < 0.0001$
$\delta^{13}C$	−18.8 a (2.1)	−16.5 a (3.6)	111.9 b (13.1)	107.5 b (11.3)	91.7 bc (18.7)	94.3 bc (3.3)	104.9 b (21.2)	91.3 bc (6.9)	100.2 bc (16.9)	71.1 c (7.1)	81.8 bc (8.9)	$F_{10,32} = 53.5; p < 0.0001$

Capital letters (E, O, and P) represent treatments with different earthworm assemblages. Single-species: E = *Estherella* spp.; O = *Onychochaeta borincana*; P = *Pontoscolex corethrurus*. Two-species: E + O = *Estherella* spp. and *O. borincana* assemblage; E + P = *Estherella* spp. and *P. corethrurus* assemblage; O + P = *O. borincana* and *P. corethrurus* assemblage. Three-species: E + O + P = *Estherella* spp., *O. borincana* and *P. corethrurus* assemblage. Value is shown as mean (S.D.) ($n = 4$, except data with: $n = 3$). Statistics shows the statistical results (F ratios and p values) from one-way ANOVA (GLM for unbalanced data). Different letters indicate significant difference among earthworm treatments (Tukey's HSD, $p < 0.05$).

Table 6. Soil microbial total nitrogen (MBN; μg N/g soil), atom percentage of ^{15}N (MBN-^{15}N; %), and delta^{15}N ($\delta^{15}N$; ‰) signature and dissolved inorganic nitrogen (DIN; μg N/g soil), and atom percentage of ^{15}N (DIN-^{15}N; %) in DIN from the control treatments (Soil Only, soil with ^{15}N-labeled grass (Grass), soil with ^{13}C-labeled leaf litter (Litter), and Control (soil with grass and leaf litter but no worms)) and earthworm treatments at the end of the 22-day mesocosm experiment with Puerto Rican soils. See Table 5 for definitions of abbreviations.

					Earthworm Treatments							
Variables	**Soil Only**	**Grass**	**Litter**	**Control**	**E**	**O**	**P**	**E + O**	**E + P**	**O + P**	**E + O + P**	**Statistics**
Microbial biomass												
MBN	124.6 (30.1)	96.8 (26.8)	110.2 (25.5)	129.1 (54.5)	190.4 (110.2)	114.5 (31.0)	162.1 (65.0)	92.3 (27.9)	111.0 (39.0)	90.4 (21.9)	136.9 (74.3)	$F_{10,31} = 1.2; p = 0.3$
MBN-^{15}N (%)	0.3691 a (0.0005)	0.3747 b (0.0017)	0.3693 a (0.0007)	0.3708 a (0.0019)	0.3709 a (0.0019)	0.3694 a (0.0012)	0.3711 a (0.0017)	0.3721 ab (0.0015)	0.3709 a (0.0004)	0.3711 a (0.0008)	0.3698 a (0.0001)	$F_{10,31} = 6.0$; $p < 0.0001$
$\delta^{15}N$	7.5 a (1.3)	23.0 b (4.8)	8.1 a (1.9)	12.3 a (5.3)	12.7 a (5.3)	8.5 a (3.4)	12.0 a (4.7)	15.7 ab (4.1)	12.5 a (1.2)	13.2 a (2.3)	9.4 a (0.1)	$F_{10,31} = 6.0$; $p < 0.0001$
Dissolved inorganic N												
DIN	62.9 a (12.8)	37.0 b (8.9)	22.6 b (8.4)	18.4 b (4.1)	40.4 ab (8.6)	38.8 ab (20.2)	23.8 b (7.7)	31.1 b (6.7)	25.7 b (6.4)	28.8 b (7.1)	33.0 b (6.2)	$F_{10,31} = 5.7$; $p < 0.0001$
DIN-^{15}N (%)	0.3692 a (0.0008)	0.3958 b (0.0149)	0.3687 a (0.0003)	0.3740 a (0.0003)	0.3770 a (0.0041)	0.3749 a (0.0025)	0.3751 a (0.0087)	0.3784 a (0.0046)	0.3749 a (0.0025)	0.3813 ab (0.0076)	0.3776 a (0.0043)	$F_{10,31} = 6.4$; $p < 0.0001$

Value is shown as mean (S.D.) ($n = 4$, except data with: $n = 3$). Statistics shows the statistical results (F ratios and p values) from one-way ANOVA (GLM for unbalanced data; significant level $\alpha = 0.05$).

3.4. Soil and Microbial Nitrogen Dynamics

There was no significant difference in microbial biomass nitrogen (MBN) between the control and earthworm treatments. However, the higher MBN-^{15}N and microbial δ^{15}N signature from the Grass treatment, compared with those in the control and earthworm treatments (except E + O treatment), indicated that the microbes did utilize and incorporate the ^{15}N-labeled grass resources (plant roots or root exudates) into the microbial biomass (both $p < 0.0001$; Table 6).

At the end of experiment (Day 21), lower soil dissolved inorganic nitrogen (DIN) was found in the control and the earthworm treatments, except native *Estherella* spp. (E) and *O. borincana* (O) treatments, as compared with the Soil Only mesocosms ($F_{10,\,31} = 5.7$, $p < 0.0001$; Table 6). Earthworms not only reduced the DIN in the experimental soil but also reduced the ^{15}N percentage in DIN (except O + P treatment) ($F_{10,\,31} = 6.4$, $p < 0.0001$; Table 6). There was no significant difference total dissolved nitrogen (TDN) between the control (10.8 $\pm$ 1.5 μg N/g soil) and the earthworm treatments (ranged from 12.0–17.1 μg N/g soil) ($F_{10,\,31} = 1.3$, $p = 0.3$).

4. Discussion

In this study, newly added ^{13}C-labeled leaf litter and ^{15}N-labeled grass were sufficiently incorporated into 10 cm of top soil, soil microbial biomass, and earthworm tissue. Natural abundance of δ^{13}C in earthworms was suggested to be 1−3‰ heavier than its dietary sources (such as leaf litter, root exudates, and microbial populations in the soil) [18,35]. In this study, earthworm δ^{13}C showed on average 1.4‰ heavier in *Estherella* spp., 3.5‰ heavier in *P. corethrurus*, and 5‰ heavier in *O. borincana*, with respect to the soil δ^{13}C, while earthworm tissue showed on average 5.9‰ heavier δ^{13}C in *Estherella* spp., 7.2‰ heavier in *P. corethrurus*, and 8.5‰ heavier in *O. borincana* than the microbial biomass δ^{13}C in which they inhabited (Tables 2–5).

We did not observe any competition exclusion among three earthworm species based on the survivorship and biomass gain among the single-species and the mixed-species treatments for each individual species. However, anecic *Estherella* spp., when cultivated alone, did show higher tissue—^{13}C (%) and δ^{13}C—compared with when it was cultivated with other earthworm species. This suggested that *Estherella* spp. might change its feeding strategy by reducing its utilization of ^{13}C-labeled litter materials and/or the microbial community that was related to the ^{13}C-labeled litter when cultivated with *P. corethrurus* or both *P. corethrurus* and *O. borincana*. Lachnicht et al. [6] observed that *P. corethrurus* and *Estherella* spp., while cultivated together, excluded each other in the bottom and upper layers of soil, respectively, in a 19-day laboratory experiment in Puerto Rican soils. The authors also found that *P. corethrurus* acquired more ^{15}N-labeled leaf litter when co-occurring with *Estherella* spp. [6]. We did not find that *P. corethrurus* changed its feeding preference in this 22-day experiment based on worm tissue ^{13}C and δ^{13}C as well as tissue ^{15}N and δ^{15}N between the single-species and mixed-species earthworm treatments, nor did *O. borincana*. In this study, cultivating live *A. glomeratus* grass plants could provide a steady, continuous supply of root exudates and rhizodeposits for soil microbes and earthworms, as compared to the one-time application of ^{13}C-labeled glucose and ^{15}N-labeled leaf litter adopted by Lachnicht et al. [6]. Such a continuous supply of food resources might relieve potential inter-specific competitive pressure derived from limited food resources in short-term experiments, especially for endogeic earthworms like *O. borincana* and *P. corethrurus* that strongly rely on rhizosphere resources.

Both endogeic *O. borincana* and *P. corethrurus* showed 5‰ or higher δ^{13}C signature than their food resources (soil organic matter and soil microbial biomass). Higher δ^{13}C signature in both endogeic earthworms could be explained by their utilization on soil microbial populations (i.e., bacteria and fungi) as food resources. Fungal species (such as mycorrhizal and saprotrophic fungi) have been reported to have a higher ^{13}C enrichment than plant foliage, fine roots, and soils because of fungal biochemical synthesis and transport between plant parts [36]. Microbial activity releases the lighter ^{12}C in respiration and gradually results in an increase of ^{13}C concentration in humified residues and its own population [37,38]. As a result, endogeic earthworms (active in rhizosphere and the mineral soils), *P. corethrurus* and *O. borincana* in this study, showed higher δ^{13}C signature and tissue ^{13}C (%)

than anecic *Estherella* spp. due to their preferential consumption of ^{13}C-enriched decayed/humified debris in the mineral soil layer, to a significant portion of ^{13}C-enriched microbial (higher microbial $\delta^{13}C$ observed in P, O + P and E + O + P earthworm treatments; Table 5) and fungal populations, or to both [6,36,37]. The possibility that both endogeic *O. borincana* and *P. corethrurus* consumed the microbial populations in the mineral soil, the rhizosphere, or both is also confirmed by their heavier $\delta^{15}N$ signatures (0.6‰ and 2.7‰ $\delta^{15}N$ heavier, respectively) compared with the soil $\delta^{15}N$ (Tables 2–4).

We found that soil microbial-$\delta^{15}N$ was on average 6.1‰ heavier than *Estherella* spp., 5.8‰ heavier than *P. corethrurus*, and 2.6‰ heavier than *O. borincana* (Tables 3, 4 and 6). The stronger ^{15}N enrichment in endogeic *O. borincana* could be derived from its utilization of ^{15}N-labeled rhizosphere (plant roots, root exudates, and rhizosphere-related microbes). Even though no study has yet investigated the feeding behavior of *O. borincana*, some endogeic earthworms (e.g., *P. corethrurus*) are often found aggregated in the root zones utilizing living root fragments and dead root cells, or as response to enhanced microbial activities in the rhizosphere [35,39]. In this study, the presence of *O. borincana* seemed to relate to higher microbial biomass ^{15}N and $\delta^{15}N$ (in the E + O earthworm mesocosms) and higher DIN and higher ^{15}N-DIN (%) in the O + P treatment (although not statistically significant), as compared with other earthworm treatments (Table 6). The potential effect of endogeic *O. borincana* on rhizospheric microbial populations and activities is a topic of interest, yet in need for further research.

Pontoscolex corethrurus showed a prolific reproduction (a total of eight juvenile *P. corethrurus*) within the 22-day soil mesocosm experiment. The stronger $\delta^{15}N$ signal observed in juvenile *P. corethrurus*, as compared with the adults, might be explained by (1) the possibility that adult *P. corethrurus* allocated its assimilated ^{15}N into cocoon reproduction, which later integrated into the tissue of juvenile *P. corethrurus*, and (2) a higher soil consumption and biomass increase in relation to overall biomass by juvenile worms than the adult worms [6]. *Pontoscolex corethrurus* has been described as one of the cosmopolitan earthworm species that has aggressively invaded many regions in the tropics, including Puerto Rico, Central Amazonian, and Peruvian soils [40–43]. Exceptional reproductive strategies of *P. corethrurus*, such as a high rate of cocoon production and hatching success, a short development time, and the ability of parthenogenesis, critically influence the local native earthworm community in the invaded soils [2]. The rapid population growth of *P. corethrurus* may increase competition pressure on food resources to the local native earthworm community [22]. The relevance of resource availability to the population growth of *P. corethrurus* and its significance in a *P. corethrurus* invasion is certainly a topic in need of future research.

Earthworms showed differential effects on soil mineralization processes in this study. All earthworm treatments along with the control (no worms) had higher soil respiration C-CO_2 at Day 21, especially in the P, E + O, and O + P treatments, as compared with other control treatments (Soil Only, Grass, and Litter mesocosms). There were higher ^{13}C-CO_2 (%) and $\delta^{13}C$ from the P mesocosms (Tukey's HSD, $p < 0.001$) and the mixed E + O mesocosms (marginally significant; $p = 0.06$) compared with those from the O + P treatments. The effects on soil microbial activities by earthworms could be explained by earthworms' direct grazing behavior on soil microbial community or indirect burrowing and casting activities [11,14]. Whether the higher soil respiration C-CO_2 from the control (no worms) mesocosms was due to the release from earthworms' grazing activity is uncertain. However, the significantly higher soil respiration ^{13}C-CO_2 (%) and $\delta^{13}C$ from *P. corethrurus* (includes P and E + P) were an indicator of facilitating effects of earthworms on the enriched soil microbial biomass $\delta^{13}C$ from the same mesocosms. *Pontoscolex corethrurus* might cause an increase in soil respiration via its simulation on the activity of the ^{13}C-labeled microbial population. However, the lower soil respiration ^{13}C-CO_2 (%) and $\delta^{13}C$ in the mixed *P. corethrurus* and *O. borincana* treatments (i.e., O + P) suggested that the presence of *O. borincana* and its interaction with *P. corethrurus* might have a negative effect on the ^{13}C-labeled microbial community and facilitate the ^{15}N-labeled microbial communities in the rhizosphere. Such a possibility is supported by the observation of the slightly increased ^{15}N (%) in the soil DIN from the increased microbial activity related to the ^{15}N-labeled rhizosphere in the O + P treatment (Table 6).

The individual stimulation on soil N mineralization by *Estherella* spp. and *O. borincana* was slightly reduced when they were incubated with other earthworm species (mixed-species earthworm treatments; Table 6). No significant change was observed in microbial biomass (C and N) between treatments, thus the changes shown in soil respiration $\delta^{13}C$ and DIN could be explained by the changed activities from the microbial population or possibly alternation of microbial community induced by the inter-specific earthworm interactions from the mixed earthworm treatments. Studies have suggested that the preference of earthworms on utilizing different food resources can reshape microbial communities in the detritusphere and the rhizosphere [44,45]. Native *Estherella* spp. and *O. borincana* may individually sustain a microbial community that specialized on N mineralization in the rhizopshere, yet the microbes switched to those which utilized a labile, newly added ^{13}C-labeled resource when sharing resources with the other species. Earthworm effect on either microbial activities or microbial community by individual species is confounded when inter-specific interactions are considered, and the individual effect on microbial activities and communities was not additive. Furthermore, changes in microbial activities and alterations to the microbial community by earthworms could gradually alter soil nutrient dynamics and availability of labile C and N over time [46], which later has an effect on habitat suitability for other species. For example, invasive *Amynthas agrestis* (Goto and Hatai, 1899) was documented to change soil microbial communities, which positively affected the habitat invasibility for another invasive species, *Lumbricus rubellus* (Hoffmeister, 1843) [47]. Many studies have focused on the earthworm effects on soil microbial biomass and soil mineralization [11,47–53]; however, research investigating the effects of earthworms with different feeding strategies (i.e., epigeic, anecic, and endogeic) on soil microbial activities and communities in terms of functional groups is still limited.

5. Conclusions

In this study, anecic *Estherella* spp. was observed to reduce its utilization on ^{13}C-labeled litter or ^{13}C-related microbial community when cultivated with *P. corethrurus* or both *P. corethrurus* and *O. borincana.* Resource utilization by different earthworms changed the activities and composition of soil microbial community and further affected soil respiration and nitrogen mineralization processes. However, the individual species effect on soil C and N dynamics was altered with mixed earthworm assemblages. *Pontoscolex corethrurus* was found to stimulate soil respiration by facilitating the activity of the ^{13}C-labeled microbial activity; however, the positive effect was suppressed when it coexisted with *O. borincana.* The stimulated N mineralization process by native *Estherella* spp. and *O. borincana* individually were reduced when each of them cultivated with other earthworm species. We concluded that the earthworm effect on soil microbial community and activity varies by species, and the individual species effect is not additive when considering multiple earthworm species assemblages. Regulation on soil nutrient dynamics by native *Estherella* sp. and *O. borincana* may potentially affect habitat suitability (e.g., resource availability) to invasive *P. corethrurus* during colonization. However, the rapid population growth of *P. corethrurus* may increase competition pressure on food resources to the local earthworm community. The relevance of resource availability to the population growth of *P. corethrurus* and its significance as an invasive species is a topic in need of future research.

Acknowledgments: We thank the staff at the Sabana Field Research Station and the International Institute of Tropical Forestry (IITF) for their help in the lab and in the field. We also thank David Kissel, the Soil Plant and Water Laboratory and the University of Georgia, for helping us with the soil transportation and sterilization process. This research was supported by the National Science Foundation grant number 0236276 to the University of Georgia Research Foundation, Inc.; the Graduate Student Research Prize, Center for Biodiversity and Ecosystem Processes, University of Georgia; grant DEB-0218039 from the NSF to the Institute of Tropical Ecosystem Studies, University of Puerto Rico, and the USDA Forest Service-IITF as part of the LTER program in the Luquillo Experimental Forest. All research at the USDA Forest Service International Institute of Tropical Forestry was performed in collaboration with the University of Puerto Rico. We also thank Ariel E. Lugo, Frank Wadsworth, and two anonymous reviewers, for their comments and suggestions on an earlier version of this manuscript.

Author Contributions: Ching-Yu Huang and Paul F. Hendrix conceived the study and designed the experiment. Grizelle González provided field site information and aided in field site selection and field sample collection.

Ching-Yu Huang processed the samples, analyzed the data, and prepared the manuscript. Paul F. Hendrix and Grizelle González provided suggestions and reviews at various stages of the manuscript.

References

1. Frelich, L.E.; Hale, C.M.; Scheu, S.; Holdsworth, A.R.; Heneghan, L.; Bohlen, P.J.; Reich, P.B. Earthworm invasion into previously earthworm-free temperate and boreal forests. *Biol. Invasions* **2006**, *8*, 1235–1245. [CrossRef]
2. Hendrix, P.F.; Baker, G.H.; Callaham, M.A., Jr.; Damoff, G.A.; Fragoso, C.; Gonzaléz, G.; Winsome, T.; Zou, X. Invasion of exotic earthworms into ecosystems inhabited by native earthworms. *Biol. Invasions* **2006**, *8*, 1287–1300. [CrossRef]
3. Hendrix, P.F.; Callaham, M.A., Jr.; Drake, J.M.; Huang, C.-Y.; James, S.W.; Snyder, B.A.; Zhang, W. Pandora's box contained bait: The global problem of introduced earthworms. *Annu. Rev. Ecol. Evol. Syst.* **2008**, *39*, 593–613. [CrossRef]
4. Callaham, M.A., Jr.; Hendrix, P.F.; Phillips, R.J. Occurrence of an exotic earthworm (*Amynthas agrestis*) in undisturbed soils of the southern Appalachian mountains, USA. *Pedobiologia* **2003**, *47*, 466–470. [CrossRef]
5. Kalisz, P.J.; Wood, H.B. Native and exotic earthworms in wildland ecosystems. In *Earthworm Ecology and Biogeography in North America*, 1st ed.; Hendrix, P.F., Ed.; Lewis Publishers: Boca Raton, FL, USA, 1995; pp. 117–126.
6. Lachnicht, S.L.; Hendrix, P.F.; Zou, X. Interactive effects of native and exotic earthworms on resource use and nutrient mineralization in a tropical wet forest soil of Puerto Rico. *Biol. Fertil. Soils* **2002**, *36*, 43–52. [CrossRef]
7. Winsome, T.; Epstein, L.; Hendrix, P.F.; Horwath, W.R. Competitive interactions between native and exotic earthworm species as influenced by habitat quality in a California grassland. *Appl. Soil. Ecol.* **2006**, *32*, 38–53. [CrossRef]
8. Bouché, M.B. Strategies Lombriciennes. In *Soil Organisms as Components of Ecosystems: Proceedings of the VI International Soil Zoology Colloquium of the International Society of Soil Science (ISSS)*; Lohm, U., Persson, T., Eds.; Swedish Natural Science Research Council: Stockholm, Sweden, 1977; pp. 122–132.
9. Lavelle, P.; Barois, I.; Cruz, I.; Fragoso, C.; Hernandez, A.; Pineda, A.; Rangel, P. Adaptive strategies of Pontoscolex corethrurus (Glossoscolecidae, Oligochaeta), a peregrine geophagous earthworm of the humid tropics. *Biol. Fertil. Soils* **1987**, *5*, 188–194. [CrossRef]
10. Lavelle, P.; Lapied, E. Endangered earthworms of Amazonia: An homage to Gilberto Righi. *Pedobiologia* **2003**, *47*, 419–417. [CrossRef]
11. Groffman, P.M.; Bohlen, P.J.; Fisk, M.C.; Fahey, T.J. Exotic earthworm invasion and microbial biomass in temperate forest soils. *Ecosystems* **2004**, *7*, 43–54. [CrossRef]
12. Hale, C.M.; Frelich, L.E.; Reich, P.B.; Pastor, J. Effects of European earthworm invasion on soil characteristics in Northern hardwood forests of Minnesota, USA. *Ecosystems* **2005**, *8*, 911–927. [CrossRef]
13. Huang, C.-Y.; Hendrix, P.F.; Fahey, T.J.; Bohlen, P.J.; Groffman, P.M. A simulation model to evaluate the impacts of invasive earthworms on soil carbon dynamics. *Ecol. Model* **2010**, *20*, 2447–2457. [CrossRef]
14. Bossuyt, H.; Six, J.; Hendrix, P.F. Rapid incorporation of carbon from fresh residues into newly formed stable microaggregates within earthworm casts. *Eur. J. Soil Sci.* **2004**, *55*, 393–399. [CrossRef]
15. Curry, J.P.; Schmidt, O. The feeding ecology of earthworms—A review. *Pedobiologia* **2007**, *50*, 463–477. [CrossRef]
16. Mummey, D.L.; Rillig, M.C.; Six, J. Endogeic earthworms differentially influence bacterial communities associated with different soil aggregate size fractions. *Soil Biol. Biochem.* **2006**, *38*, 1608–1614. [CrossRef]
17. Hendrix, P.F.; Lachnicht, S.L.; Callaham, M.A., Jr.; Zou, X. Stable isotopic studies of earthworm feeding ecology in tropical ecosystems of Puerto Rico. *Rapid Commun. Mass Sp.* **1999**, *13*, 1295–1299. [CrossRef]
18. Neilson, R.; Boag, B.; Simth, M. Earthworm $\delta^{13}C$ and $\delta^{15}C$ analyses suggest that putative functional classifications of earthworms are site-specific and may also indicate habitat diversity. *Soil Biol. Biochem.* **2000**, *32*, 1053–1061. [CrossRef]
19. Schmidt, O.; Scrimgeour, C.M.; Handley, L.L. Natural abundance of ^{15}N and ^{13}C in earthworms from a wheat and a wheat-clover field. *Soil Biol. Biochem.* **1997**, *29*, 1301–1308. [CrossRef]

20. Zhang, W.; Hendrix, P.F.; Snyder, B.A.; Molina, M.; Li, J.; Rao, X.; Siemann, E.; Fu, S. Dietary flexibility aids Asian earthworm invasion in North American forests. *Ecology* **2010**, *91*, 2070–2079. [CrossRef] [PubMed]
21. Gonzaléz, G.; Zou, X.; Borges, S. Earthworm abundance and species composition in abandoned tropical croplands: Comparison of tree plantations and secondary forests. *Pedobiologia* **1996**, *40*, 385–391.
22. Sánchez-de León, Y.; Zou, X.; Borges, S.; Ruan, H. Recovery of native earthworms in abandoned tropical pastures. *Conserv. Biol.* **2003**, *17*, 999–1006. [CrossRef]
23. Gonzaléz, G.; Zou, X.; Sabat, A.; Fetcher, N. Earthworm abundance and distribution pattern in contrasting plant communities within a tropical wet forest in Puerto Rico. *Caribb. J. Sci.* **1999**, *35*, 93–100.
24. Huang, C.-Y.; Gonzaléz, G.; Hendrix, P.F. The re-colonization ability of native earthworm, Estherella spp., in Puerto Rican forests and pastures. *Caribb. J. Sci.* **2006**, *42*, 386–396.
25. Scatena, F.N. An Introduction to the Physiography and History of the Bisley Experimental Watersheds in the Luquillo Mountains of Puerto Rico. Available online: www.srs.fs.usda.gov/pubs/gtr/gtr_so072.pdf (accessed on 19 September 2016).
26. Schmidt, O.; Scrimgeour, C.M. A simple urea leaf-feeding method for the production of ^{13}C and ^{15}N labelled plant material. *Plant Soil* **2001**, *229*, 197–202. [CrossRef]
27. Abelleira, O.J. Ecology of Novel Forests Dominated by the African Tulip Tree (*Spathodea campanulata* Beauv.) in Northcentral Puerto Rico. Master's Thesis, University of Puerto Rico, Rio Piedras, Puerto Rico, 2009.
28. Liu, Z.G.; Zou, X.M. Exotic earthworms accelerate plant litter decomposition in a Puerto Rican pasture and a wet forest. *Ecol. Appl.* **2002**, *12*, 1406–1417. [CrossRef]
29. Joergensen, R.G. The fumigation-extraction method to estimate soil microbial biomass: Calibration of the k_{EC} value. *Soil Biol. Biochem.* **1996**, *28*, 25–31. [CrossRef]
30. Sparling, G.P.; West, A.W. A direct extraction method to estimate soil microbial C: Calibration *in situ* using microbial respiration and ^{14}C labelled cells. *Soil Biol. Biochem.* **1988**, *20*, 337–343. [CrossRef]
31. Cabrera, M.L.; Beare, M.H. Alkaline persulfate oxidation for determining total nitrogen in microbial biomass extracts. *Soil Sci. Soc. Am. J.* **1993**, *57*, 1007–1012. [CrossRef]
32. Stark, J.M.; Hart, S.C. Diffusion technique for preparing salt solutions, Kjeldahl digests, and persulfate digests for nitorgen-15 analysis. *Soil Sci. Soc. Am. J.* **1996**, *60*, 1846–1855. [CrossRef]
33. Coleman, D.C.; Fry, B. *Carbon Isotope Techniques*, 1st ed.; Academic Press, Inc.: San Diego, CA, USA, 1991.
34. SAS Institute Inc. *SAS Technical Report, SAS/STAT Software: The GLM Procedure*; Version 6; SAS Institute Inc.: Cary, NC, USA, 1991; p. 217.
35. Spain, A.V.; Saffigna, P.G.; Wood, A.W. Tissue carbon sources for *Pontoscolex corethrurus* (Oligochaeta: Glossoscolecidae) in a sugarcane ecosystem. *Soil Biol. Biochem.* **1990**, *22*, 703–706. [CrossRef]
36. Hobbie, E.A.; Macko, S.A.; Shugart, H.H. Insights into nitrogen and carbon dynamics of ectomycorrhizal and saprotrophic fungi from isotopic evidence. *Oecologia* **1999**, *118*, 353–360. [CrossRef]
37. Pollierer, M.M.; Langel, R.; Scheu, S.; Maraun, M. Compartmentalization of the soil animal food web as indicated by dual analysis of stable isotope ratios ($^{15}N/^{14}N$ and $^{13}C/^{12}C$). *Soil Biol. Biochem.* **2009**, *41*, 1221–1226. [CrossRef]
38. Dijkstra, P.; Ishizu, A.; Doucett, R.; Hart, S.C.; Schwartz, E.; Menyailo, O.V.; Hungate, B.A. ^{13}C and ^{15}N natural abundance of the soil microbial biomass. *Soil Biol. Biochem.* **2006**, *38*, 3257–3266. [CrossRef]
39. Binet, F.; Hallaire, V.; Curmi, P. Agricultural practices and the spatial distribution of earthworms in maize fields. Relationships between earthworm abundance, maize plants and soil compaction. *Soil Biol. Biochem.* **1997**, *29*, 577–583. [CrossRef]
40. Chauvel, A.; Grimaldi, M.; Barros, E.; Blanchart, E.; Desjardins, T.; Sarrazin, M.; Lavelle, P. Pasture damage by an Amazonian earthworm. *Nature* **1999**, *398*, 32–33. [CrossRef]
41. Fragoso, C.; Kanyonyo, J.; Moreno, A.; Senapati, B.K.; Blanchart, E.; Rodríguez, C. A survey of tropical earthworms: Taxonomy, biogeography and environmental plasticity. In *Earthworm Management in Tropical Agroecosystems*; Lavelle, P., Brussaard, L., Hendrix, P., Eds.; CABI: New York, NY, USA, 1999; pp. 1–26.
42. Gonzaléz, G.; Huang, C.-Y.; Zou, X.; Rodriguez, C. Earthworm invasions in the tropics. *Biol. Invasions* **2006**, *8*, 1247–1256. [CrossRef]
43. Hallaire, V.; Curmi, P.; Duboisset, A.; Lavelle, P.; Pashanasi, B. Soil structure changes induced by the tropical earthworm *Pontoscolex corethrurus* and organic inputs in a Peruvian ultisol. *Euro. J. Soil Biol.* **2000**, *36*, 35–44. [CrossRef]

44. Butenschoen, O.; Marhan, S.; Scheu, S. Response of soil microorganisms and endogeic earthworms to cutting of grassland plants in a laboratory experiment. *Appl. Soil Ecol.* **2008**, *38*, 152–160. [CrossRef]
45. Sheehan, C.; Kirwan, L.; Connolly, J.; Bolger, T. The effects of earthworm functional diversity of microbial biomass and the microbial community level physiological profile of soils. *Euro. J. Soil Biol.* **2008**, *44*, 65–70. [CrossRef]
46. Bohlen, P.J.; Edwards, C.A.; Zhang, Q.; Parmelee, R.W.; Allen, M. Indirect effects of earthworms on microbial assimilation of labile carbon. *App. Soil Ecol.* **2002**, *20*, 255–261. [CrossRef]
47. Zhang, B.-G.; Li, G.-T.; Shen, T.-S.; Wang, J.-K.; Sun, Z. Changes in microbial C, N, and P and enzyme activities in soil incubated with the earthworms *Metaphire guillelmi* or *Eisenia fetida*. *Soil Biol. Biochem.* **2000**, *32*, 2055–2062. [CrossRef]
48. Bohlen, P.J.; Scheu, S.; Hale, C.M.; McLean, M.A.; Migge, S.; Groffman, P.M.; Parkinson, D. Non-native invasive earthworms as agents of change in northern temperate forests. *Front. Ecol. Environ.* **2004**, *2*, 427–435. [CrossRef]
49. Eisenhauer, N.; Partsch, S.; Parkinson, D.; Scheu, S. Invasion of a deciduous forest by earthworms: Changes in soil chemistry, microflora, microarthopods and vegetation. *Soil Biol. Biochem.* **2007**, *39*, 1099–1110. [CrossRef]
50. Fisk, M.C.; Fahey, T.J.; Groffman, P.M.; Bohlen, P.J. Earthworm invasion, fine-root distributions, and soil respiration in North temperate forests. *Ecosystems* **2004**, *7*, 55–62. [CrossRef]
51. Lachnicht, S.L.; Hendrix, P.F. Interaction of the earthworm *Diplocardia mississippiensis* (Megascolecidae) with microbial and nutrient dynamics in a subtropical Spodosol. *Soil Biol. Biochem.* **2001**, *33*, 1411–1417. [CrossRef]
52. Li, X.; Fisk, M.C.; Fahey, T.J.; Bohlen, P.J. Influence of earthworm invasion on soil microbial biomass and activity in a northern hardwood forest. *Soil Biol. Biochem.* **2002**, *34*, 1929–1937. [CrossRef]
53. Wolters, V.; Joergensen, R.G. Microbial carbon turnover in beech forest soils worked by *Aporrectodea caliginosa* (Savigny) (Oligochaeta: Lumbricidae). *Soil Biol. Biochem.* **1992**, *24*, 171–177. [CrossRef]

3

Understory Plant Functional Types Alter Stoichiometry Correlations between Litter and Soil in Chinese Fir Plantations with N and P Addition

Junyi Xie, Haifu Fang, Qiang Zhang, Mengyun Chen, Xintong Xu, Jun Pan, Yu Gao, Xiangmin Fang, Xiaomin Guo and Ling Zhang *

Jiangxi Provincial Key Laboratory of Silviculture, Jiangxi Agricultural University, Nanchang 330045, China
* Correspondence: lingzhang@jxau.edu.cn or lingzhang09@126.com

Abstract: Research Highlights: This study identifies the effect of nitrogen (N) and phosphorus (P) addition on stoichiometry correlations between understory plants and soil in subtropical Chinese fir plantations. Background and Objectives: Nitrogen and P are two nutrients limiting forest ecosystem production. To obtain more wood production, N and P are usually applied in plantation management. Changes in soil N and P will generally alter the stoichiometric characteristics of understory plants, which control carbon (C) and nutrient cycles between plants and soil. However, different correlations between plant and soil stoichiometry among functional groups of understory plants have not been investigated, which also impacted element cycling between plants and soil. Materials and Methods: Subtropical Chinese fir plantations were selected for N (100 kg ha^{-1} $year^{-1}$) and P (50 kg ha^{-1} $year^{-1}$) addition study. We collected fresh litter and the corresponding soil of four understory plants (*Lophatherum gracile* Brongn., *Woodwardia japonica* (L.f.) Sm., *Dryopteris atrata* (Kunze) Ching and *Dicranopteris dichotoma* (Thunb.) Berhn.) for study of C, N, and P stoichiometric ratios. Results: Nitrogen and P addition affected C, N, and P concentrations and stoichiometric ratios in litter and soil as well as correlations between litter and soil stoichiometric ratios. Understory plant species with different functional types impacted the correlations between plants and soil in C, N, and P stoichiometric ratios, especially correlations between litter C and soil C and N. Conclusions: Changes in soil N and P affect the stoichiometric ratios of understory plants. Functional groups impacted the correlation in C, N, and P stoichiometric ratios between plants and soil, indicating functional groups varied in their impacts on element cycling between plants and soil in plantations with exogenous nutrient addition, which should be considered in future management of plantations with intensive fertilization practice.

Keywords: nitrogen and phosphorus addition; understory plants; stoichiometric ratio; litter decomposition

---

1. Introduction

Combining stoichiometry with biology, chemistry, and physics, ecological stoichiometry addresses the balance of carbon (C), nitrogen (N), and phosphorus (P) [1]. Carbon, N, and P, which act on nutrient cycling and utilization, play a very important role in the nutrient cycling process of the ecosystem and ecological stoichiometry [2]. Thus, numerous scholars focus on ecological stoichiometry [3]. Studying the stoichiometric characteristics of C, N, and P of the forest ecosystem can provide insight into the stability of the forest ecosystem in a certain region; in addition, it can clarify the proportion of the demand for environmental nutrients by plants and the drivers of change. Moreover, studying the stoichiometric characteristics in forest ecosystems can uncover the amount of nutrients returned to the soil from plant litter, as well as the nutrient supply of soil to plants in the ecosystem [4,5]. The C:N:P ratio of forest ecosystems has been shown to be significantly different with different vegetation types, climatic factors, and environmental factors [3,6,7]. The input of N and P directly changes the environmental factors of the forest ecosystem, and the addition of N and P affects the structure, function, and nutrient cycles of forest ecosystem [8]. Nitrogen inputs affect forest ecosystems in different ways. For example, N addition may cause an increase in leaf biomass and photosynthetic efficiency, and reduce the number of thin roots [9,10]. Nitrogen and P addition acts directly on leaves. Nitrogen and P addition leads to changes in the correlation of the stoichiometric ratio between plants and soil, affecting plant productivity and species diversity in communities, finally resulting in a change in the forest ecosystem [11–14]. Nitrogen addition reduced the C:N ratio and increased the N:P ratio of leaves, thus affecting the N and P balance in plants, resulting in the restriction of plant growth by P or N and P, thus reducing the productivity of the ecosystem [14,15]. The addition of P to the soil generally increases the P content in plant tissues and reduces the N:P ratio [16,17]. Nitrogen addition promotes P absorption in subtropical plants and accelerates plant growth, but the results are different from those in some temperate forests [18–20]. Therefore, to a large extent, the nutrient status of leaves can reflect a plant's environmental status; there is a close relationship between leaf nutrient status and soil stoichiometric ratio [21].

Research has shown that the effect of N and P addition on vegetation type was related to ecosystem type, time, and dose of N and P addition [22–24]. At present, research is increasing on the nutrient circulation of forest ecosystems with a focus on ecological stoichiometry. Due to the improvement of forest quality and the advantage of high canopy productivity, there are many studies on the C, N, and P ecological stoichiometry of the plant-litter-soil system [25,26]. However, most previously published studies have focused on overstory plant species, largely ignoring understory plant species, which have a faster C and N turnover rate than the overstory plants. Moreover, understory species tend to be more sensitive to environmental changes, which limits our understanding of the nutrient cycles of ecosystems [27]. Under the current state of global climate change, understory vegetation plays a very important role in the C and N cycle of forest ecosystems and helps to maintain forest biodiversity and ecosystem stability [28]. Understory vegetation plays an important ecological role in nutrient cycling, soil physical and chemical properties, and community succession [29–31]. Previous research has shown that the addition of N and P to soil changed the species richness of understory vegetation and reduced plant diversity, highlighting that more research is needed on common understory plant species [32–34].

Chinese fir (*Cunninghamia lanceolata* (Lamb.) Hook.) is widely distributed in 14 provinces in China, including Zhejiang, Jiangxi, and Hunan [35–37]. Chinese fir has a history of more than 1000 years of planting because it is one of the most important fast-growing timber species in southern China; therefore, it plays an irreplaceable role in economic and ecological benefits [38]. In recent years, the management of understory vegetation in plantations and the application of fertilizer have significantly influenced the productivity of forest ecosystems [39,40]. However, the stoichiometric responses of subtropical fir plantation litter and soil to persistent exogenous N and P input are still not clear. Therefore, under the global climate change, in order to improve our understanding for the interaction between plant and soil in plantation forests ecosystem, we studied the response of C, N, and P concentrations and stoichiometric characteristics of four representative plants (two different functional types) in the lower layer of Chinese fir plantation to the addition of N and P in the subtropical red land area. We try to solve the following problems: (1) What is the reaction of C, N, P concentration and stoichiometric ratio in plant litter and soil to N and P addition; (2) How does N or P addition affect the stoichiometric ratios of different functional plants litter and soil?

2. Materials and Methods

2.1. Study Sites

This study was conducted at the Qianyanzhou Ecological Research station in Taihe County, Ji'an City, Jiangxi Province, China (115°03′29.2″ E, 26°44′29.1″ N). The research site is a typical red hilly region with subtropical monsoonal climate. The annual average precipitation is up to 1600 mm and the annual average temperature is approximately 18 °C [41]. The experimental site is within the native distribution of Chinese fir. The study was conducted in an experimental Chinese fir plantation that was established approximately 15 years ago. The coverage of the sample plot was between 75 and 85%, and the slope was between 20 and 30°. The natural atmospheric N deposition in this area was 49 kg N ha^{-1} $year^{-1}$ [42].

2.2. Experimental Design

The experiment began in November 2011, and four treatments (control: CK, N addition, P addition, and N and P addition together) were established. These four treatments were arranged in a random complete block design, with a total of four blocks. There were 16 of 20 × 20 m plots. To prevent influence between the plots, the spacing between plots was greater than 20 m. Nitrogen and P were applied first in December 2011 and then four times each year (March, June, September, and December). Nitrogen (100 kg ha^{-1} $year^{-1}$) was added in the form of NH_4NO_3, and P (50 kg ha^{-1} $year^{-1}$) was added in the form of NaH_2PO_4 [43]. To more evenly distribute N and P addition plots in the study site, we added sand (8 kg $plot^{-1}$) in the N plots, phosphate plots, and control (CK) plots. For fertilization we chose a date where the forecast predicted no rain for four consecutive days [27].

2.3. Sample Collection

In November 2018, we investigated the presence or absence of understory species in the Chinese fir plantation in each plot. From this preliminary data, we selected four dominant species (*Lophatherum gracile* Brongn., *Woodwardia japonica* (L.f.) Sm., *Dryopteris atrata* (Kunze) Ching and *Dicranopteris dichotoma* (Thunb.) Berhn.), which appeared in all 16 plots and formed communities, as sampling targets. Four communities were randomly selected in each plot. To collect plant samples, five individual plants and the corresponding soils were collected from each plot. These samples were immediately brought back to the laboratory in cooler. After sampling, the four collected plants were divided into different functional types (Gramineae and ferns) for nutritional analysis.

The leaves of each plant from all plots were collected and combined to produce one leaf sample. Leaf samples were air-dried, and then ground to pass through 0.149 mm sieve for determination of C, N, and P. Sub-sample of air-dried leaves were oven dried at 60 °C to obtain dry mass of measured leaves without residual water content. Soil samples were also processed by removing visible plant debris and rocks, passed through a 0.149 mm sieve to determine elements, or passed through a 2 mm sieve for determination of available nutrients. Sub-samples of soils were oven dried to a constant weight to obtain water content in the air-dried soil.

2.4. Litter and Soil Nutrient Measurement

Fresh soil samples through a 2 mm sieve were used for the determination of soil dissolved organic C (DOC) (VarioTOC, Elementar, Germany) and available N (AN, including NH_4^+ and NO_3^-) [44]. Soil/water (2:5) was used to determine the pH value of soil [44]. Air-dried soil samples that had been passed through a 2 mm sieve were used for the determination of soil available P (AP) [44]. We measured AN and AP with a flow injection auto analyzer (Smartchem 200 Alliance Corp. France). Before analyzing total organic C (TOC), total N (TN), and total P (TP) in litter and soil samples, we ground the litter and air-dried soil samples and passed them through a 0.149 mm sieve. Then, we used a digester (FOSS. Tecator Digestor 20) to digest plant litter and soil and used a flow injection auto analyzer to analyze the TN and TP of the digested products. The stoichiometric ratio of C, N, and P of litter and soil was reported as mass ratio.

2.5. Data Analysis

Using the One-way analysis of variance (ANOVA) for four kinds of plants soil DOC, NH_4^+, NO_3^-, AN, AP, and pH difference analysis, when the difference is significant, using LSD (least significant difference) comparison ($p = 0.05$). Nested ANOVAs were used to analyze the effects of N addition (N), phosphorous addition (P), function type plant (f), and understory species (S, nested within Function type plant) on litter and soil C, N, and P and their stoichiometry. The variation trends of C, N, and P concentrations in the litter of two functional plants and the soil were fitted by one-dimensional regression equation. All statistical analyses were conducted by JMP 9.0 (SAS Institute, Cary, NC, USA).

3. Results

3.1. *Availability of Soil Nutrients and Carbon, Nitrogen, and Phosphorus (CNP) Concentration of Plant Litter and Soil*

Seven years after the annual addition of N and P to the plantation, we found that the soil of *L. gracile* has a higher NO_3^--N content compared with the other three understory species. Moreover, the AP content in the soil of *Dicranopteris dichotoma* is significantly lower than that of the other plants ($p < 0.01$). DOC, $NH4^+$-N, AN, and pH do not change among the four understory plant species (Table 1). Among the four species, *L. gracile* has higher concentrations of C and N in litter and N and P in associated soil (Table A1). The litter C, N, and P and soil C concentrations of *W. japonica* are all low (Table A1). The litter and soil C concentrations of *Dryopteris atrata* are all higher than the other species ($p < 0.01$). The litter and soil P concentrations of *Dicranopteris dichotoma* are all low (Table 1).

Table 1. Litter and soil nutrients of four understory plant species (means ± SE). Superscript letters in the same line indicate a significant difference at $p = 0.05$ based on LSD (least significant difference) comparison. Lines without letters indicate that the difference is not significant. LC: litter total organic carbon (C); LN: litter total nitrogen (N); LP: litter total phosphorus (P); SC: soil total organic C; SN: soil total N; SP: soil total P.

Factors	*Lophatherum gracile* Brongn. (Gramineae)	*Woodwardia japonica* (L.f.) Sm. (Fern)	*Dryopteris atrata* (Kunze) Ching (Fern)	*Dicranopteris dichotoma* (Thunb.) Berhn.) (Fern)
DOC (mg kg^{-1})	88.05 ± 6.88	100.79 ± 8.23	103.53 ± 10.59	85.67 ± 4.64
NH_4^+-N (mg kg^{-1})	5.26 ± 0.40	5.01 ± 0.40	5.23 ± 0.32	5.45 ± 0.51
NO_3^--N (mg kg^{-1})	1.92 ± 0.19 ^{A}	1.41 ± 0.09 ^{B}	1.40 ± 0.09 ^{B}	1.42 ± 0.15 ^{B}
AN (mg kg^{-1})	7.18 ± 0.50	6.42 ± 0.44	6.63 ± 0.37	6.87 ± 0.64
AP (mg kg^{-1})	1.17 ± 0.15 ^{A}	1.15 ± 0.18 ^{A}	1.34 ± 0.20 ^{A}	0.56 ± 0.09 ^{B}
pH	3.32 ± 0.05	3.31 ± 0.04	3.37 ± 0.04	3.33 ± 0.04
LC (g kg^{-1})	465.09 ± 8.61 ^{A}	419.44 ± 7.18 ^{B}	459.67 ± 6.32 ^{A}	475.82 ± 6.62 ^{A}
LN (g kg^{-1})	24.39 ± 0.69 ^{A}	17.07 ± 0.57 ^{C}	20.10 ± 0.77 ^{B}	15.05 ± 0.92 ^{C}
LP (g kg^{-1})	1.39 ± 0.12 ^{B}	1.18 ± 0.06 ^{BC}	1.94 ± 0.18 ^{A}	1.02 ± 0.07 ^{C}
SC (g kg^{-1})	31.57 ± 2.10 ^{BC}	27.87 ± 0.93 ^{C}	37.78 ± 1.39 ^{A}	35.88 ± 2.00 ^{AB}
SN (g kg^{-1})	1.30 ± 0.04 ^{A}	1.10 ± 0.05 ^{B}	0.62 ± 0.05 ^{C}	1.22 ± 0.10 ^{AB}
SP (g kg^{-1})	0.43 ± 0.02 ^{A}	0.39 ± 0.02 ^{A}	0.29 ± 0.02 ^{B}	0.32 ± 0.01 ^{B}

3.2. *Effects of N and P Addition on Carbon, Nitrogen, and Phosphorus (C, N, and P) Concentrations and Stoichiometric Ratios of Different Functional Groups*

Litter C, N, and the C:N ratio were significantly affected by N addition, the plant functional group, understory species, and the randomized complete block design. Litter P and the C:P ratio depended on P addition, understory species, and the randomized complete block design (Table 2). Litter P also depended on understory species × P addition ("S × P[f]"). The N:P ratio of litter depended on P, N × P, and the plant functional group. However, the results varied with blocks except L N/P (Table 2).

Table 2. ANOVAs on the effects of nitrogen (N) and phosphorus (P) deposition, plant functional group (F), understory species (Sp; nested within plant functional group, f), and their interactions on litter (raw decomposed leaves) C, N, P, and C/N/P. B&R indicates randomized complete block design. The six indicators include litter total organic C (LC), litter total N (LN), litter total P (LP), litter C to N ratio (L C/N), litter C to P ratio (L C/P), and litter N to P ratio (L N/P) along the first row of the table. df is the degrees of freedom. *: $p < 0.05$; **: $p < 0.01$; ***: $p < 0.001$.

Factor	df	LC (g kg^{-1})		LN (g kg^{-1})		LP (g kg^{-1})		L C/N		L C/P		L N/P	
		MS	F	MS	F	MS	F	MS	F	MS	F	MS	F
N	2	8256	5.92 ***	113	10.10 ***	0.70	2.30	376	5.19 **	323	0.02	90.33	2.47
P	1	26.78	0.02	17.18	1.54	10.37	34.22 ***	56.65	0.78	618,936	37.03 ***	982	26.84 ***
N*P	2	2440	26.71	28.54	2.56	0.81	2.66	70.56	0.98	27,073	1.62	156	4.26 *
f	1	23,734	17.00 ***	2066	185.37 ***	0.01	0.04	2777	38.41 ***	2941	0.18	1472	40.26 ***
N*f	2	10,323	7.40 ***	21.97	1.97	0.12	0.40	49.92	0.69	28,014	1.68	58.89	1.61
P*f	1	497.30	0.36	31.06	2.79	0.11	0.38	64.55	0.89	2928	0.18	71.35	1.95
N*P*f	2	3603	2.58	2.88	0.26	0.02	0.05	36.05	0.50	884	0.05	23.27	0.64
Sp[f]	2	47,217	33.83 ***	398	35.69 ***	13.89	45.85 ***	2517	34.81 ***	660,744	39.53 ***	174	4.76 ***
Sp*N[f]	4	2545	1.82	11.58	1.04	0.26	0.86	68.92	0.95	4379	0.26	1.59	0.04
Sp*P[f]	2	2714	1.94	19.89	1.79	3.35	11.04 ***	7.05	0.10	38,456	2.30	80.97	2.21
B&R	3	5465	3.92 ***	201	18.01 ***	1.16	3.82 **	447	6.19 ***	59,442	3.56*	43.52	1.19

3.3. *Correlation between the Litter of Different Functional Groups and Soil Carbon, Nitrogen, and Phosphorus (C, N, and P) with the Addition of N and P*

Based on the unitary regression analysis, the litter N of the fern functional group was positively correlated with soil P ($p = 0.051$, Figure A1f) compared with the CK. In addition, for the fern functional group, the litter P was positively correlated with soil C ($p = 0.002$, Figure A1g) and negatively correlated with soil P ($p = 0.019$, Figure A1i). With the addition of N, the litter C of Gramineae was positively correlated with soil P ($p = 0.029$, Figure A2c); the litter N was positively correlated with soil N ($p = 0.015$, Figure A2e). The litter P and soil P were positively correlated in the Gramineae group but negatively correlated in the fern group (Figure A2i). With the addition of P, there was a negative correlation between the C and soil P of the litter of the fern group ($p = 0.051$, Figure A3c); the litter N and soil N were negatively correlated ($p < 0.001$, Figure A3e) in ferns but positively correlated ($p = 0.008$, Figure A3f) in Gramineae.

Litter P is positively correlated with soil P in both the Gramineae and fern groups (Figure A3i). With the application of both N and P, the change of the litter C concentration in the fern group was affected by the change of soil C, soil N, and soil P concentration. The litter C concentration increased with the increase of soil C but decreased with the increase of soil N and soil P concentration (Figure 1a–c). The litter C concentration was negatively correlated with soil N ($p = 0.059$, Figure 1b) in the fern group, whereas soil N was positively correlated with litter C ($p = 0.003$, Figure 1b) in the Gramineae group. The litter N concentration of the fern group decreased with the increase of the soil N concentration ($p < 0.001$, Figure 1e).

There was no correlation between litter N, soil C, and soil P of the two different plant functional groups (Figure 1d, f). Moreover, the litter P concentration was positively correlated with both soil N and soil P (Figure 1h, i) in the fern functional group.

3.4. *Effects of N and P Addition on Carbon, Nitrogen, and Phosphorus (C, N, and P) Concentrations and Stoichiometric Ratios of the Soils Associated with Different Plant Functional Groups*

A nested ANOVA showed that soil C was significantly affected by the understory species. Soil N and soil P were significantly affected by N addition, plant functional group, understory species, and randomized complete block design. The soil N also depended on P addition*plant functional group (Table 3). The soil C/N ratio and C:P ratio depended on the plant functional group, understory species, understory species*P addition, and randomized complete block design (Table 3; Figure 2). The soil C/P ratio depended on P addition*plant functional group. The soil N:P ratio depended on N addition, understory species, and randomized complete block design (Figure 2). Similarly, results of SN, SP, S C/N, SC/P, and SN/P varied with blocks (Table 3).

Table 3. ANOVAs on the effects of N, P, and Species (Sp, nested within f) and their interactions with soil carbon, nitrogen, and phosphorus (C, N, P), soil C/N, soil C/P, and soil N/P. Terms are the same as in Table 2. B&R is randomized complete block design. df is the degrees of freedom. *: $p < 0.05$; **: $p < 0.01$; ***: $p < 0.001$.

Factor	df	SC (g kg^{-1})		SN (g kg^{-1})		SP (g kg^{-1})		S C/N		S C/P		S N/P	
		MS	F	MS	F	MS	F	MS	F	MS	F	MS	F
N	2	182	1.41	0.90	5.59 **	0.05	3.49 *	914	0.24	5964	2.92	18.93	7.39 ***
P	1	109	0.84	0.12	0.73	0.05	3.29	1178	0.31	1865	0.91	3.38	1.32
N*P	2	16.63	0.13	0.33	2.03	0.01	0.64	1479	0.39	13.27	0.01	5.32	2.08
f	1	186	1.44	3.75	23.39 ***	0.30	21.79 ***	31,687	8.31 **	41,992	20.54 ***	5.85	2.28
N*f	2	278	2.15	0.05	0.31	0.02	1.78	821	0.22	4381	2.14	2.97	1.16
P*f	1	38.80	0.30	0.62	3.86 *	0.05	3.58	7680	2.01	11,111	5.43 *	0.55	0.22
N*P*f	2	356	2.76	0.02	0.12	0.01	0.92	1074	0.28	2776	1.36	2.39	0.93
Sp[f]	2	1327	10.28 ***	4.77	29.77 ***	0.13	9.57 ***	90,281	23.68 ***	68,901	33.70 ***	20.70	8.08 ***
Sp*N[f]	4	176	1.36	0.12	0.78	0.01	1.07	5364	1.41	3802	1.86	2.05	0.80
Sp*P[f]	2	22.99	0.18	0.12	0.76	0.04	2.91	15,940	4.18 *	8820	4.31 *	5.02	1.96
B&R	3	325	2.52	1.41	8.83 ***	0.17	12.44 ***	26,816	7.04 ***	6256	3.06 *	33.72	13.16 ***

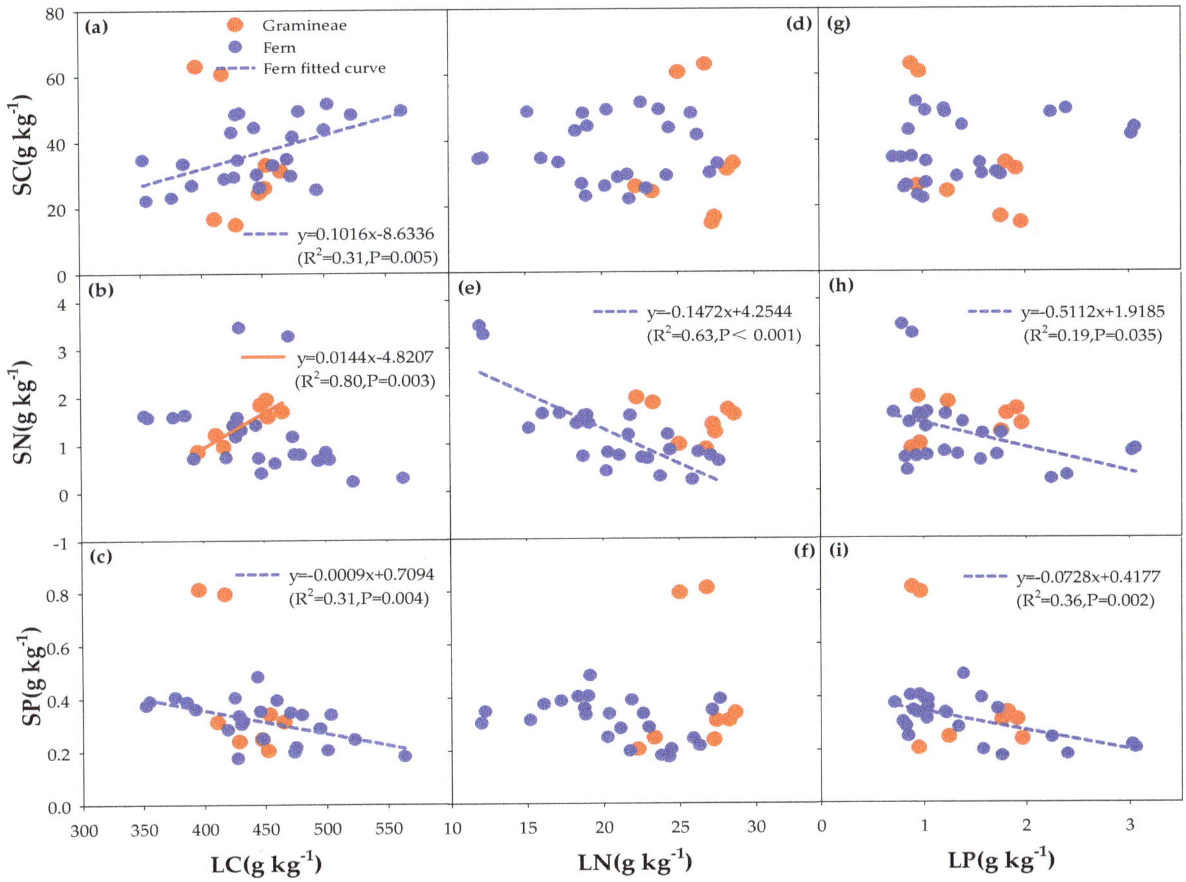

Figure 1. Correlations between litter and soil carbon, nitrogen, and phosphorus (C, N, and P) of four dominate species with two different functional types (Gramineae *Lophatherum gracile* and fern *Woodwardia japonica*, *Dryopteris atrata* and *Dicranopteris dichotoma*) with N (100 kg ha^{-1} year^{-1}) and P (50 kg ha^{-1} year^{-1}) addition.

4. Discussion

Ecological stoichiometry plays an important role in the study of plants and soils within an entire ecosystem. Especially under global climate change, it is valuable to study the understory plant community and the nutrient cycle [45,46]. In the Chinese fir plantation, this study identified the differences in N and P nutrients availability in the soil associated with four understory species. Soil pH was lower than that of other study [43], and the soil of understory plants has a very low pH (3.31–3.37, soil/water, 2:5), indicating that these four species were acid-tolerant. In addition, there were significant differences in the concentrations of C, N, and P in litter and soil among the four species, indicating that species may be the factors contributing to this difference [2,47]. After the continuous addition of N and P, the C, N, and P concentration and stoichiometric ratio of the litter of different plant functional groups showed different degrees of variation compared to the CK. This change may be caused by changes in the C, N, and P cycles in the understory ecosystem after the addition of N and P [12,48,49]. With the addition of N and P, the C, N, and P concentrations of different plant functional groups and soil showed different degrees of correlation, and the correlation would change with different rates of N and P addition. However, most results were significantly affected by blocks, indicating potential variations in both litter and soil C and nutrients with spatial distribution of plots, which should be further considered in future studies.

4.1. *Effects of Different Understory Vegetation Types on Stoichiometric Ratio*

Understory vegetation plays an important role in maintaining biodiversity in forest ecosystems and soil nutrient cycling [50,51]. Most ecological indicators are correlated with different understory vegetation types [52]. Understory vegetation often participates in soil C and N nutrient cycling processes via multiple methods [53]. There are different correlations between water-soluble organic C and microbial biomass C of different understory vegetation types [54]. In this study, litter C, N, and C:N and soil N, P, C:N, and C:P in species of different functional groups are significantly correlated (Tables 2 and 3). The litter from different species is significantly correlated with soil C, N, and P and the stoichiometric ratio (Tables 2 and 3).

This is consistent with Güsewell's conclusion that plants have different C, N, and P concentrations in leaves due to the species and nutritional conditions [2]. These findings are consistent with Cheng et al.'s conclusions that indicated that the soil N utilization efficiency of Gramineae is higher than ferns [55]. In addition, litter from the Gramineae group has a lower C:N ratio, which is similar to the results of a study on the understory species of a *Pinus massoniana* plantation. Gramineae may be more conducive to improving the nutrient cycling rate in its dominant area [54].

4.2. Effect of Nitrogen (N) and Phosphorus (P) Addition on Stoichiometric Ratio

Nitrogen and P are the main limiting elements of plant growth in terrestrial ecosystems. It is generally believed that a N:P mass ratio lower than 14 indicates that plant growth is limited by N, while a N:P mass ratio higher than 16 indicates that plant growth is limited by P [47]. An unbalanced input of N and P will seriously affect the ecological stoichiometry, ultimately affecting the function of the ecosystem [15,56,57]. We found that N:P ratio of Gramineae litter (>16) was higher than ferns, After the addition of P, the N:P ratio of Gramineae decreased, but it was not significantly different from the CK (Figure 2i). Plants regulate their growth rate by adjusting the C:N:P ratio [58–60].

Because of]. the growth dilution effect, the available plant P concentration would decrease with increasing N [60 However, our study did not find that N addition significantly reduced the P concentration of plant litter (Figure 2c, Table 2), which was in agreement with a study on the effect of long-term application of N and P fertilizer on plant N:P in a Tibetan alpine meadow [23]. It is possible that this effect is species specific or depends on the N application rate [61,62]. In comparison, the addition of N increased the N:P of litter and soil in the fern group (Figure 2i,l). The addition of P reduced the C:P and N:P of litter and soil in the fern group (Figure 2h,i,k,l). The separate addition of N and P had no significant effect on the CNP stoichiometric ratio of Gramineae, which indicates that the fern group was affected by the changes of a single environmental factor.

This suggests that the fern population could rapidly expand if the environment was conducive to their favored growth conditions. Gramineae species have a good adaptability with the single change of N and P [53]. The co-addition of N and P reduced the C:N and C:P of litter (Figure 2g,h) of two of the different functional groups, increased the soil N:P (Figure 2i) of the fern group, and reduced the C:N ratio of understory species, which indicated that the addition of N and P together increased the N concentration in plants.

Coupling between N and P was observed, which plays an important role in the regulation of]. nutrient limitations and the strategies for plants to obtain nutrients in the changing environment [63,64 These results show that the addition of N and P affects the correlation between litter and soil C, N, and P, and has different degrees of influence in different plant functional groups.

With the increasing deposition of N in southern China, the availability of soil P is limited, impacting the balance of C, N, and P in the ecosystem. In the context of global climate change, long-term experiments adding N and P are beneficial to understanding the nutrient cycle of understory vegetation in forest ecosystems and to cope with environmental changes. Future studies on understory vegetation management should focus on the effects of N and P addition and understory vegetation types on soil C and N cycles in plantations.

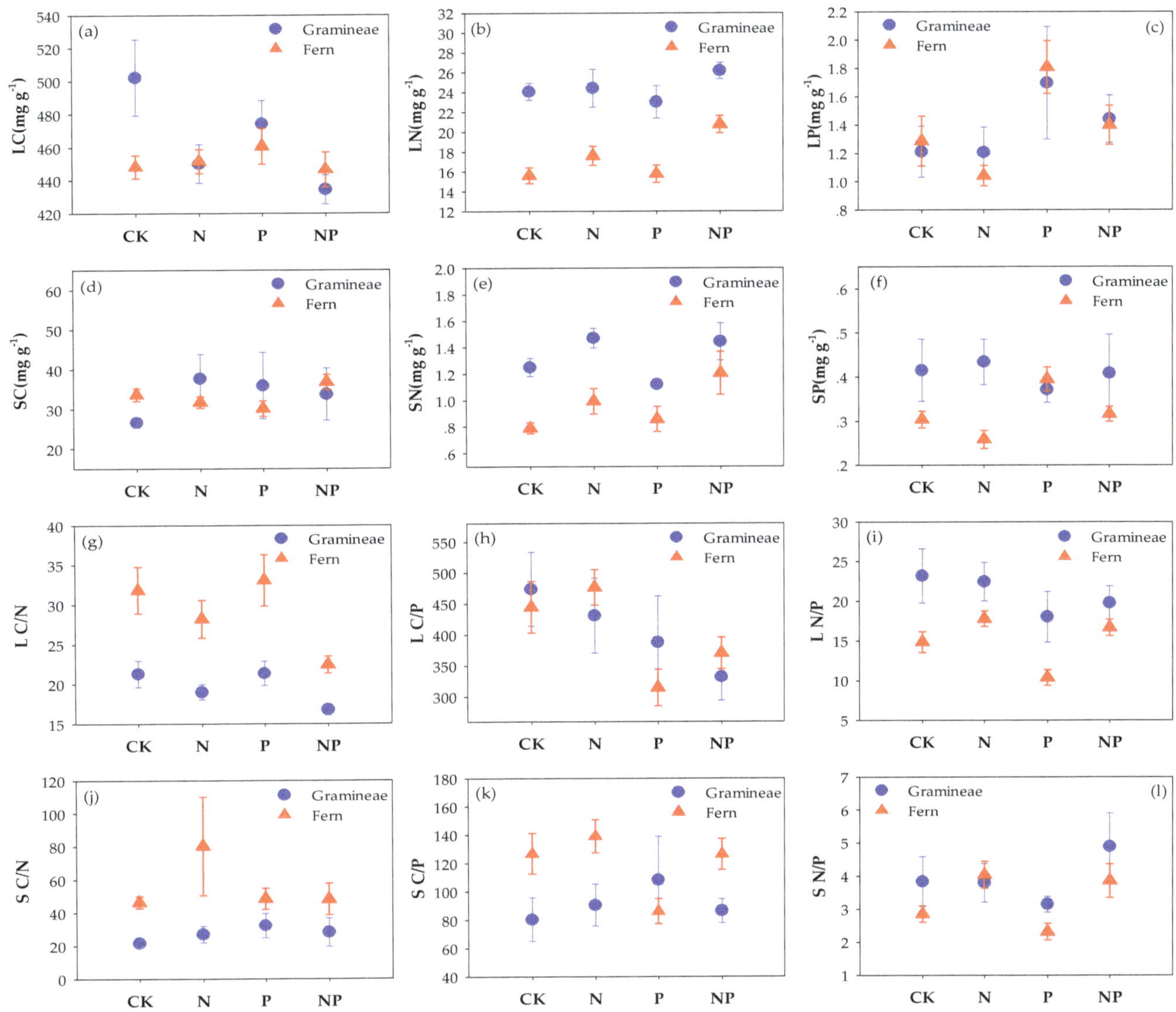

Figure 2. Litter and soil carbon, nitrogen, and phosphorus (C, N, P), C/N, C/P and N/P of four dominate species with two different functional types (Gramineae *Lophatherum gracile* and fern *Woodwardia japonica*, *Dryopteris atrata* and *Dicranopteris dichotoma*) under four nutrient addition modes (CK, N, P and NP). Terms are the same as in Table 2.

5. Conclusions

The C, N, and P concentrations of plants and soil vary greatly among different species, and these differences may be species dependent. The input of exogenous N and P changes the stoichiometric characteristics of understory plants to different degrees; thus, we believe that different species have different response mechanisms to changes of environmental factors. Different functional plant groups show different changes with the addition of N and P. The addition of N and P had an interactive effect on the N:P of litter, indicating that the growth of understory plants was dually restricted by N and P. Over the long-term, the addition of N and P may change the competitiveness of different functional plant groups in the forest, leading to a change in the community of understory plant species. Therefore, the response of the stoichiometric characteristics of understory plants to different N and P inputs should be studied in future research on understory vegetation.

Author Contributions: Conceptualization, L.Z. and J.X.; methodology, L.Z. and J.X.; software, J.X. and M.C.; investigation, J.X., H.F., Q.Z. and Y.G.; resources, L.Z.; data curation, J.P. and J.X.; writing—original draft preparation, J.X.; writing—review and editing, L.Z., X.X., X.F. and J.X.; supervision, X.G. and L.Z.; funding acquisition, X.F. and L.Z.

Acknowledgments: The authors thank the following people for their help in this research. Li-Li Ma, Shu-Li Wang, Xi Yuan provided field assistance. Jia-Shun Peng assisted with lab analyses. Jiao Wu assisted with syntax modification.

Abbreviations

C	carbon
N	nitrogen
P	phosphorus
DOC	dissolved organic carbon
TOC	total organic carbon
TN	total nitrogen
TP	total phosphorus
AN	available nitrogen
AP	available phosphorus
LC	total litter organic carbon
LN	total litter nitrogen
LP	total litter phosphorus
SC	total soil organic carbon
SN	total soil nitrogen
SP	total soil phosphorus
Sp	understory species
F	plant functional group
B&R	block as random effect
L C/N	litter carbon to nitrogen ratio
L C/P	litter carbon to phosphorus ratio
L N/P	litter nitrogen to phosphorus ratio
S C/N	soil carbon to nitrogen ratio
S C/P	soil carbon to phosphorus ratio
S N/P	soil nitrogen to phosphorus ratio

Appendix A

Table A1. The values of organic C (C), total N (N) and total P (P) in litter and soil of four plants under the addition of N and P (means ± SE).

Species			No P Addition		P Addition	
			CK	N	CK	N
Lophatherum gracile (Gramineae)	Litter	C	502.10 ± 22.99	449.88 ± 11.72	474.09 ± 13.99	434.28 ± 8.70
		N	24.04 ± 0.85	24.38 ± 1.89	23.00 ± 1.63	26.14 ± 0.82
		P	1.21 ± 0.18	1.21 ± 0.18	1.69 ± 0.40	1.44 ± 0.17
	soil	C	26.61 ± 1.37	37.68 ± 6.08	35.88 ± 8.32	33.71 ± 6.54
		N	1.25 ± 0.07	1.47 ± 0.07	1.12 ± 0.03	1.44 ± 0.14
		P	0.41 ± 0.07	0.43 ± 0.05	0.37 ± 0.03	0.41 ± 0.09
Woodwardia japonica (Fern)	Litter	C	420.27 ± 12.05	421.31 ± 9.00	442.22 ± 19.48	393.95 ± 11.78
		N	15.36 ± 1.39	17.56 ± 1.23	16.43 ± 0.93	18.95 ± 0.67
		P	1.23 ± 0.08	1.13 ± 0.18	1.33 ± 0.13	1.05 ± 0.08
	soil	C	29.38 ± 2.08	29.88 ± 2.15	25.22 ± 1.75	32.07 ± 2.97
		N	0.96 ± 0.07	1.18 ± 0.10	1.00 ± 0.13	1.34 ± 0.14
		P	0.38 ± 0.03	0.29 ± 0.05	0.40 ± 0.03	0.39 ± 0.02

Table A1. *Cont.*

Species			No P Addition CK	No P Addition N	P Addition CK	P Addition N
Dryopteris atrata (Fern)	Litter	C	443.67 ± 6.65	457.00 ± 6.40	454.05 ± 15.97	483.93 ± 15.52
		N	16.62 ± 0.68	19.82 ± 1.65	18.77 ± 1.14	25.17 ± 0.69
		P	1.88 ± 0.45	1.16 ± 0.09	2.57 ± 0.42	2.18 ± 0.22
	soil	C	40.89 ± 1.94	35.70 ± 3.45	31.08 ± 1.68	38.16 ± 3.02
		N	0.64 ± 0.05	0.60 ± 0.17	0.40 ± 0.03	0.74 ± 0.13
		P	0.24 ± 0.03	0.19 ± 0.04	0.40 ± 0.08	0.25 ± 0.03
Dicranopteris dichotoma (Fern)	Litter	C	480.50 ± 6.65	476.03 ± 13.68	485.45 ± 19.59	461.29 ± 10.67
		N	14.74 ± 1.94	15.37 ± 1.90	12.01 ± 1.54	18.08 ± 1.55
		P	0.75 ± 0.04	0.84 ± 0.04	1.53 ± 0.17	0.97 ± 0.06
	soil	C	30.64 ± 2.27	29.51 ± 0.70	34.01 ± 4.97	39.89 ± 3.83
		N	0.77 ± 0.07	1.19 ± 0.144	1.18 ± 0.15	1.53 ± 0.42
		P	0.30 ± 0.02	0.29 ± 0.02	0.38 ± 0.03	0.32 ± 0.01

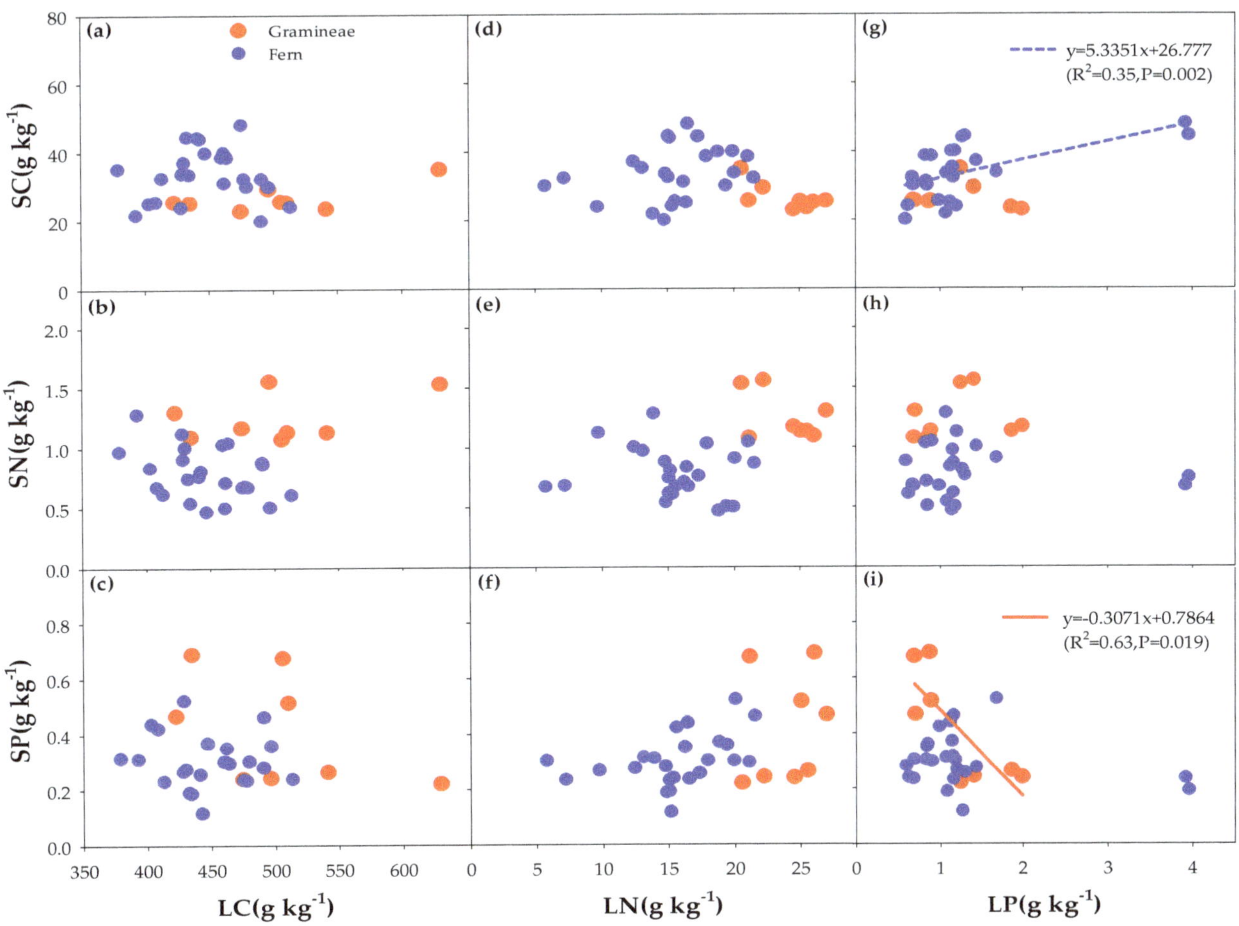

Figure A1. Correlations between litter and soil C, N and P of four dominate species with two different functional types (Gramineae *Lophatherum gracile* and fern *Woodwardia japonica*, *Dryopteris atrata* and *Dicranopteris dichotoma*) in CK treatment.

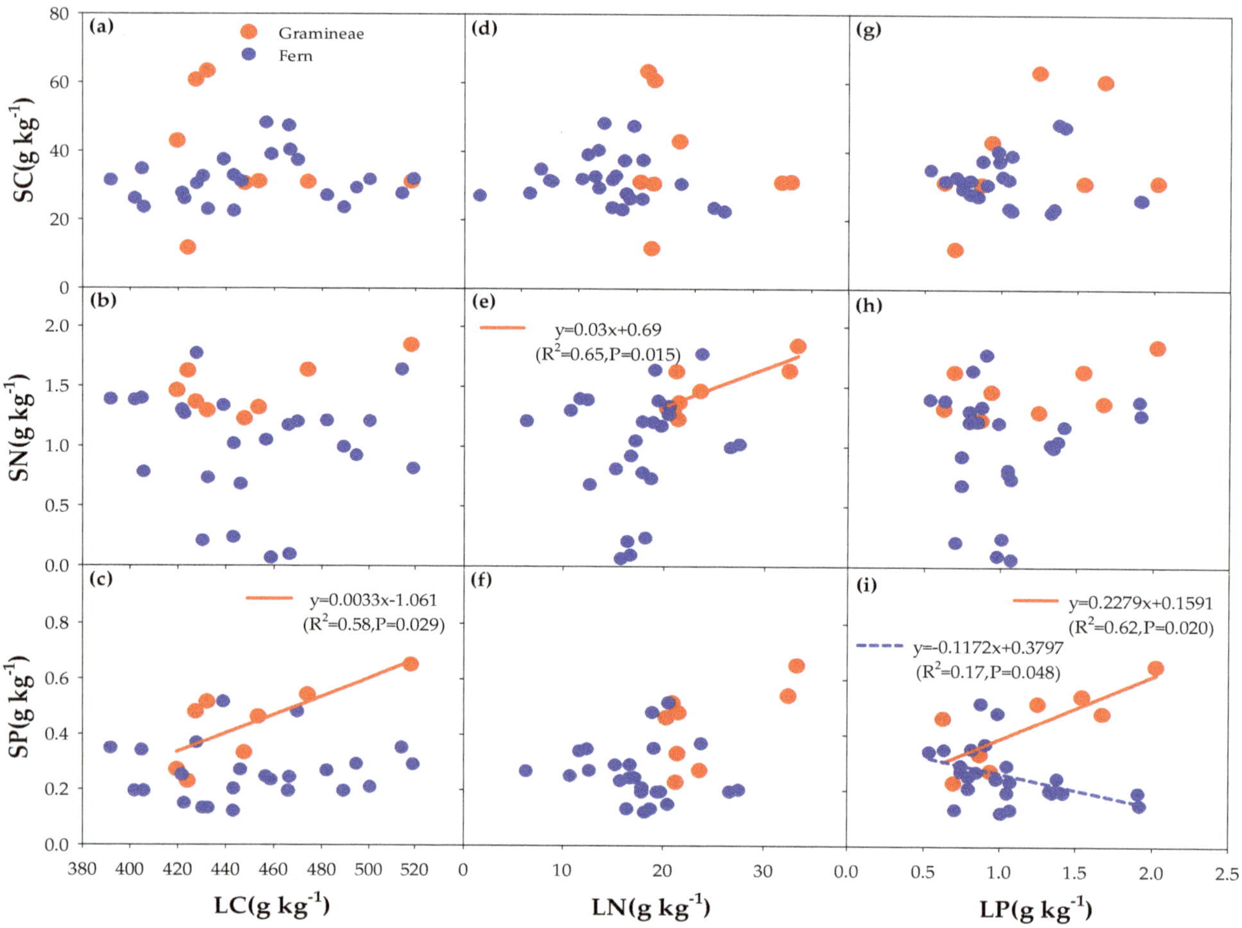

Figure A2. Correlations between litter and soil C, N and P of four dominate species with two different functional types (Gramineae *Lophatherum gracile* and fern *Woodwardia japonica*, *Dryopteris atrata* and *Dicranopteris dichotoma*) in N addition treatment.

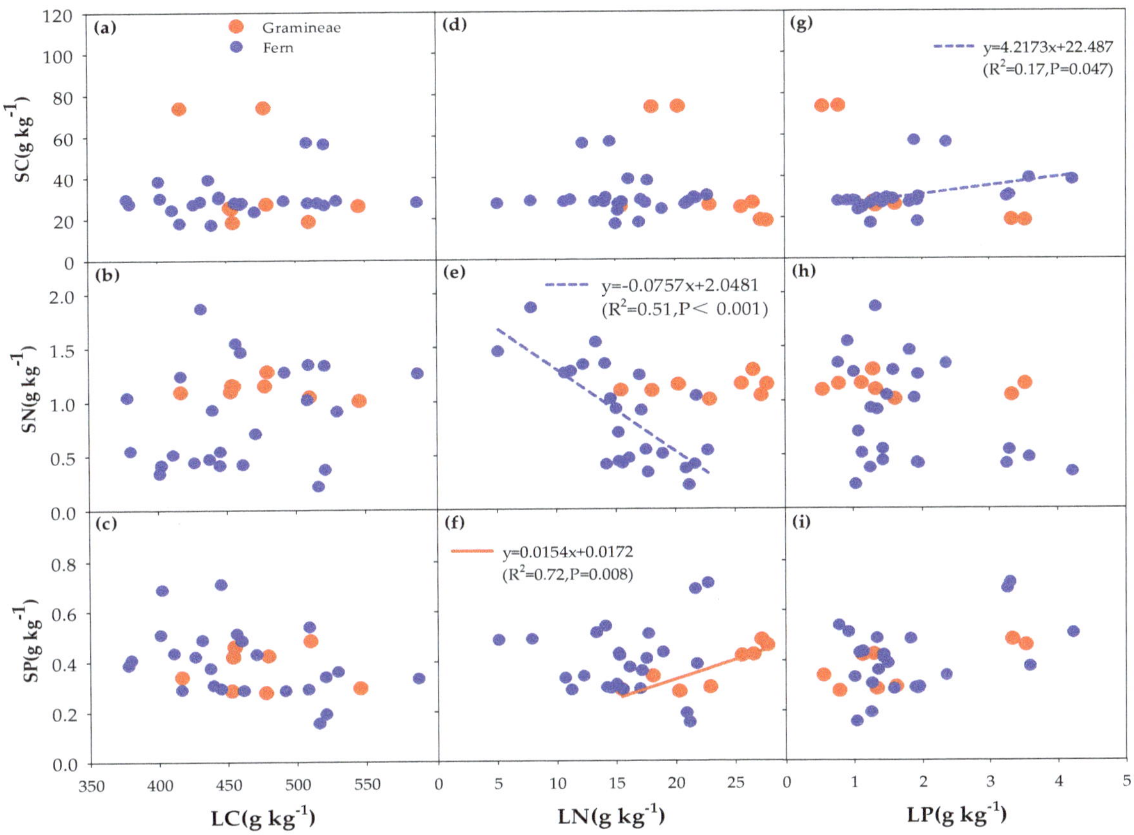

Figure A3. Correlations between litter and soil C, N and P of four dominate species with two different functional types (Gramineae *Lophatherum gracile* and fern *Woodwardia japonica*, *Dryopteris atrata* and *Dicranopteris dichotoma*) in P addition treatment.

References

1. Elser, J.; Sterner, R.; Fagan, W.; Markow, T.; Cotner, J.; Hobbie, S.; Odell, G.; Weider, L.; Gorokhova, E.; Harrison, J. Biological stoichiometry from genes to ecosystems. *Ecol. Lett.* **2000**, *3*, 540–550. [CrossRef]
2. Güsewell, S. N:P ratios in terrestrial plants: Variation and functional significance. *New Phytol.* **2004**, *164*, 243–266. [CrossRef]
3. Han, W.; Fang, J.; Guo, D.; Zhang, Y. Leaf nitrogen and phosphorus stoichiometry across 753 terrestrial plant species in China. *New Phytol.* **2005**, *168*, 377–385. [CrossRef] [PubMed]
4. Reich, P.B.; Oleksyn, J. Global patterns of plant leaf N and P in relation to temperature and latitude. *Proc. Natl. Acad. Sci. USA* **2004**, *101*, 11001–11006. [CrossRef] [PubMed]
5. Ladanai, S.; Ågren, G.I.; Olsson, B.A. Relationships between tree and soil properties in *picea abies* and *pinus sylvestris* forests in sweden. *Ecosystems* **2010**, *13*, 302–316. [CrossRef]
6. Bui, E.N.; Henderson, B.L. C:N:P stoichiometry in Australian soils with respect to vegetation and environmental factors. *Plant Soil* **2013**, *373*, 553–568. [CrossRef]
7. Tian, H.; Chen, G.; Zhang, C.; Melillo, J.M.; Hall, C.A.S. Pattern and variation of C: N:P ratios in China's soils: A synthesis of observational data. *Biogeochemistry* **2010**, *98*, 139–151. [CrossRef]
8. Siddique, I.; Vieira, I.C.G.; Schmidt, S.; Lamb, D.; Carvalho, C.J.R.; Figueiredo, R.D.O.; Blomberg, S.; Davidson, E.A. Nitrogen and phosphorus additions negatively affect tree species diversity in tropical forest regrowth trajectories. *Ecology* **2010**, *91*, 2121–2131. [CrossRef]
9. Wei, X.; Blanco, J.A.; Jiang, H.; Kimmins, J.H. Effects of nitrogen deposition on carbon sequestration in Chinese fir forest ecosystems. *Sci. Total Environ.* **2012**, *416*, 351–361. [CrossRef]
10. Litton, C.M.; Raich, J.W.; Ryan, M.G. Carbon allocation in forest ecosystems. *Glob. Chang. Biol.* **2007**, *13*, 2089–2109. [CrossRef]
11. Clark, C.M.; Tilman, D. Loss of plant species after chronic low-level nitrogen deposition to prairie grasslands. *Nature* **2008**, *451*, 712–715. [CrossRef] [PubMed]

12. Peñuelas, J.; Sardans, J.; Rivas-Ubach, A.; Janssens, I.A. The human-induced imbalance between C, N and P in Earth's life system. *Glob. Chang. Biol.* **2012**, *18*, 3–6. [CrossRef]
13. Strengbom, J.; Walheim, M.; Näsholm, T.; Ericson, L. Regional differences in the occurrence of understorey species reflect nitrogen deposition in Swedish forests. *Ambio* **2003**, *32*, 91–97. [CrossRef] [PubMed]
14. Menge, D.N.L.; Field, C.B. Simulated global changes alter phosphorus demand in annual grassland. *Glob. Chang. Biol.* **2007**, *13*, 2582–2591. [CrossRef]
15. Sardans, J.; Rivas-Ubach, A.; Peñuelas, J. The elemental stoichiometry of aquatic and terrestrial ecosystems and its relationships with organismic lifestyle and ecosystem structure and function: A review and perspectives. *Biogeochemistry* **2012**, *111*, 1–39. [CrossRef]
16. Mayor, J.R.; Wright, S.J.; Turner, B.L. Species-specific responses of foliar nutrients to long-term N and P additions in a lowland tropical forest. *J. Ecol.* **2014**, *102*, 36–44. [CrossRef]
17. Huang, W.; Zhou, G.; Liu, J.; Zhang, D.; Xu, Z.; Liu, S. Effects of elevated carbon dioxide and nitrogen addition on foliar stoichiometry of nitrogen and phosphorus of five tree species in subtropical model forest ecosystems. *Environ. Pollut.* **2012**, *168*, 113–120. [CrossRef] [PubMed]
18. Koller, E.K.; Press, M.C.; Callaghan, T.V.; Phoenix, G.K. Tight coupling between shoot level foliar N and P, leaf area, and shoot growth in Arctic dwarf shrubs under simulated climate change. *Ecosystems* **2016**, *19*, 326–338. [CrossRef]
19. Zhang, Q.; Xie, J.; Lyu, M.; Xiong, D.; Wang, J.; Chen, Y.; Li, Y.; Wang, M.; Yang, Y. Short-term effects of soil warming and nitrogen addition on the N:P stoichiometry of *Cunninghamia lanceolata* in subtropical regions. *Plant Soil* **2016**, *411*, 395–407. [CrossRef]
20. Braun, S.; Thomas, V.F.; Quiring, R.; Flückiger, W. Does nitrogen deposition increase forest production? The role of phosphorus. *Environ. Pollut.* **2010**, *158*, 2043–2052. [CrossRef]
21. Chen, F.S.; Niklas, K.J.; Zeng, D.H. Important foliar traits depend on species-grouping: Analysis of a remnant temperate forest at the Keerqin sandy lands, China. *Plant Soil* **2011**, *340*, 337–345. [CrossRef]
22. Heuck, C.; Smolka, G.; Whalen, E.D.; Frey, S.; Gundersen, P.; Moldan, F.; Fernandez, I.J.; Spohn, M. Effects of long-term nitrogen addition on phosphorus cycling in organic soil horizons of temperate forests. *Biogeochemistry* **2018**, *141*, 167–181. [CrossRef]
23. Zhang, J.; Yan, X.; Su, F.; Li, Z.; Wang, Y.; Wei, Y.; Ji, Y.; Yang, Y.; Zhou, X.; Guo, H.; et al. Long-term N and P additions alter the scaling of plant nitrogen to phosphorus in a Tibetan alpine meadow. *Sci. Total Environ.* **2018**, *625*, 440–448. [CrossRef] [PubMed]
24. Craft, C.B.; Vymazal, J.; Richardson, C.J. Response of everglades plant communities to nitrogen and phosphorus additions. *Wetlands* **1995**, *15*, 258–271. [CrossRef]
25. Yang, Y.; Liu, B.R.; An, S.S. Ecological stoichiometry in leaves, roots, litters and soil among different plant communities in a desertified region of Northern China. *Catena* **2018**, *166*, 328–338. [CrossRef]
26. Wang, W.Q.; Wang, C.; Sardans, J.; Zeng, C.S.; Tong, C.; Peñuelas, J. Plant invasive success associated with higher N-use efficiency and stoichiometric shifts in the soil–plant system in the Minjiang River tidal estuarine wetlands of China. *Wetl. Ecol. Manag.* **2015**, *23*, 865–880. [CrossRef]
27. Wang, F.; Chen, F.; Wang, G.G.; Mao, R.; Fang, X.; Wang, H.; Bu, W. Effects of experimental nitrogen addition on nutrients and nonstructural carbohydrates of dominant understory plants in a Chinese fir plantation. *Forests* **2019**, *10*, 155. [CrossRef]
28. Takafumi, H.; Hiura, T. Effects of disturbance history and environmental factors on the diversity and productivity of understory vegetation in a cool-temperate forest in Japan. *For. Ecol. Manag.* **2009**, *257*, 843–857. [CrossRef]
29. Dobson, A.M.; Blossey, B.; Richardson, J.B. Invasive earthworms change nutrient availability and uptake by forest understory plants. *Plant Soil* **2017**, *421*, 175–190. [CrossRef]
30. Wang, F.; Zou, B.; Li, H.; Li, Z. The effect of understory removal on microclimate and soil properties in two subtropical lumber plantations. *J. For. Res.* **2014**, *19*, 238–243. [CrossRef]
31. Jules, M.J.; Sawyer, J.O.; Jules, E.S. Assessing the relationships between stand development and understory vegetation using a 420-year chronosequence. *For. Ecol. Manag.* **2008**, *255*, 2384–2393. [CrossRef]
32. Lu, X.; Mo, J.; Gilliam, F.S.; Zhou, G.; Fang, Y. Effects of experimental nitrogen additions on plant diversity in an old-growth tropical forest. *Glob. Chang. Biol.* **2010**, *16*, 2688–2700. [CrossRef]
33. Hurd, T.M.; Brach, A.R.; Raynal, D.J. Response of understory vegetation of Adirondack forests to nitrogen additions. *Can. J. For. Res.* **1998**, *28*, 799–807. [CrossRef]

34. Rainey, S.M.; Nadelhoffer, K.J.; Silver, W.L.; Downs, M.R. Effects of chronic nitrogen additions on understory species in a red pine plantation. *Ecol. Appl.* **1999**, *9*, 949–957. [CrossRef]
35. Bi, J.; Blanco, J.A.; Seely, B.; Kimmins, J.P.; Ding, Y.; Welham, C. Yield decline in Chinese-fir plantations: A simulation investigation with implications for model complexity. *Can. J. For. Res.* **2007**, *37*, 1615–1630. [CrossRef]
36. Lu, Z.H.; Wu, G.; Ma, X.; Bai, G.X. Current situation of Chinese forestry tactics and strategy of sustainable development. *J. For. Res.* **2002**, *13*, 319–322.
37. Chen, G.S.; Yang, Z.J.; Gao, R.; Xie, J.S.; Guo, J.F.; Huang, Z.Q.; Yang, Y.S. Carbon storage in a chronosequence of Chinese fir plantations in southern China. *For. Ecol. Manag.* **2013**, *300*, 68–76. [CrossRef]
38. Tang, X.; Pérez-Cruzado, C.; Fehrmann, L.; Álvarez-González, J.G.; Lu, Y.; Kleinn, C. Development of a compatible taper function and stand-level merchantable volume model for Chinese fir plantations. *PLoS ONE* **2016**, *11*, e0147610. [CrossRef]
39. Peng, Y.; Tian, D.; Thomas, S.C. Forest management and soil respiration: Implications for carbon sequestration. *Environ. Rev.* **2008**, *16*, 93–111. [CrossRef]
40. Powers, R.F.; Busse, M.D.; McFarlane, K.J.; Zhang, J.W.; Young, D.H. Long-term effects of silviculture on soil carbon storage: Does vegetation control make a difference? *Forestry* **2013**, *86*, 47–58. [CrossRef]
41. Chen, F.S.; Niklas, K.J.; Liu, Y.; Fang, X.M.; Wan, S.Z.; Wang, H. Nitrogen and phosphorus additions alter nutrient dynamics but not resorption efficiencies of Chinese fir leaves and twigs differing in age. *Tree Physiol.* **2015**, *35*, 1106–1117. [CrossRef] [PubMed]
42. Lu, C.; Tian, H. Spatial and temporal patterns of nitrogen deposition in China: Synthesis of observational data. *J. Geophys. Res. Space Phys.* **2007**, *112*, 10–15. [CrossRef]
43. Tang, Y.; Zhang, X.; Li, D.; Wang, H.; Chen, F.; Fu, X.; Fang, X.; Sun, X.; Yu, G. Impacts of nitrogen and phosphorus additions on the abundance and community structure of ammonia oxidizers and denitrifying bacteria in Chinese fir plantations. *Soil Biol. Biochem.* **2016**, *103*, 284–293. [CrossRef]
44. Bao, S.D. *Soil and Agricultural Chemistry Analysis*, 3rd ed.; Agriculture Press: Beijing, China, 2008. (In Chinese)
45. Zechmeister-Boltenstern, S.; Keiblinger, K.M.; Mooshammer, M.; Peñuelas, J.; Richter, A.; Sardans, J.; Wanek, W. The application of ecological stoichiometry to plant–microbial–soil organic matter transformations. *Ecol. Monogr.* **2015**, *85*, 133–155. [CrossRef]
46. Metcalfe, D.B.; Fisher, R.A.; Wardle, D.A. Plant communities as drivers of soil respiration: Pathways, mechanisms, and significance for global change. *Biogeosciences* **2011**, *8*, 2047–2061. [CrossRef]
47. Meuleman, A.F.M.; Koerselman, W. The Vegetation N:P Ratio: A new tool to detect the nature of nutrient limitation. *J. Appl. Ecol.* **1996**, *33*, 1441–1450.
48. Marklein, A.R.; Houlton, B.Z. Nitrogen inputs accelerate phosphorus cycling rates across a wide variety of terrestrial ecosystems. *New Phytol.* **2011**, *193*, 696–704. [CrossRef]
49. Lu, M.; Yang, Y.; Luo, Y.; Fang, C.; Zhou, X.; Chen, J.; Yang, X.; Li, B. Responses of ecosystem nitrogen cycle to nitrogen addition: A meta-analysis. *New Phytol.* **2011**, *189*, 1040–1050. [CrossRef]
50. Wu, J.; Liu, Z.; Chen, D.; Huang, G.; Zhou, L.; Fu, S. Understory plants can make substantial contributions to soil respiration: Evidence from two subtropical plantations. *Soil Biol. Biochem.* **2011**, *43*, 2355–2357. [CrossRef]
51. Wu, J.; Liu, Z.; Wang, X.; Sun, Y.X.; Zhou, L.; Lin, Y.; Fu, S. Effects of understory removal and tree girdling on soil microbial community composition and litter decomposition in two Eucalyptus plantations in South China. *Funct. Ecol.* **2011**, *25*, 921–931. [CrossRef]
52. Yimer, F.; Ledin, S.; Abdelkadir, A. Soil organic carbon and total nitrogen stocks as affected by topographic aspect and vegetation in the Bale Mountains, Ethiopia. *Geoderma* **2006**, *135*, 335–344. [CrossRef]
53. Zhao, J.; Wan, S.; Li, Z.; Shao, Y.; Xu, G.; Liu, Z.; Zhou, L.; Fu, S. Dicranopteris-dominated understory as major driver of intensive forest ecosystem in humid subtropical and tropical region. *Soil Biol. Biochem.* **2012**, *49*, 78–87. [CrossRef]
54. Pan, P.; Zhao, F.; Ning, J.; Zhang, L.; Ouyang, X.; Zang, H. Impact of understory vegetation on soil carbon and nitrogen dynamic in aerially seeded *Pinus massoniana* plantations. *PLoS ONE* **2018**, *13*, e0191952. [CrossRef] [PubMed]
55. Chen, J.; Stark, J.M. Plant species effects and carbon and nitrogen cycling in a sagebrush–crested wheatgrass soil. *Soil Biol. Biochem.* **2000**, *32*, 47–57. [CrossRef]

56. Finzi, A.C.; Austin, A.T.; Cleland, E.E.; Frey, S.D.; Houlton, B.Z.; Wallenstein, M.D. Responses and feedbacks of coupled biogeochemical cycles to climate change: Examples from terrestrial ecosystems. *Front. Ecol. Environ.* **2011**, *9*, 61–67. [CrossRef]
57. Peñuelas, J.; Poulter, B.; Sardans, J.; Ciais, P.; Van Der Velde, M.; Bopp, L.; Boucher, O.; Godderis, Y.; Hinsinger, P.; Llusià, J.; et al. Human-induced nitrogen–phosphorus imbalances alter natural and managed ecosystems across the globe. *Nat. Commun.* **2013**, *4*, 2934. [CrossRef] [PubMed]
58. Klausmeier, C.A.; Litchman, E.; Daufresne, T.; Levin, S.A. Optimal nitrogen-to-phosphorus stoichiometry of phytoplankton. *Nature* **2004**, *429*, 171–174. [CrossRef] [PubMed]
59. Moe, S.J.; Stelzer, R.S.; Forman, M.R.; Harpole, W.S.; Daufresne, T.; Yoshida, T. Recent advances in ecological stoichiometry: Insights for population and community ecology. *Oikos* **2005**, *109*, 29–39. [CrossRef]
60. Van Heerwaarden, L.M.; Toet, S.; Aerts, R. Nitrogen and phosphorus resorption efficiency and proficiency in six sub-arctic bog species after 4 years of nitrogen fertilization. *J. Ecol.* **2003**, *91*, 1060–1070. [CrossRef]
61. Gusewell, S. Variation in nitrogen and phosphorus concentrations of wetland plants. *Perspect. Plant Ecol. Evol. Syst.* **2002**, *5*, 37–61. [CrossRef]
62. Bubier, J.L.; Smith, R.; Juutinen, S.; Moore, T.R.; Minocha, R.; Long, S.; Minocha, S. Effects of nutrient addition on leaf chemistry, morphology, and photosynthetic capacity of three bog shrubs. *Oecologia* **2011**, *167*, 355–368. [CrossRef] [PubMed]
63. Niklas, K.J.; Cobb, E.D. N, P, and C stoichiometry of *Eranthis hyemalis* (Ranunculaceae) and the allometry of plant growth. *Am. J. Bot.* **2005**, *92*, 1256–1263. [CrossRef] [PubMed]
64. Elser, J.J.; Bracken, M.E.; Cleland, E.E.; Gruner, D.S.; Harpole, W.S.; Hillebrand, H.; Ngai, J.T.; Seabloom, E.W.; Shurin, J.B.; Smith, J.E. Global analysis of nitrogen and phosphorus limitation of primary producers in freshwater, marine and terrestrial ecosystems. *Ecol. Lett.* **2007**, *10*, 1135–1142. [CrossRef] [PubMed]

4

Estimation of Nutrient Exports Resulting from Thinning and Intensive Biomass Extraction in Medium-Aged Spruce and Pine Stands in Saxony, Northeast Germany

Christine Knust *, Karoline Schua and Karl-Heinz Feger

TU Dresden, Institute of Soil Science and Site Ecology, 01737 Tharandt, Germany; karoline.schua@tu-dresden.de (K.S.); karl-heinz.feger@tu-dresden.de (K.-H.F.)

* Correspondence: christine.knust@tu-dresden.de

Academic Editors: Scott X. Chang and Xiangyang Sun

Abstract: A growing interest in using forest biomass for bioenergy generation may stimulate intensive harvesting scenarios in Germany. We calculated and compared nutrient exports of conventional stem only (SO), whole tree without needles (WT excl. needles), and whole tree (WT) harvesting in two medium aged Norway spruce (*Picea abies* L. KARST.) and Scots pine (*Pinus sylvestris* L.) stands differing in productivity, and related them to soil nutrient pools and fluxes at the study sites. We established allometric biomass functions for each aboveground tree compartment and analyzed their nutrient contents. We analyzed soil nutrient stocks, estimated weathering rates, and obtained deposition and seepage data from nearby Level II stations. WT (excl. needles) and WT treatments cause nutrient losses 1.5 to 3.6 times higher than SO, while the biomass gain is only 1.18 to 1.25 in case of WT (excl. needles) and 1.28 to 1.30 in case of WT in the pine and spruce stand, respectively. Within the investigated 25-year period, WT harvesting would cause exports of N, K^+, Ca^{2+}, and Mg^{2+} of 6.6, 8.8, 5.4, and 0.8 kg·ha^{-1} in the pine stand and 13.9, 7.0, 10.6, and 1.8 kg·ha^{-1} in the spruce stand annually. The relative impact of WT and WT (excl. needles) on the nutrient balance is similar in the pine and spruce stands, despite differences in stand productivities, and thus the absolute amount of nutrients removed. In addition to the impact of intensive harvesting, both sites are characterized by high seepage losses of base cations, further impairing the nutrient budget. While intensive biomass extraction causes detrimental effects on many key soil ecological properties, our calculations may serve to implement measures to improve the nutrient balance in forested ecosystems.

Keywords: spruce; pine; thinning; aboveground biomass; energetic use; stand growth; nutrient contents; nutrient accumulation

---

1. Introduction

In order to mitigate fossil carbon dioxide emissions for energy generation, renewable energies such as bioenergy are being promoted as an alternative to fossil fuels. In Germany in 2013, renewable sources accounted for 12.3% of total energy consumption, of which biomass accounted for 7.6% [1]. Within the biomass sector, forest residues have been identified as a large underused source for potentially increasing the raw material supply. Already, about 11 million·m^3 of wood has been used directly for energy purposes, which accounts for about one fifth of the total annual harvest from forests. The potential for increasing forest biomass for energy (at a sustainable rate) is estimated to be between 12 and 19 million·m^3·year^{-1}, which could be achieved by increasing the utilization of forest residues and currently underused hardwood stands [2].

The process of extracting harvesting residues from forests, or whole tree harvesting, has a number of technological and ecological constraints. On the technological side, there is a high expenditure for the logistics due to the scattered location of the biomass and its low density, and thus transportability [3,4]. Advanced technologies and management practices could help to improve both feedstock quality and cost efficiency in the future [3,5]. The biomass market situation, and thus the actual price paid by biomass combustion facilities per unit of feedstock, will eventually determine economic feasibility of the utilization of forest residues for bioenergy.

On the ecological side, there is a threat of potentially high nutrient exports when extracting the nutrient-rich crown material, and thus a loss of productivity [6–10]. Also, if large parts of the harvesting residues are exported from the site, there will be fewer habitats for decomposer fauna and flora and less input material to refill the site-specific humus stock [11,12]. The nutrient issue could be attenuated by returning nutrients into the forests from sources such as wood ash from biomass combustion facilities [13–16]. Implementing wood ash recycling systems into practice would require thorough knowledge regarding the actual amounts of nutrients removed in management scenarios of various intensity [14]. Additionally needed is an estimation of the available soil nutrient pool, which plays a major role in assessing the impact of harvest-induced nutrient losses on site productivity, and thus on management sustainability in terms of maintaining nutrient reserves. Finally, site specific nutrient balances are needed to evaluate the impact of different harvesting intensities and to draw conclusions for adapted forest management.

Estimating nutrient fluxes associated with the extraction of aboveground biomass compartments under intensive management scenarios requires knowledge and estimation methods about the distribution of biomass and nutrients of all tree compartments. Yet yield tables and forest growth models typically used in forest management and planning in Germany concentrate on just the volume of the marketable round wood, which is typically the tree trunk up to a certain diameter (e.g., 7 cm over bark), depending on the current market situation. Information about the distribution of the biomass to above-ground tree compartments, such as the stem with a diameter smaller than 7 cm, branches, foliage, and dead branches is needed in order to estimate the profitability and the impact on the nutrient budget of going from classic stem-only (SO) harvesting to more intensive scenarios, such as whole tree without foliage (WT excl. needles) or whole tree (WT) harvesting [17,18]. Some studies on the impact of nutrient removals in intensive scenarios focus on the final felling and harvesting at the end of a stand's rotation time and its impact on the growth of the next stand generation [7,10]. The rationale for this approach is that the biomass and nutrient fluxes in final fellings, such as clear cuts or shelterwood cuttings, are especially large. Alternately, other studies have concentrated on the impact of intensive biomass exports in thinning operations of medium aged stands [10,19–22], which happen at a stage of stand development when the remaining trees exhibit high productivity, and thus require large amounts of nutrients for the buildup of aboveground biomass.

In our study, we aimed at quantifying the nutrient exports of thinning operations with variable intensities of biomass removal in two medium-aged coniferous stands representing typical site and stand conditions in the region of Saxony in northeast (NE) Germany and evaluating the impact of such treatments on the nutrient budget of the stands. We hypothesize that intensive harvesting scenarios (1) impose a strong negative effect on the nutrient balance of forest stands, even in the thinning stage; and (2) will lead to a more negative nutrient balance in the highly productive stands than in less productive stands, due to larger absolute amounts of biomass and thus nutrients removed. Therefore, we developed single-tree biomass equations to predict the dry mass of all aboveground tree compartments based on the tree's diameter at breast height (DBH) and analyzed nutrient contents of these compartments. Based on this knowledge, we estimated the biomass and nutrient extraction by thinning with three different management intensities: conventional SO, WT (excl. needles), and WT harvesting, which refers to aboveground biomass only. In all scenarios, stumps and roots are left in the stand. Also, we calculated the nutrients required to build up aboveground biomass within the examined time span. Finally, we set the amount of extracted and stored nutrients in relation to the site-specific nutrient stocks and fluxes in order to estimate the sustainability of intensive management scenarios.

2. Materials and Methods

2.1. Study Area and Sites

The study area was the Oberlausitz in Saxony, NE Germany. In the northern lowlands, the typically nutrient poor pleistocene soils are often covered with Scots pine (*Pinus sylvestris* L.) stands. Depending on the site potential which varies on a small scale, forests could be dominated by Scots pine, sessile oak (*Quercus petraea* (Mattuschka) Liebl.), pedunculated oak (*Q. robur* L.), silver birch (*Betula pendula* Roth), European beech (*Fagus sylvatica* L.), and various mixtures of these species. Of these species, pines are the most dominant in the study region, covering 69% of the forested area. In the hills in southern Oberlausitz, site conditions are generally more suitable for forestry, as soils are derived from weathered granodiorite bedrock overlain with a loess layer of variable thickness (often 10 to 40 cm). Many forest stands here are pure Norway spruce (*Picea abies* (L.) Karst.) plantations or spruce dominated, and spruce accounts for 13% of the forest cover, while under natural conditions most sites consist of beech dominated mixed forests. We selected two study stands to represent these contrasting stand types: a pine stand on a nutrient-poor Podzol-type soil with a low water holding capacity derived from pleistocene gravel and sand in the lowlands, and a spruce stand on a deep Cambisol-type soil in the southern hills (Table 1). Although the soil conditions at the spruce site were nominally more favorable than at the pine site, the base saturation of the mineral soil was substantially lower (Table 1), which was caused by atmospheric acid inputs from recent decades (notably from 1970 to 1990). Both stands were 38 years-old in 2013. At this age, thinning operations are carried out once or twice in ten years to increase stand stability and promote growth of the most favorable trees by reducing stand density.

Table 1. Site and stand characteristics of the studied Scots pine and Norway spruce stand in Saxony.

Ecosystem Component	Parameter	Unit	Scots Pine (Laußnitz)	Norway Spruce (Neusalza-Spremberg)
Site characteristics	Mean annual temperature	°C	9.4	8.1
	mean annual rainfall	mm	757	910
	altitude	m above sealevel	190	405
Soil characteristics	Soil type		Dystric Arenosol on pleistocenic sediment	Cambisol on weathered granodiorite overlain with loess
	Profile depth (root zone)	cm	35	60
	field capacity	$mm{\cdot}dm^{-1}$ ***	8.5	18.8
	Base saturation in the organic layer (mean ± SD)	%	30.4 ± 14.9	68.4 ± 43.5
	Base saturation in the mineral soil	%	8.2–3.4	4.5–1.7 ± 5.2–1.2
Stand characteristics in 2013	Age	Year	38	38
	Stand density	$N{\cdot}ha^{-1}$	1850	725
	Stock density **		0.89	0.97
	Growing stock	$m^3{\cdot}ha^{-1}$	199.0	400.5
	Average DBH *	cm	14.3	24.5
	Average height	m	13.9	23.8
	Mean annual increment	$m^3{\cdot}ha^{-1}{\cdot}year^{-1}$	10.2	16.2
	Basal area	$m{\cdot}ha^{-1}$	29.9	34.1
	Average slenderness coefficient		0.95	0.97

* DBH: diameter at breast height; ** ratio of actual growing stock to potential maximum growing stock (according to yield table); *** mm or water per dm of soil depth.

2.2. Stand Biomass and Treatment

In 2013, inventories of the pine stand and the spruce stand were carried out on representative 0.2 ha subplots, determining the number of trees, the DBH, and the height of 10 trees representing the largest 10% of each stand. This data was used to model current stand properties and stand development for 25 years, using the individual tree based forest growth model BWinPro-S [23]. In this time span, forest management will focus on the process known as 'stand qualification', which is selective cutting to promote the growth of the healthiest trees. According to current management practices, we assumed moderate to heavy thinning from above in the spruce stand, and heavy thinning from below in the pine stand. The thinning intensity in terms of removed volume per hectare was adjusted in such a way to keep the degree of stocked area as well as the current increment constant, which resulted in the removal of about 60 m^3 every 5 years in the spruce stand and 45 m^3 every 10 years in the pine stand. This was done in order to maximize stand stability, and to achieve a high growth rate at the stand level as well as a reasonable DBH increment of the single trees.

Based on the inventories, nine to ten trees were selected across the range of DBH values in order to investigate the relative mass distribution as well as nutrient contents in the aboveground biomass compartments. The selected trees were felled and separated into the following compartments in the field: (1) stem wood (diameter > 7 cm) including bark; (2) tree top wood (diameter < 7 cm) including bark; (3) branches including bark and needles; (4) dead branches (Table 2). The fresh mass of these compartments was determined using a hanging scale mounted in the stand. Subsamples of each compartment were taken to the lab and dried at 60 °C until weight was constant. For the stem wood including bark compartment, we collected three 5 cm thick discs from the following tree heights: 1.3 m, 1/2 of the length of the stem wood, top of the stem wood. For the branches including needles, we collected one entire branch from each third whorl, starting at the youngest whorl. For the tree top wood we collected one 5 cm thick slice from the middle of the section, and for the dead branches we randomly collected three to five dead branches per tree. This resulted in relatively large amounts of sample material for the most important tree compartments in terms of mass and nutrient contents (i.e. stem wood, bark of the stem wood, branches and needles). In the lab, the subsamples were further divided into the following target compartments: stem wood, bark of the stem wood, tree top wood including bark, branches including bark, needles and dead branches (Table 2). During the drying and separating process, the water content of each compartment was determined as well as the wood/bark ratio and the branches/needles ratio. Thus, we were able to determine the dry mass of each compartment of each tree by simple ratio calculations. This data was used to fit allometric biomass models of the dry mass of each biomass compartment (M_{bc}) of the type $M_{bc} = a \times DBH\ (\text{cm})^b$, where a and b are specific coefficients and the explaining variable is the DBH (cm), by using non-linear regressions of the original data. As our emphasis was to estimate nutrient stocks in each biomass component, we used the described independent biomass models for each component, instead of an approach focusing more on the additivity of the biomass equation.

Table 2. Partitioning of the aboveground biomass into tree compartments in the field and in the lab, number of samples taken per tree and compartment.

Compartment		Number of Samples
Field	**Lab**	**Samples/Tree**
Stem wood incl. bark	Stem wood > 7 cm	3
	Bark of stem wood	3
Tree top wood incl. bark	tree top wood incl. bark	1
Branches incl. bark and needles	Branches incl. bark	4–7, depending on tree size
	Needles	4–7, depending on tree size
Dead branches	Dead branches	3–5, depending on tree size

2.3. Nutrient Contents

Prior to chemical analysis, one composite sample per tree and compartment was created and ground to 0.25 mm (needles: 0.08 mm). For C and N analysis, aliquots of 10 mg of sample material were analyzed using a Vario EL III (Elementar Analysensysteme GmbH, Hanau, Germany). Nitrogen contents of low N biomass compartments (stem wood, bark, tree top wood, branches, dead branches) were determined with a Vario Max cube (Elementar Analysensysteme GmbH, Hanau,

Germany) using 300 mg aliquots of sample material. The contents of P, K, Ca, and Mg contents were analyzed after HNO_3 digestion using a CCD-ICP Spectrometer CIROS (Spectro Analytical Instruments, Kleve, Germany).

2.4. Soil Nutrient Stocks and Nutrient Balance

Soil profiles were dug and characterized in both stands, and each mineral soil horizon was sampled using metal rings, so that bulk density could be determined in addition to chemical analyses. We calculated the stock of exchangeable Ca^{2+}, K^+, and Mg^{2+} based on the cation exchange capacity (CEC_{eff}) after percolating 5 g subsamples with 1 M NH_4Cl and subsequent analysis in the CCD-ICP spectrometer CIROS (Spectro Analytical Instruments, Kleve, Germany). Total concentrations of N and P were determined after digestion in HNO_3, HF, and $HClO_4$, using the CCD-ICP spectrometer CIROS (Spectro Analytical Instruments, Kleve Germany). Based on horizon-wise nutrient contents and bulk densities, the exchangeable and total nutrient stocks per hectare were calculated for the root zone (35 cm at the pine site and 60 cm at the spruce site).

For the nutrient balance, data from the EU-monitoring Level II sites were used to estimate atmospheric deposition inputs and seepage outputs. For the pine stand, the Level II station "Laußnitz" was used, which is only 500 m away from the study site, while for the spruce stand data from the Level II station "Bautzen/Neukirch" was used, which is 15 km away from the study site, but is comparable in terms of site, soil, and stand conditions [24]. The weathering inputs were roughly estimated according to a method by [25]. First, a total weathering rate of 0.4–1.0 $kmol \cdot ha^{-1} \cdot year^{-1}$ was estimated according to literature values [26]. Then, the share of K^+, Ca^{2+}, and Mg^{2+} was determined in relation to their share in base saturation. Due to data scarcity, we could not set up nutrient budgets for P.

3. Results

3.1. Forest Development

Both the pine and the spruce stand are very productive. Thus, they exhibited a relatively high current increment which decreased slightly during the modeled period from 14.4 to 13.1 and 27.5 to 26.7 $m^3 \cdot ha \cdot year^{-1}$ at the pine and spruce site, respectively. The high productivity can be attributed both to proper forest treatments before the beginning of the study period, and to the favorable growing conditions. Figure 1 shows the development of the total aboveground biomass and the impact of the thinnings that were modeled to be carried out once in 10 years in the pine stand and once in 5 years in the spruce stand.

In both stands, the volume increment over the study period was slightly lower than the volume harvested in thinnings (132.5 vs. 141.1 $m^3 \cdot ha^{-1}$ in the pine stand and 277.1 vs. 318.9 $m^3 \cdot ha^{-1}$ in the spruce stand, Table 3). With regard to the further stand development after the study period, we set a slightly lower stocking density (Table 3). After the qualification phase focused in this paper, the pine stand will reach the dimensioning phase, in which no more thinnings are carried out and the trees are left to grow until they reach the target DBH (40 cm). Due to its high productivity, the spruce stand will more or less skip the dimensioning phase, because the largest trees will have already reached the target DBH (40 cm) after the qualification phase. Once they have reached it, forest management will focus on harvesting the mature trees and at the same time initiating stand regeneration.

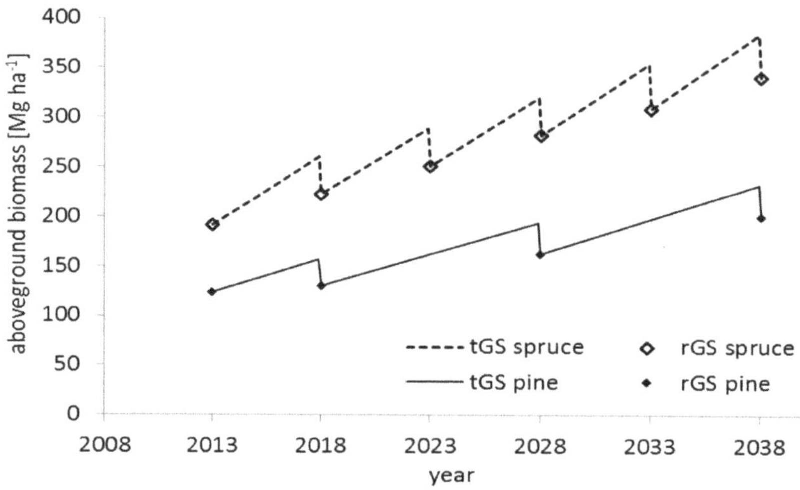

Figure 1. Development of total aboveground biomass (t·ha^{-1}) of the pine and spruce stand. tGS: total growing stock, rGS: remaining growing stock after thinning. Thinning of the pine stand in 2018, 2028, and 2038; thinning of the spruce stand is carried out in 2018, 2023, 2028, 2033, and 2038.

Table 3. Development of the remaining stand and sum of extracted volume in thinnings. Scots pine stand: three thinnings, Norway spruce stand: five thinnings during a 25-year period.

Date / Period	Current Increment	Stocking Density	Growing Stock	Extracted Volume
Pine	m^3·ha^{-1}·a^{-1}		m^3·ha^{-1}	m^3·ha^{-1}
2013	14.4	0.89	200.2	
2038	13.1	0.84	332.7	
Σ thinnings 2018–2038				141.1
Spruce				
2013	27.5	0.97	400	
2038	26.7	0.93	677.1	
Σ thinnings 2018–2038				318.9

3.2. Biomass and Nutrient Distribution in Tree Compartments

We achieved a satisfactory fit of the allometric biomass equations to our data for most of the biomass compartments. As an example, Figure 2 shows the biomass equation of the total aboveground biomass of the spruce stand. The equation parameters and the coefficients of determination are given in Table 4.

The coefficients of determination of the biomass functions were generally high, indicating a strong correlation of the DBH and the mass of the biomass compartments, especially in the case of stem wood, bark, branches, and needles, which are the most important aboveground biomass compartments with respect to their share of total aboveground biomass (Figure 3) and nutrient storage.

The development of the share of the biomass compartments in total aboveground biomass is shown in Figure 3, based on the biomass equations in Table 4. Obviously, the tree top wood incl. bark is relevant only for trees with a DBH < 20 cm, because a relatively large part of the stem is smaller than the defined 7 cm diameter. In larger trees, the stem wood with a diameter > 7 cm including bark accounts for 79% in both spruce and pine. The most important difference between the pine and spruce is the share of branches and needles, with pine trees having a larger share of branches at 14% and a smaller share of needles at only 4.4%, compared to 11.0% and 8.2% in spruce trees, respectively, due to their species-specific crown architecture and physiology. Figure 3 clearly shows the larger relative importance of low quality crown material in small trees, and thus in young stands. This probably makes whole tree harvesting for bioenergy purposes especially attractive in younger stands in order to increase the overall economic outcome of thinning operations.

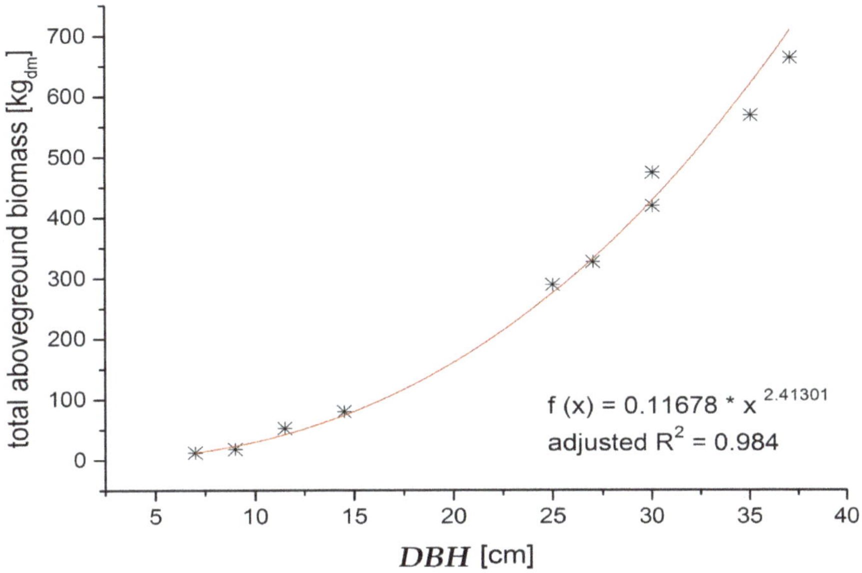

Figure 2. Relation of total aboveground biomass of the sampled spruce trees to DBH and biomass equation.

Table 4. Equation coefficients (a, b) and coefficients of determination (R^2) of the biomass equations $M_{bc} = a \times DBH\ (\text{cm})^b$ for the aboveground biomass compartments of the pine and spruce stand.

Compartment	Pine Stand			Spruce Stand		
	a	*b*	R^2	*a*	*b*	R^2
total aboveground biomass	0.0786	2.539	0.98	0.11678	2.41301	0.98
stem wood	0.0075	3.2281	0.98	0.033	2.68382	0.98
bark of the stem wood	0.0052	2.5924	0.98	0.00926	2.3432	0.98
tree top wood incl. bark	174.34	−1.526	0.89	50.69505	−1.02053	0.78
branches	0.0008	3.3992	0.99	0.00542	2.66276	0.96
needles	0.0015	2.8154	0.94	0.01253	2.35013	0.94
dead branches	0.0002	3.3465	0.88	0.01722	1.91842	0.86

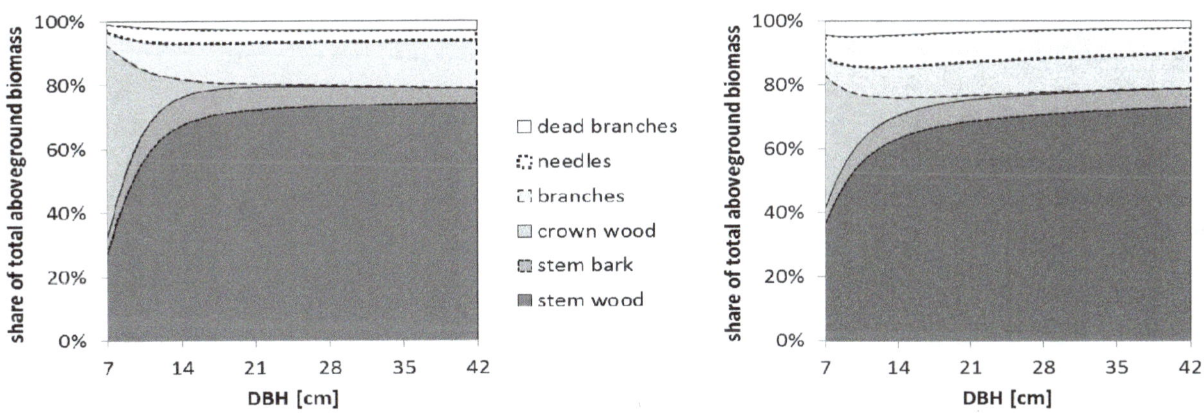

Figure 3. Share of aboveground biomass compartments of the examined Scots pine (**left**) and Norway spruce (**right**) stand at a DBH-range of 7 to 42 cm.

3.3. Nutrient Contents in Aboveground Biomass Compartments

Although nutrient contents for aboveground tree compartments were measured in various other studies, the results are often only partly comparable, because of differences in the defined compartments, in stand age and soil conditions [27]. Thus, with respect to the purpose of this study, we performed our own measurements of nutrient contents (Table 5).

Table 5. Nutrient contents (g·kg^{-1}) of aboveground biomass compartments of pine and spruce trees at the trial sites in East Saxony. Means and standard deviation, n = 5.

	Stem Wood		Stem Bark		Tree Top Wood		Branches		Needles		Dead Branches	
						Scots Pine						
N	0.72	(0.3)	2.76	(0.5)	1.98	(0.6)	4.24	(1.1)	16.4	(2.6)	2.24	(0.7)
P	0.04	(0.4)	0.26	(1.3)	0.17	(1)	0.40	(1.7)	1.41	(7.2)	0.09	(1.1)
K	0.36	(0.4)	1.25	(1.6)	0.91	(1.2)	1.78	(2.1)	6.06	(7.8)	0.26	(1.3)
Ca	0.72	(0.4)	6.86	(1.6)	1.34	(1)	3.11	(2.3)	4.03	(6.8)	2.06	(0)
Mg	0.17	(0.3)	0.46	(1.2)	0.35	(0.9)	0.50	(1.9)	0.74	(5.4)	0.21	(0)
						Norway Spruce						
N	0.80	(0.3)	3.78	(0.8)	2.07	(0.3)	4.49	(0.7)	13.2	(1.6)	1.86	(0.3)
P	0.06	(0)	0.54	(0.1)	0.31	(0.1)	0.60	(0.1)	1.36	(0.2)	0.10	(0)
K	0.46	(0.1)	2.70	(0.5)	1.23	(0.3)	2.73	(0.5)	5.10	(0.7)	0.28	(0.1)
Ca	0.90	(0.1)	7.55	(0.9)	1.27	(0.2)	3.16	(0.3)	4.16	(0.4)	2.98	(0.4)
Mg	0.14	(0)	0.98	(0.1)	0.34	(0.1)	0.63	(0.1)	0.90	(0.1)	0.29	(0.1)

The highest concentrations of all nutrients are generally found in the needles, with the exception of Ca, which shows the highest concentration in the bark. Magnesium concentrations differ between pine and spruce. In the pine trees, Mg concentrations are highest in the needles, followed by branches and bark; while in spruce trees, Mg is distributed similar to Ca with the highest concentrations in the bark, followed by needles and branches. The nutrient contents of the dead branches range somewhat in between those of branches and stem wood.

3.4. Biomass, Nutrient Extraction, Uptake, and Storage

Relative to the increment of the remaining stand, the harvested biomass in the thinnings of the 25-year study period are larger in the spruce stand compared to the pine stand (Table 6). While the biomass extraction in the WT scenario in the pine stand is lower than the biomass ingrowth (129 vs. 114 t·ha^{-1}), it exceeds the ingrowth in the spruce stand (150 vs. 201 t·ha^{-1}) due to a higher overall productivity of the spruce stand, and thus heavier and more frequent thinnings. When intensifying the biomass extraction from conventional stem only (SO) harvesting to whole tree harvesting (WT), the factor of increase is slightly higher in the pine stand compared to the spruce stand (factor of increase 1.30 in the pine stand compared to 1.28 in the spruce stand), due to the smaller dimension of the harvested pine trees, and thus a relatively larger proportion of crown material (see Figure 3).

The difference in the factor of increase of WT (excl. needles), with a higher biomass extraction in case of pine compared to spruce (factor of increase 1.25 vs. 1.18), results from the higher share of branches in pine trees compared to spruce trees.

The relative loss of nutrients in intensive scenarios far exceeds the biomass gain, with factors of increase in whole tree harvesting between 1.7 (Ca, pine and spruce) and 3.6 (P, pine) to 3.4 (P, spruce), respectively.

If needles are allowed to fall off before extracting crown material (WT excl. needles), the increase in nutrient exports is generally higher in pine trees than in spruce, which goes along with the relatively larger biomass extraction due to the higher share of branches in pine trees in the WT (excl. needles) scenario.

Table 6. Aboveground biomass increment and nutrient uptake/storage in the remaining stand.

Scots Pine Stand		Biomass	N	P	K	Ca	Mg
		t·ha^{-1}			kg·ha^{-1}		
Remaining stand Δ 2013–2038 (WT)		129	237	18	99	189	32
Σ thinnings	SO *	88	80	5	39	113	17
	WT (excl. needles) *	110	156	12	69	169	26
	WT *	114	230	18	96	187	30
Ratio Compared to SO							
thinnings	WT (excl. needles) *	1.25	2.0	2.4	1.8	1.5	1.5
	WT *	1.30	2.9	3.6	2.5	1.7	1.8
		Biomass	**N**	**P**	**K**	**Ca**	**Mg**
Norway Spruce Stand		t·ha^{-1}			kg·ha^{-1}		
remaining stand Δ 2013–2038 (WT) *		150	335	35	172	261	44
Σ thinnings	SO *	157	162	15	99	220	33
	WT (excl. needles) *	185	271	29	161	304	48
	WT *	201	486	51	245	372	63
Ratio Compared to SO							
thinnings	WT (excl. needles) *	1.18	1.7	1.9	1.6	1.4	1.5
	WT *	1.28	3.0	3.4	2.5	1.7	1.9

* SO: stem only, WT (exl. needles): whole tree without needles, WT: whole tree.

3.5. Evaluation of the Nutrient Budget as Influenced by Intensive Harvesting

For evaluation of the impact of the thinning scenarios of varying intensity on the nutrient regime of the study sites, annual rates of nutrient inputs and outputs were estimated and compared to the nutrient pools in the organic layer and the mineral soil in the rooting zone (Table 7). Both sites are characterized by relatively high atmospheric N inputs, which far exceed seepage loss and uptake by the trees. The result is a positive N balance on both stands regardless of the biomass extraction intensity, showing that the studied forest stands currently function as nitrogen sinks. The situation is fundamentally different for the base cations. Moderate atmospheric and weathering inputs are hardly sufficient to compensate for seepage loss and nutrient fixation in the biomass increment of the remaining stand. Even conventional SO harvesting would lead to negative nutrient balances and the more intensive biomass extraction scenarios would impair the nutrient balance even further. The only exception is K in the spruce stand, which would have a negative balance only under the WT scenario. Regardless of the intensity of biomass extraction, both sites are characterized by high seepage losses of base cations, in relation to the atmospheric and weathering inputs. In the case of Ca, soil available pools and seepage losses are both twice as high at the pine site compared to the spruce site, indicating that seepage losses of Ca are coupled to availability. On the other hand, this relation is much less pronounced in the case of Mg, and K does not show such a relation at all.

Table 7. Pools of total N and exchangeable K^+, Ca^{2+}, and Mg^{2+}, annual rates of nutrient inputs and outputs, as well as the nutrient balance under different biomass extraction scenarios (SO, WT excl. needles, WT) of the Scots pine stand and the Norway spruce stand.

	Scots Pine Stand				Norway Spruce Stand			
	N	K^+	Ca^{2+}	Mg^{2+}	N	K^+	Ca^{2+}	Mg^{2+}
Soil pools		kg·ha^{-1}				kg·ha^{-1}		
	2955	53	263	13	3599	172	127	34
	N	K	Ca	Mg	N	K	Ca	Mg

Table 7. *Cont.*

		Scots Pine Stand				Norway Spruce Stand		
Inputs		kg·ha^{-1}·year^{-1}				kg·ha^{-1}·year^{-1}		
Deposition	19.3	2.1	3.3	0.7	36	4.8	4.9	1.3
Weathering		4	9.8	0.5		7.6	7.5	1.2
Outputs		kg·ha^{-1}·year^{-1}				kg·ha^{-1}·year^{-1}		
Seepage	1.5	3.1	15.5	3.2	1.9	2.9	7	3.9
Uptake and storage in ingrowth	6.8	2.8	5.4	0.9	9.6	4.9	7.5	1.3
Thinning SO	2.3	1.1	3.2	0.5	4.6	2.8	6.3	0.9
WT excl. needles	4.5	2.0	4.8	0.8	7.7	4.6	8.7	1.4
WT	6.6	2.8	5.4	0.8	13.9	7.0	10.6	1.8
Balance		kg·ha^{-1}·year^{-1}				kg·ha^{-1}·year^{-1}		
SO	8.7	−0.9	−11.0	−3.4	19.9	1.8	−8.3	−3.6
WT excl. needles	6.6	−1.8	−12.6	−3.7	16.8	0.0	−10.8	−4.0
WT	4.4	−2.6	−13.2	−3.8	10.6	−2.4	−12.7	−4.5

4. Discussion

In this study we simulated relatively heavy thinnings in order to sustain stand stability to adapt the stands to climate change in the long term, and make use of the high forest productivity [28,29], which is accelerated by high atmospheric N inputs of about 19.3 and 36.0 kg·ha·year^{-1} at the pine and spruce site, respectively. The large difference in productivity between pine and spruce is both a species and a site effect, because the less demanding pine stands are usually grown on the least productive forest sites [19].

The biomass functions we established in order to model the mass of each aboveground compartment were easy to apply and fit our data very well. Only for the tree top wood of spruce and the dead branches, the R^2 was slightly lower than for the other compartments, probably because the tree top wood is a somewhat artificially defined compartment and the retention of dead branches on a tree depends on factors other than the tree's DBH, such as the actual stand density in the area surrounding the respective tree. Nevertheless, since these two compartments play a minor role for addressing the questions raised in this paper, and the R^2 was still sufficient, we retained the equations for use. The drawback of the biomass functions we established is that they are stand specific, and can therefore only be applied for the DBH range for which they were calibrated. As a result, we concentrated our study on just the thinning phase. In order to create more widely applicable biomass functions, more trees with a wider range of DBH, and from stands representing a wider range of productivity and stand density, would need to be measured. The biomass equations would then need to account for this greater variability by incorporating further explanatory variables such as height and stand density, in addition to the DBH [30].

One option proposed to increase the harvested biomass, and keep nutrient exports relatively low at the same time, is to leave tree tops in the forest for one year to allow the needles to fall off before collecting the tree tops (WT excl. needles scenario) [7,20]. The different pattern of crown biomass allocation between pine and spruce, with a greater share of branches in pine trees versus needles in spruce trees, shows that this method would lead to a greater biomass loss in spruce than in pine trees, when compared to the WT scenario. With respect to biomass feedstock quantities, this procedure would be more viable in pine than in spruce stands. If biomass feedstock quality is also considered, it would be preferable in any case to exclude the needles, as wood is a higher quality feedstock in terms of particle size distribution when chipping at the forest industries [3], as well as the behavior in the burning process.

The nutrient contents in the aboveground biomass compartments match with the findings of previous studies, as far as the compartments are comparable. Jacobsen et al. [31] performed nutrient analyses that are partly comparable, and found similar nutrient contents in the stem wood, bark, needles, and branches of pine and spruce trees. The nutrient contents reported by Weis and Göttlein [9] in the needles, bark, and wood of spruce trees are also within the range of our findings.

Due to the rough estimation method of weathering and the utilization of deposition and seepage values from similar and close-by sites (but not exactly the study sites), the nutrient balance should be interpreted with some caution. Nevertheless, it is still useful for evaluating the impact of biomass, and thus nutrient extraction, in the different thinning scenarios.

Our findings confirm our first hypothesis, that intensive harvesting scenarios impose a strong negative effect on the nutrient balance of forest stands even in the thinning stage. We demonstrated that intensive WT and WT (exc. needles) harvesting causes fairly high nutrient exports in the thinning phase of the trial sites. The biomass gain is far lower than the nutrient expenses, as the biomass gain is only 1.18 to 1.30 times compared to SO, but the ratio of nutrient losses is 1.4 to 3.6 compared to SO. The increase of nutrient losses from SO to WT harvesting are greater for N and P compared to the base cations, reflecting the over proportional abundance of N and P in the crown material, especially in the physiologically active needles. Yet the impact on the site-specific nutrient sustainability is far greater for the base cations than for N, because of excess atmospheric N inputs. We unfortunately have no data for the nutrient balance of P, but as P inputs into the forest ecosystem by weathering and deposition are usually low, a rather negative P balance similar to that of the base cations can be assumed. High seepage losses contribute strongly to the negative nutrient balance of the base cations. The seepage losses of base cations result from sulfur depositions during the 1970s to 1990s, which were greatly reduced by the end of the last century [32]. However, SO_4^{2-} from accumulated S pools still gets washed out of the soil profile, coupled with base cations to neutralize the charge [33]. This leads to negative nutrient balances for Ca^{2+} and Mg^{2+} already at the SO scenario, which decreases further in the WT (excl. needles) and the WT scenario, when K^+ also becomes negative.

According to our findings, we reject our second hypothesis, according to which the negative impact of intensive harvesting is greater in highly productive stands than in less productive stands, due to larger absolute amounts of removed biomass, and thus nutrients. In our study, the higher stand productivity of the spruce stand corresponded with a higher nutrient demand for uptake and storage, as well as higher nutrient losses due to biomass extraction, compared to the less productive pine stand. At the same time, these differences in productivity, and thus biomass-related nutrient fluxes, are balanced out by higher nutrient stocks and inputs through deposition (except for Ca), while seepage losses are equal to or even lower than the pine site. Thus, the relative impact of nutrient exports through intensive biomass extraction is almost equally severe in the spruce and pine stand, despite different absolute levels of biomass extraction.

Our findings should, however, be interpreted with some caution, as they are based on static assumptions regarding the development of forest growth, the nutrient concentrations in the biomass, and the weathering rate. In reality, increased nutrient scarcity due to intensive biomass extraction may cause reduced growth of the remaining stand, as has been shown in studies conducted in North American and Scandinavian forests [7,8,10], while other studies found no clear effect on tree growth [34,35]. A meta-analysis by Achat et al. [36] indicates an average growth reduction of 3%–7% after intensive biomass extraction. Declining pools of available nutrients may also cause reduced nutrient concentrations in the foliage of the remaining trees, which would also alter the nutrient balance. The nutrient balance is also known to be strongly influenced by the rate of mineral weathering [37], which is in practice very difficult to determine and could only be roughly estimated in our study. Furthermore, mineral weathering rates are not static, but mycorrhizal weathering may be accelerated if there are nutrient deficiencies [38]. Despite these uncertainties, our findings underline the possible nutrient balance risk of intensive biomass extraction from thinnings, which brings into question the nutrient balance sustainability of these intensive management practices.

Our study highlights that intensive harvesting scenarios like WT (excl. needles) and WT harvesting can have severe consequences for the nutrient balance of forest sites. Furthermore, intensive biomass removal is known to negatively affect the soil as the basis for biomass production. Reduced residue inputs reduce the organic carbon content [39] and mineralization rates and cause soil compaction and thus a reduced water holding capacity [36]. Based on the analysis of detrimental forest management practices (i.e., litter raking, fuelwood collection), the influence of biomass extraction on soil acidification was highlighted, because the uptake and storage of cations exceeds that of anions during tree growth and biomass buildup [40,41]. In order to mitigate the negative effects of intensive biomass extraction on the nutrient budget, the redistribution of wood ash is frequently discussed and has proven to be suitable to compensate for nutrient losses in forests, with the exception of N [13,42–45], and was previously found to be applicable in the form of bark-ash-pellets at the study sites [14].

5. Conclusions

Our study provides evidence that intensive biomass extraction may cause substantial losses of nutrients in pine and spruce stands, even in the thinning stage of stand development. This is amplified by the high productivity of both investigated stands, which is accelerated by high atmospheric N inputs. Despite the fact that the nutrient balance established in this study contains only a rough estimation of the weathering rate, it suggests that the budgets of base cations are already stressed due to high seepage losses caused by the still ongoing effects of very high S-driven acid deposition which occurred previously in the region. Therefore, the nutrient balance became negative even for SO harvesting in the case of Ca^{2+} and Mg^{2+}. Further intensification of biomass removal clearly impairs the nutrient budget, also for K^+, while the N balance remains positive even in the most intensive biomass removal scenario (WT). Even if a further development of the wood energy sector will increase the demand for forest biomass in the future, forest management should take the nutrient balance into account. Intelligent concepts to limit forest biomass extraction to a sustainable level need to be developed. At the same time, additional measures, such as wood ash recycling, may help to improve the nutrient balance of the base cations and P, without burdening the ecosystems with ever more N.

Acknowledgments: We acknowledge support by the Federal Ministry of Education and Research and the Open Access Publication Funds of the TU Dresden. The authors would like to thank Clemens Kurth for advice concerning the forest treatment and Henning Andreae, Frank Jacob, and Hendrik Horn for help with the biomass inventories and functions. We are very grateful for the valuable improvements of the manuscript provided by two anonymous reviewers.

Author Contributions: Christine Knust and Karl-Heinz Feger conceived and designed the experiments; Christine Knust and Karoline Schua performed the experiments; Christine Knust analyzed the data; Christine Knust wrote the paper.

References

1. Fachagentur Nachwachsende Rohstoffe e.V. (FNR). *Basisdaten Bioenergie Deutschland 2014: Festbrennstoffe, Biokraftstoffe, Biogas*; FNR: Gülzow, Germany, 2014.
2. BMU; BMELV. *Nationaler Biomasseaktionsplan für Deutschland: Beitrag der Biomasse für Eine Nachhaltige Energieversorgung*; BMU, BMELV: Berlin, Germany, 2010.
3. Kizha, A.R.; Han, H.-S. Processing and sorting forest residues: Cost, productivity and managerial impacts. *Biomass Bioenergy* **2016**, *93*, 97–106. [CrossRef]
4. Wolfsmayr, U.J.; Rauch, P. The primary forest fuel supply chain: A literature review. *Biomass Bioenergy* **2014**, *60*, 203–221. [CrossRef]
5. Cremer, T. Bereitstellung von Holzhackschnitzeln Durch Die Forstwirtschaft: Produktivitätsmodelle als Entscheidungsgrundlage über Verfahren und Aushaltungsvarianten, Entwickelt auf der Basis Einer Metaanalyse. Ph.D. Thesis, University of Freiburg i.Br., Freiburg, Germany, 2008.
6. Vangansbeke, P.; de Schrijver, A.; de Frenne, P.; Verstraeten, A.; Gorissen, L.; Verheyen, K. Strong negative impacts of whole tree harvesting in pine stands on poor, sandy soils: A long-term nutrient budget modelling approach. *For. Ecol. Manag.* **2015**, *356*, 101–111. [CrossRef]

7. Egnell, G. Is the productivity decline in Norway spruce following whole-tree harvesting in the final felling in boreal Sweden permanent or temporary? *For. Ecol. Manag.* **2011**, *261*, 148–153. [CrossRef]
8. Walmsley, J.; Jones, D.; Reynolds, B.; Price, M.; Healey, J. Whole tree harvesting can reduce second rotation forest productivity. *For. Ecol. Manag.* **2009**, *257*, 1104–1111. [CrossRef]
9. Weis, W.; Göttlein, A. Nährstoffnachhaltige Biomassenutzung: Bei der Nutzung von Biomasse ist Vorsicht geboten: Nicht jeder Waldstandort verträgt den erhöhten Nährstoffentzug. *LWF Aktuell* **2012**, *90*, 44–47.
10. Helmisaari, H.-S.; Hanssen, K.H.; Jacobson, S.; Kukkola, M.; Luiro, J.; Saarsalmi, A.; Tamminen, P.; Tveite, B. Logging residue removal after thinning in Nordic boreal forests: Long-term impact on tree growth. *For. Ecol. Manag.* **2011**, *261*, 1919–1927. [CrossRef]
11. Clarke, N.; Gundersen, P.; Jönsson-Belyazid, U.; Kjønaas, O.J.; Persson, T.; Sigurdsson, B.D.; STUPAK, I.; Vesterdal, L. Influence of different tree-harvesting intensities on forest soil carbon stocks in boreal and northern temperate forest ecosystems. *For. Ecol. Manag.* **2015**, *351*, 9–19. [CrossRef]
12. Lassauce, A.; Lieutier, F.; Bouget, C. Woodfuel harvesting and biodiversity conservation in temperate forests: Effects of logging residue characteristics on saproxylic beetle assemblages. *Biol. Conserv.* **2012**, *147*, 204–212. [CrossRef]
13. Pitman, R.M. Wood ash use in forestry—A review of the environmental impacts. *Forestry* **2006**, *79*, 563–588. [CrossRef]
14. Knust, C.; Schua, K.; Göttlein, A.; Ettl, R.; Wolferstetter, T.; Feger, K.-H. Compensation of nutrient losses resulting from the intensified use of woody biomass using bark-ash-pellets. In *Bioenergy from Dendromass for the Sustainable Development of Rural Areas*; Butler Manning, D., Bemmann, A., Bredemeier, M., Lamersdorf, N., Ammer, C., Eds.; Wiley (Imprint), Inc.: Weinheim, Germany, 2015; pp. 269–286.
15. Demeyer, A.; Voundi Nkana, J.C.; Verloo, M. Characteristics of wood ash and influence on soil properties and nutrient uptake: An overview. *Bioresour. Technol.* **2001**, *77*, 287–295. [CrossRef]
16. Ingerslev, M.; Hansen, M.; Pedersen, L.B.; Skov, S. Effects of wood chip ash fertilization on soil chemistry in a Norway spruce plantation on a nutrient-poor soil. *For. Ecol. Manag.* **2014**, *334*, 10–17. [CrossRef]
17. Rademacher, P.; Schönfelde, E.; Meiwes, K.J. Elementgehalte in Baumkompartimenten von Fichte (*Picea abies* [L.] Karst), Kiefer (*Pinus sylvestris* [L.] Karst), Buche (*Fagus sylvatica* [L.] Karst), Eiche (*Quercus petraea und robur* [L.] Karst) und Douglasie (*Pseudotsuga menziesii* [L.] Karst). In *Möglichkeiten und Grenzen der Vollbaumnutzung: Ergebnisbericht (FKZ: 22015407)*; Nordwestdeutsche Forstliche Versuchsanstalt, Universität Göttingen: Göttingen, Germany, 2011; pp. 125–154.
18. Merino, A.; Balboa, M.A.; Rodríguez Soalleiro, R.; González, J.Á. Nutrient exports under different harvesting regimes in fast-growing forest plantations in southern Europe. *For. Ecol. Manag.* **2005**, *207*, 325–339. [CrossRef]
19. Tveite, B.; Hanssen, K.H. Whole-tree thinnings in stands of Scots pine (Pinus sylvestris) and Norway spruce (Picea abies): Short- and long-term growth results. *For. Ecol. Manag.* **2013**, *298*, 52–61. [CrossRef]
20. Zetterberg, T.; Olsson, B.A.; Löfgren, S.; Hyvönen, R.; Brandtberg, P.-O. Long-term soil calcium depletion after conventional and whole-tree harvest. *For. Ecol. Manag.* **2016**, *369*, 102–115. [CrossRef]
21. Jacobson, S.; Kukkola, M.; Mälkönen, E.; Tveite, B. Impact of whole-tree harvesting and compensatory fertilization on growth of coniferous thinning stands. *For. Ecol. Manag.* **2000**, *129*, 41–51. [CrossRef]
22. Jacobson, S.; Kukkola, M.; Mälkönen, E.; Tveite, B.; Möller, G. Growth response of coniferous stands to whole-tree harvesting in early thinnings. *Scand. J. For. Res.* **1996**, *11*, 50–59. [CrossRef]
23. Döbbeler, H.; Alber, M.; Schmidt, M.; Nagel, J.; Schröder, J. *BWINPro—Programm zur Bestandesanalyse und Prognose: Handbuch zur Gemeinsamen Version von BWINPro und BWINPro-S*; Version 6.3; Nordwestetsche Forstliche Versuchsanstalt, Göttingen, Germany and TU Dresden, Lehrstuhl für Holzmeßkunde und Waldwachstum: Tharandt, Germany, 2011.
24. Andreae, H. (Staatsbetrieb Sachsenforst, Kompetenzzentrum Wald und Forstwirtschaft, Saxony, Germany). Personal communication, 2014.
25. Von Wilpert, K.; Bösch, B.; Bastian, P.; Zirlewagen, D.; Hepperle, F.; Holzmann, S.; Puhlmann, H.; Schäffer, J.; Kändler, G.; Sauter, U.H. Biomasse-Aufkommensprognose und Kreislaufkonzept für den Einsatz von Holzaschen in der Bodenschutzkalkung in Oberschwaben. *Ber. Freibg. Forstl. Forsch.* **2011**, *87*, 155.
26. Hildebrand, E.E.; Wilpert, K. von; Kohler, M. *Ionenspeicher- und Mobilisierungspotentiale im Skelett und Feinboden des Standortes Conventwald bei Unterschiedlichen Depositionsszenarien: Abschlussbericht, FKZ PEF196009 und PEF 196010, BWC 99 009 und BWC 99010*; Universität Freiburg und Forstliche Versuchs- und Forschungsanstalt: Freiburg, Germany, 2000.

27. Rademacher, P. Contents of nutrient elements in tree components of economical important species in relation to their residual utilisation. *Holz Werkst.* **2005**, *63*, 285–296. [CrossRef]
28. Eisenhauer, D.-R.; Sonnemann, S. Waldbaustrategien unter sich ändernden Umweltbedingungen: Leitbilder, Zielsysteme und Waldentwicklungstypen. *Waldökol. Landsch. Nat.* **2009**, *8*, 71–88.
29. Fischer, H.; Wagner, S. Silvicultural responses to predicted climate change scenarios. *J. For. Ecol. For. Manag.* **2009**, *2*, 12–23.
30. Rumpf, S.; Nagel, J.; Schmidt, M. Biomasseschätzfunktionen von Fichte (*Picea abies* [L.] Karst), Kiefer (*Pinus sylvestris* [L.] Karst), Buche (*Fagus sylvatica* [L.] Karst), Eiche (*Quercus robur und petraea* [L.] Karst) und Douglasie (*Pseudotsuga menziesii* [L.] Karst) für Nordwestdeutschland. In *Möglichkeiten und Grenzen der Vollbaumnutzung: Ergebnisbericht (FKZ: 22015407)*; Nordwestdeutsche Forstliche Versuchsanstalt, Universität Göttingen: Göttingen, Germany, 2011; pp. 25–123.
31. Jacobsen, C.; Rademacher, P.; Meesenburg, H.; Meiwes, K.J. Gehalte chemischer Elemente in Baumkompartimenten: Literaturstudie und Datensammlung. In *Berichte des Forschungszentrums Waldökosysteme, Reihe B*; Universität Göttingen: Göttingen, Germay, 2003.
32. Oulehle, F.; Hofmeister, J.; Cudlín, P.; Hruška, J. The effect of reduced atmospheric deposition on soil and soil solution chemistry at a site subjected to long-term acidification, Načetín, Czech Republik. *Sci. Total Environ.* **2006**, *370*, 532–544. [CrossRef] [PubMed]
33. Wunderlich, S.; Raben, G.; Andreae, H.; Feger, K.-H. Schwefel-Vorräte und Sulfat-Remobilisierungspotential in Böden der Level-II-Standorte Sachsens. *AFZ-Der. Wald* **2006**, *60*, 762–765.
34. Kaarakka, L.; Tamminen, P.; Saarsalmi, A.; Kukkola, M.; Helmisaari, H.-S.; Burton, A.J. Effects of repeated whole-tree harvesting on soil properties and tree growth in a Norway spruce (*Picea abies* (L.) Karst.) stand. *For. Ecol. Manag.* **2014**, *313*, 180–187. [CrossRef]
35. Roxby, G.E.; Howard, T.E. Whole-tree harvesting and site productivity: Twenty-nine northern hardwood sites in central New Hampshire and western Maine. *For. Ecol. Manag.* **2013**, *293*, 114–121. [CrossRef]
36. Achat, D.L.; Deleuze, C.; Landmann, G.; Pousse, N.; Ranger, J.; Augusto, L. Quantifying consequences of removing harvesting residues on forest soils and tree growth—A meta-analysis. *For. Ecol. Manag.* **2015**, *348*, 124–141. [CrossRef]
37. Klaminder, J.; Lucas, R.W.; Futter, M.N.; Bishop, K.H.; Köhler, S.J.; Egnell, G.; Laudon, H. Silicate mineral weathering rate estimates: Are they precise enough to be useful when predicting the recovery of nutrient pools after harvesting? *For. Ecol. Manag.* **2011**, *261*, 1–9. [CrossRef]
38. Vadeboncoeur, M.A.; Hamburg, S.P.; Yanai, R.D.; Blum, J.D. Rates of sustainable forest harvest depend on rotation length and weathering of soil minerals. *For. Ecol. Manag.* **2014**, *318*, 194–205. [CrossRef]
39. Kreutzer, K. Ecological problems of full tree utilization. *Forstwiss. Cent.* **1979**, *98*, 298–308. [CrossRef]
40. Glatzel, G. The impact of historic land use and modern forestry on nutrient relations of Central European forest ecosystems. *Fertil. Res.* **1991**, *27*, 1–8. [CrossRef]
41. Feger, K.H. Bedeutung von ökosysteminternen Umsätzen und Nutzungseingriffen für den Stoffhaushalt von Waldlandschaften (Habilitationsschrift)—*Freiburger Bodenkundliche. Abhandlungen* **1993**, *31*, 237.
42. Ingerslev, M.; Skov, S.; Sevel, L.; Pedersen, L.B. Element budgets of forest biomass combustion and ash fertilisation—A Danish case-study. *Biomass Bioenergy* **2011**, *35*, 2697–2704. [CrossRef]
43. Kikamägi, K.; Ots, K.; Kuznetsova, T. Effect of wood ash on the biomass production and nutrient status of young silver birch (Betula pendula Roth) trees on cutaway peatlands in Estonia. *Ecol. Eng.* **2013**, *58*, 17–25. [CrossRef]
44. Saarsalmi, A.; Smolander, A.; Kukkola, M.; Moilanen, M.; Saramäki, J. 30-year effects of wood ash and nitrogen fertilization on soil chemical properties, soil microbial processes and stand growth in a Scots pine stand. *For. Ecol. Manag.* **2012**, *278*, 63–70. [CrossRef]
45. Solla-Gullón, F.; Santalla, M.; Rodríguez-Soalleiro, R.J.; Merino, A. Nutritional status and growth of a young Pseudotsuga menziesii plantation in a temperate region after application of wood-bark ash. *For. Ecol. Manag.* **2006**, *237*, 312–321. [CrossRef]

5

Stocks and Stoichiometry of Soil Organic Carbon, Total Nitrogen and Total Phosphorus after Vegetation Restoration in the Loess Hilly Region, China

Hongwei Xu [1,3], Qing Qu [2], Peng Li [2], Ziqi Guo [2], Entemake Wulan [2] and Sha Xue [1,2,3,*]

1 State Key Laboratory of Soil Erosion and Dryland Farming on Loess Plateau, Institute of Soil and Water Conservation, Chinese Academy of Sciences and Ministry of Water Resources, Yangling 712100, China; xuhongwei16@mails.ucas.ac.cn

2 Institute of Soil and Water Conservation, Northwest A&F University, Yangling 712100, China; ylxnqq@nwafu.edu.cn (Q.Q.); lipenglipeng1900@gmail.com (P.L.); Guoziqi100@outlook.com (Z.G.); 5141509166@mail.sjtu.edu.cn (E.W.)

3 Institute of College of Natural Resources and Environment, University of Chinese Academy of Sciences, Beijing 100049, China

* Correspondence: xuesha100@163.com

Abstract: The Loess Plateau is an important region for vegetation restoration in China; however, changes in soil organic carbon (SOC), soil nutrients, and stoichiometry after restoration in this vulnerable ecoregion are not well understood. Typical restoration types, including orchardland, grassland, shrubland, and forestland, were chosen to examine changes in the stocks and stoichiometry of SOC, soil total nitrogen (TN), and soil total phosphorus (TP) at different soil depths and recovery times. Results showed that SOC stocks first increased and then stabilized in orchardland, grassland, and shrubland at 0–30 cm depths, while in forestland, SOC stocks gradually increased. Soil TN stocks first increased and then decreased in orchardland, shrubland, and forestland with restoration age at 0–30 cm depths, while soil TP stocks showed little variation between restoration types; at the same time, the overall C:N, C:P, and N:P ratios increased with restoration age. In the later stages of restoration, the stocks of SOC and soil TN at 0–30 cm soil depths were still lower than those in natural grassland and natural forest. Additionally, the SOC, soil TN, and soil TP stocks and the C:N, C:P, and N:P ratios decreased with soil depth. The forestland had the highest rate of change in SOC and soil TN stocks, at 0–10 cm soil depth. These results indicate a complex response of SOC, soil TN, and soil TP stocks and stoichiometry to vegetation restoration, which could have important implications for understanding C, N, and P changes and nutrient limitations after vegetation restoration.

Keywords: soil stoichiometry; soil nutrient; nutrient limitations; natural grassland; natural forest

---

1. Introduction

Soil is an important component of terrestrial ecosystems and the main source of the nutrients required for plant growth and development. Soil organic carbon (SOC), soil total nitrogen (TN), and soil total phosphorus (TP) are the main structural and nutritional components of soil and are also the main limiting factors in terrestrial ecosystems [1]. Soil organic C, soil TN, and soil TP stocks reflect the potential of the soil to provide nutrients to vegetation. These elements continuously circulate between the layers of the soil (the biogeochemical cycles of C, N, and P), which ensures a smooth flow of energy and maintains the stability of ecosystems [2]. The availabilities of soil TN and soil TP are major factors regulating the carbon balance of the ecosystem. Elemental stoichiometry is an important indicator reflecting the C, N, and P cycles in soil and the accumulation and balance of nutrients in ecosystems, and it can help to determine the responses of ecological processes to global changes [3].

Vegetation restoration has received intensive interest because of its potential influence on global C and N cycling, soil quality improvement, land management, and regional economic development [4]. Land use change in the form of vegetation restoration plays an important role in improving the ecological environment and function of ecosystems and can also improve soil quality and soil nutrient cycling. Improved soil quality will, in turn, affect plant production and ecosystem function.

A large number of related studies have shown that stocks and stoichiometry of SOC, soil TN, and soil TP are closely related to land use type [5,6], and nutrient inputs and outputs are considered to be the main factors affecting soil nutrient content [7–9]. Some studies have found that vegetation restoration can promote photosynthesis, soil nutrient accumulation, and microbial activity [10–12], and increase the stoichiometry of SOC, soil TN, and soil TP [9]. However, other studies have indicated that land use change can lead to decreases in soil nutrient contents [13]. Studies estimating the impact of land use on the stocks and stoichiometry of SOC, TN, and TP have mainly focused on the topsoil (0–20 cm), as this is considered to be the most active soil layer in terms of natural and anthropogenic disturbances [14]. Recent studies have shown that the nutrient content of deep soils may also vary greatly with land use [14,15]. Therefore, understanding how C, N, and P stocks and stoichiometry change in soil with land use changes can clarify soil nutrient availability and nutrient cycling and balance mechanisms, and is of great significance for regional ecosystem health assessments.

The Loess Plateau, China, is located in a semi-arid/semi-humid climate zone which has undergone serious soil erosion. It is an ecologically fragile area and a key area for soil and water conservation efforts in China. Before the 1950s, extreme weather such as droughts, heavy rain, hailstorms, strong winds, and dust storms occurred frequently in this area, resulting in serious soil erosion. In addition, as a result of long-term and unsustainable land use, vegetation has been destroyed over large areas due to grazing and farming. The large-scale cultivation of sloping cropland further aggravated the soil erosion. The amount of nitrogen, phosphorus, and potassium lost from slope farmland has been estimated at 12.7 million tons per year [16]. Vegetation restoration was implemented in the 1970s in this region. In order to control soil erosion and improve ecosystem function, ecological restoration and environmental reconstruction work has been carried out in which slope cropland (slope > 25°) has been converted into orchardland, grassland, shrubland, and forestland. After decades of continuous efforts, vegetation coverage has increased, and the ecological environment has been greatly improved [17]. The sequestration and stoichiometry of SOC, soil TN, and soil TP varies among different vegetation types and restoration ages; therefore, its effect on soil physicochemical properties varies as well. It is important, therefore, to clarify annual and vertical variations in the SOC, soil TN, and soil TP stocks and stoichiometry in soils with different vegetation types in the Loess Hilly Region, China.

In order to better understand the SOC processes, carbon budget of the soil, and soil fertility issues after afforestation, we addressed the following questions: (1) How have stocks of SOC, soil TN, and soil TP, and their ratios, changed across the Loess Hilly Region, China, after decades of vegetation restoration? (2) Are these changes associated with soil depth? (3) How do these changes vary with restoration type? We further hypothesized that (1) as litter inputs to the soil increase with restoration age, stocks of SOC and soil TN increase, whereas soil TP stocks do not significantly change since P stocks are primarily affected by parent minerals. The change rate will be greatest for SOC, followed by soil TN, and then soil TP, causing an increase in the C:N, C:P, and N:P ratios with restoration age; (2) vegetation restoration affects stocks of SOC, soil TN, and soil TP at the soil surface more than at greater soil depths; and (3) due to differences in the litter produced by different vegetation, root secretions and soil microorganisms will also vary between restoration types, resulting in differences in stocks and ratios of SOC, soil TN, and soil TP.

2. Materials and Methods

2.1. Study Area

The study area was located in Ansai County, Shanxi Province, China in the center of the Loess Plateau (Figure 1). This region has a warm temperate and semi-arid climate with an annual average temperature of 8.8 °C. Annual precipitation is approximately 500 mm, 60% of which falls between July and September, and the frost-free period is 157 days. The soil is mainly composed of Huangmian soil developed on wind-deposited loessial parent material. This type of soil is characterized by weak cohesion, which has made it prone to severe soil erosion. The sand (2.00–0.02 mm grain size), silt (0.02–0.002 mm), and clay (<0.002 mm) contents are 65%, 24%, and 11%, respectively. The soil bulk density (BD) and soil pH of the tillage layer range from 1.15 to 1.35 g cm^{-3} and 8.4 to 8.6, respectively.

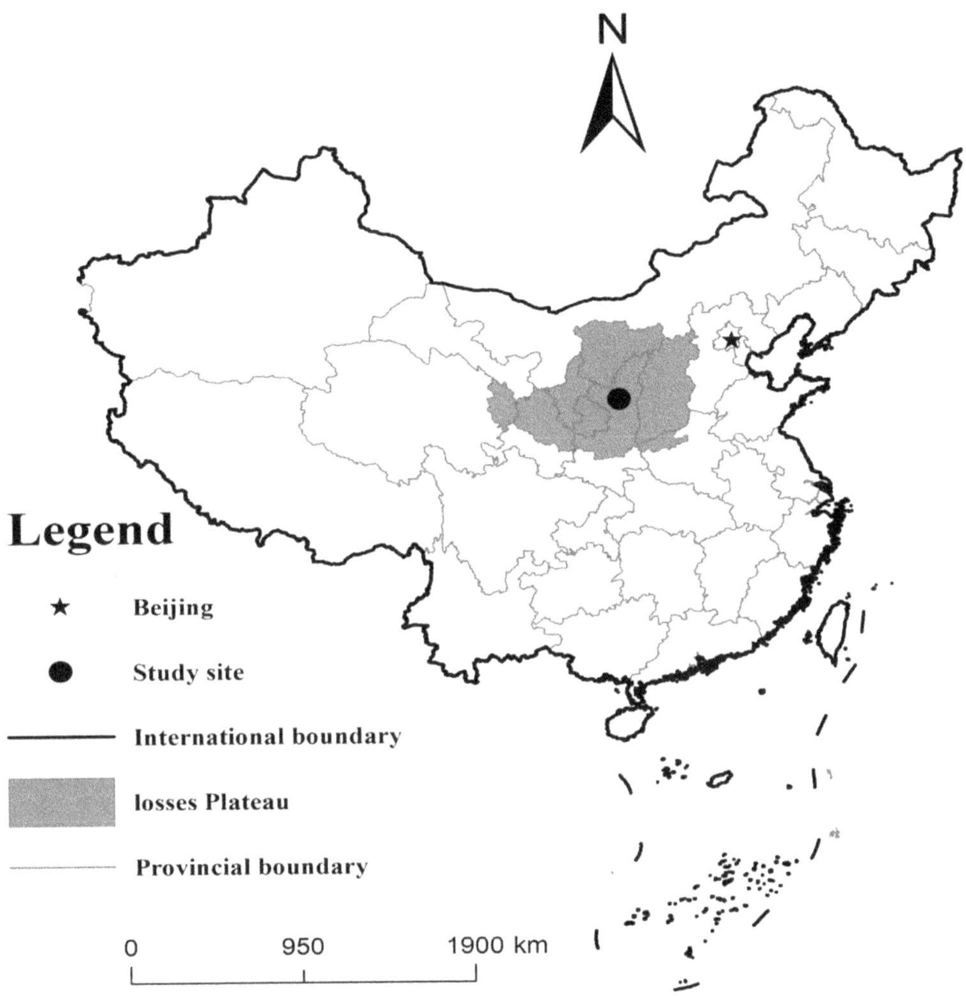

Figure 1. The locations of study sites in Ansai County, Shanxi Province, China.

2.2. Soil Sampling and Laboratory Analyses

With the aim to examine changes in the stocks and stoichiometry of soil organic carbon (SOC), soil total nitrogen (TN), and soil total phosphorus (TP) after vegetation restoration in different restoration types, this study adopted a "space for time" approach. A total of 82 sites representing four restoration types were selected based on vegetation type, topographic features, and restoration age, comprising 9 sites of orchardland (5, 10, and 20 years), 34 sites of grassland (2, 5, 8, 11, 15, 18, 26, and 30 years), 24 shrubland (5, 10, 20, 30, 36, and 47 years), and 15 forestland (5, 10, 20, 37, and 56 years). In addition, we selected three slope cropland sites, which were studied at 0 years. Because the four restoration areas were transformed from croplands, four natural grassland sites (age > 50 years) and nine natural forest sites (age > 100 years) were selected as controls (Table S1). Three 10 × 10 m plots were chosen in each orchardland, three 2 × 2 m plots in each grassland (grassland and natural grassland), three 10 × 10 m plots in each shrubland, and three 20 × 20 m plots in each forestland type (forestland and natural forest). Each plot was at least 50 m from the other plots. A total of 15 soil samples were collected from five soil depths (0–10, 10–20, 20–30, 30–50, and 50–100 cm) in a random sampling design using a soil drilling sampler (4 cm inner diameter). Soil samples from each plot from the same soil depth were mixed to form one sample. These soil samples were brought back to the laboratory and then divided into two parts. One part of the sample was naturally air-dried, plant roots and other impurities were removed, and then the SOC, soil TN, and soil TP were measured. The other part was stored in a refrigerator at 4 °C until further analysis of other indicators, which are not presented in this paper.

The soil bulk density (BD) of each depth was measured using the cutting ring method. The SOC was determined using the H_2SO_4–$K_2Cr_2O_7$ method [18]. The soil TN was measured using the Kjeldahl method [19], and the soil TP was determined colorimetrically using the ammonium molybdate method [20].

2.3. Calculation of SOC, Soil TN, and Soil TP Stocks

The stocks of SOC, soil TN, and soil TP from five soil depths of 0–10, 10–20, 20–30, 30–50, and 50–100 cm in different restoration types were selected in our study to research the carbon budget of the soil and soil fertility issues after afforestation. The SOC, soil TN, and soil TP stocks (Mg ha^{-1}) were calculated as follows:

$$\mathrm{SOC_i\ stock} = \mathrm{SOC_i} \times \mathrm{BD_i} \times \mathrm{D_i}/10, \quad (1)$$

$$\mathrm{Soil\ TN_i\ stock} = \mathrm{soil\ TN_i} \times \mathrm{BD_i} \times \mathrm{D_i}/10, \quad (2)$$

$$\mathrm{Soil\ TP_i\ stock} = \mathrm{soil\ TP_i} \times \mathrm{BD_i} \times \mathrm{D_i}/10, \quad (3)$$

where SOC_i is the soil organic carbon content of the *i*th layer of soil (g kg^{-1}), soil TN_i is the soil total N content of the *i*th layer of soil (g kg^{-1}), soil TP_i is the soil total P content of the *i*th layer of soil (g kg^{-1}), BD_i is the soil bulk density of the *i*th layer of soil (g cm^{-3}), and D_i is the soil depth of the *i*th layer of soil (cm).

2.4. Statistical Analyses

Two-way ANOVAs were used to determine the effects of restoration age, soil depth, and their interaction on SOC, soil TN, and soil TP stocks and C:N, C:P, and N:P ratios. An independent samples *t*-test was used to compare the SOC, soil TN, and soil TP stocks and the C:N, C:P, and N:P ratios from sites grassland30, natural grassland, shrubland47, forestland56, and natural forest ($p < 0.05$). Before ANOVA analyses, we performed tests for normality and homogeneity of variance. In order to compare the effects of vegetation type, we selected SOC, soil TN, and soil TP contents and stoichiometry for the same or similar years from the four restoration types: orchardland (5 years, 10 years, 20 years), grassland (5 years, 11 years as 10 years, 18 as 20 years), shrubland (5 years, 10 years), and forestland (5 years, 10 years). In addition, we selected the SOC, soil TN, and soil TP contents from different soil depths as the rate of change with recovery years and compared the SOC, soil TN, and soil TP sequestration rates between different restoration types. All statistical analyses were conducted using SPSS 21.0 (SPSS Inc., Chicago, IL, USA). Figures were drawn using Origin 9.0.

3. Results

3.1. SOC, Soil TN, and Soil TP Stocks in Different Restoration Types

Restoration age had a significant effect on SOC and soil TN stocks ($p < 0.01$) (Table S2). In orchardland, the SOC stocks of the soil at 0–30 cm depth first increased and then stabilized after

5 years; the soil TN stock at the same depth first increased and then decreased with restoration age. The soil TP stocks at 0–100 cm depth showed no significant change with restoration age (Figure 2a–c). In grassland, the SOC and soil TN stocks at 0–30 cm depth first increased and then stabilized after 18 years, and the soil TP stocks at 0–100 cm depth showed no significant change with different restoration ages (Figure 2d–f). In shrubland, the SOC stocks at 0–50 cm depth and soil TN at 0–10 cm increased at first, peaked after 5 years, and then stabilized after 30 years; the soil TP stocks at 0–100 cm depth showed no significant change (Figure 2h–j). In forestland, the SOC at 0–50 cm gradually increased, and the soil TN stocks increased before peaking after 37 years, then decreased with restoration age; the soil TP stock at 0–100 cm depth showed no significant change (Figure 2k–m). In addition, SOC and soil TN stocks from the 0–30 cm soil layer of grassland30 were significantly lower than those in natural grassland (Figure 3a,b). SOC and soil TN stocks at 0–30 cm soil depth in shrubland47, as well as SOC stocks at 0–100 cm depth and soil TN stocks at 0–20 cm depth in forestland56, were significantly lower than those in natural forest (Figure 3d,e). The soil TP stocks at all soil depths showed no significant differences between grassland30 and natural grassland, shrubland47 and natural forest, or forestland56 and natural forest (Figure 3c,f).

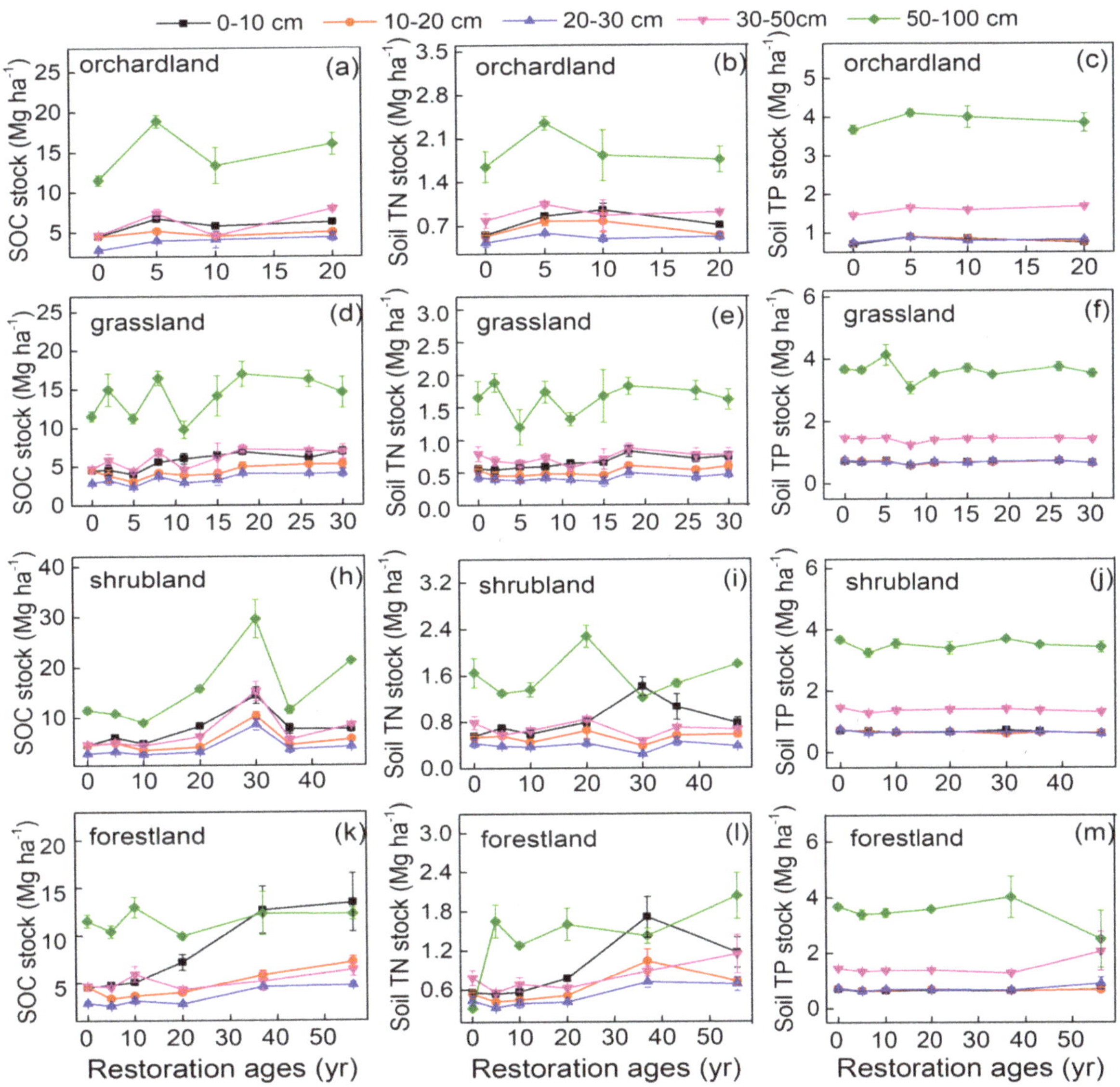

Figure 2. Changes in stocks of soil organic carbon (SOC), soil total nitrogen (TN), and soil total phosphorus (TP) with restoration age. Note: values are mean ± standard error.

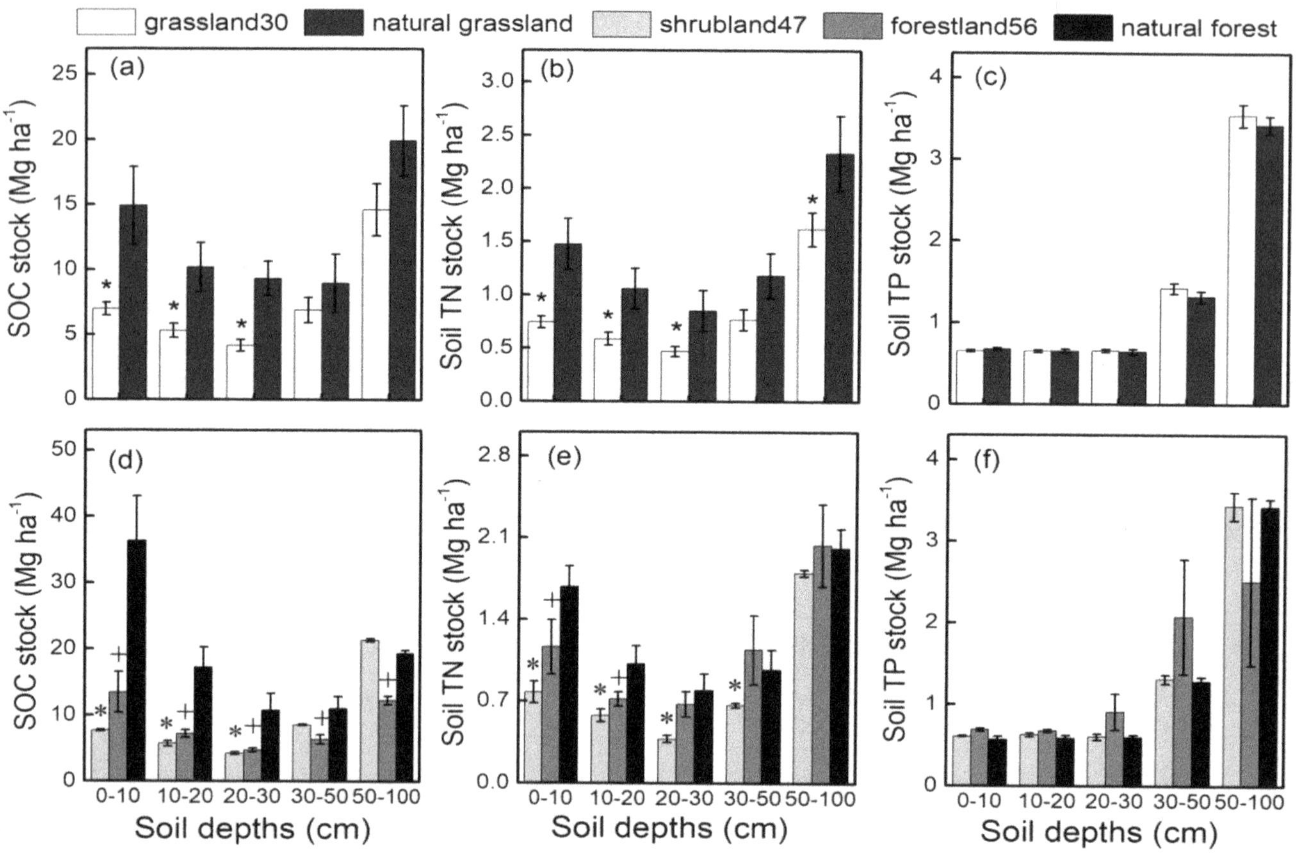

Figure 3. Stocks of soil organic carbon (SOC), soil total nitrogen (TN), and soil total phosphorus (TP) at different soil depths in grassland at 30 years compared to those of natural grassland, and those of shrubland at 47 years and forestland at 56 years compared to those of natural forest. Note: values are mean ± standard error. (**a**–**c**) means the SOC stock, Soil TN stock and Soil TP stock at different soil depths in grassland30 compared to those of natural grassland. (**d**–**f**) means the SOC stock, Soil TN stock and Soil TP stock at different soil depths in shrubland47 and forestland56 compared to those of natural forest * denotes significant differences between the natural grassland and grassland at 30 years after vegetation restoration ($p < 0.05$). * denotes significant differences between the natural forest and shrubland at 47 years after vegetation restoration ($p < 0.05$). + denotes significant differences between the natural forest and forestland at 56 years after vegetation restoration ($p < 0.05$).

3.2. Changes in SOC, Soil TN, and Soil TP Stoichiometry

In orchardland, the C:N ratio at 0–100 cm depth showed no obvious changes with restoration age, but there was an overall increasing trend (Figure 4a, Table S3). The C:P ratio at 0–30 cm increased, while the N:P ratio at 0–20 cm first increased and then decreased after 10 years of restoration (Figure 4b,c). In grassland, the C:N ratio at 0–100 cm showed an overall increasing trend with restoration age (Figure 4d, Table S3); the C:P and N:P ratios in soils of 0–30 cm depth gradually increased with restoration age (Figure 4e,f). In shrubland, the C:N ratio at 0–100 cm showed little variation with restoration age, but the overall trend was an increase (Figure 4h, Table S3); the C:P and N:P ratios at 0–10 cm depth first increased, then peaked at 30 years, before decreasing again with restoration age (Figure 4i,j). In forestland, the C:N ratio at 0–100 cm depth showed little change with restoration age, but the overall trend was an increase (Figure 4k, Table S3). The C:P ratio in soils of 0–100 cm depth gradually increased with restoration age (Figure 4l), and the N:P ratio at 0–50 cm first increased, then peaked at 37 years before decreasing again with restoration age (Figure 4m).

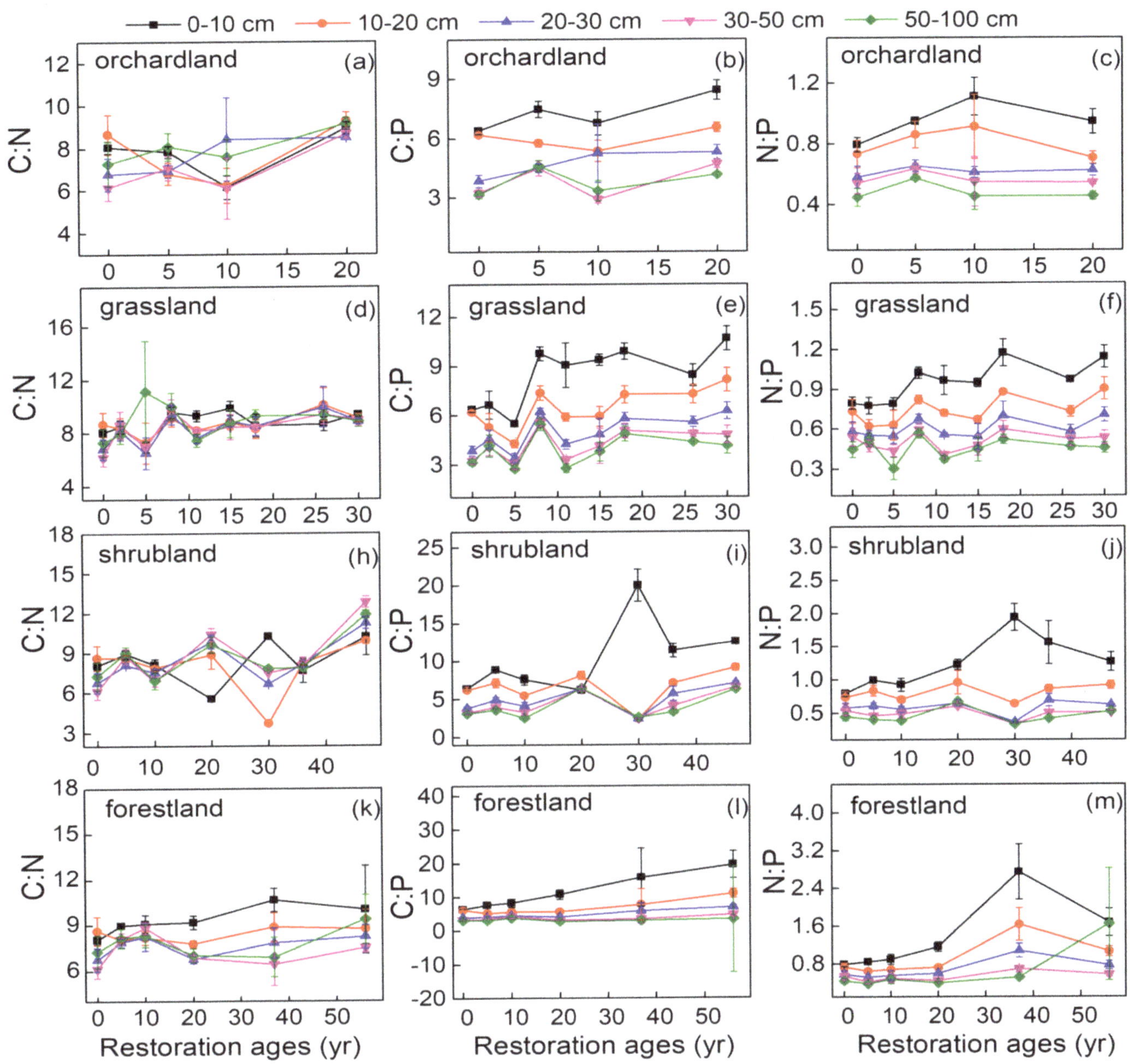

Figure 4. Stoichiometric characteristics of soil organic carbon (SOC), soil total nitrogen (TN), and soil total phosphorus (TP) with restoration age. Note: values are mean ± standard error.

The C:P and N:P ratios at 0–30 cm in grassland30 were significantly lower than those in natural grassland (Figure 5b,c), and the C:P and N:P ratios at 0–30 cm depth in shrubland47 and forestland56 were significantly lower than those in natural forest (Figure 5e,f). In addition, the C:N ratio decreased with soil depth in forestland (Figure 4k). Overall, the C:P and N:P ratios gradually decreased with soil depth in the three land types (Figure 4b,c,e,f,i,j,l,m).

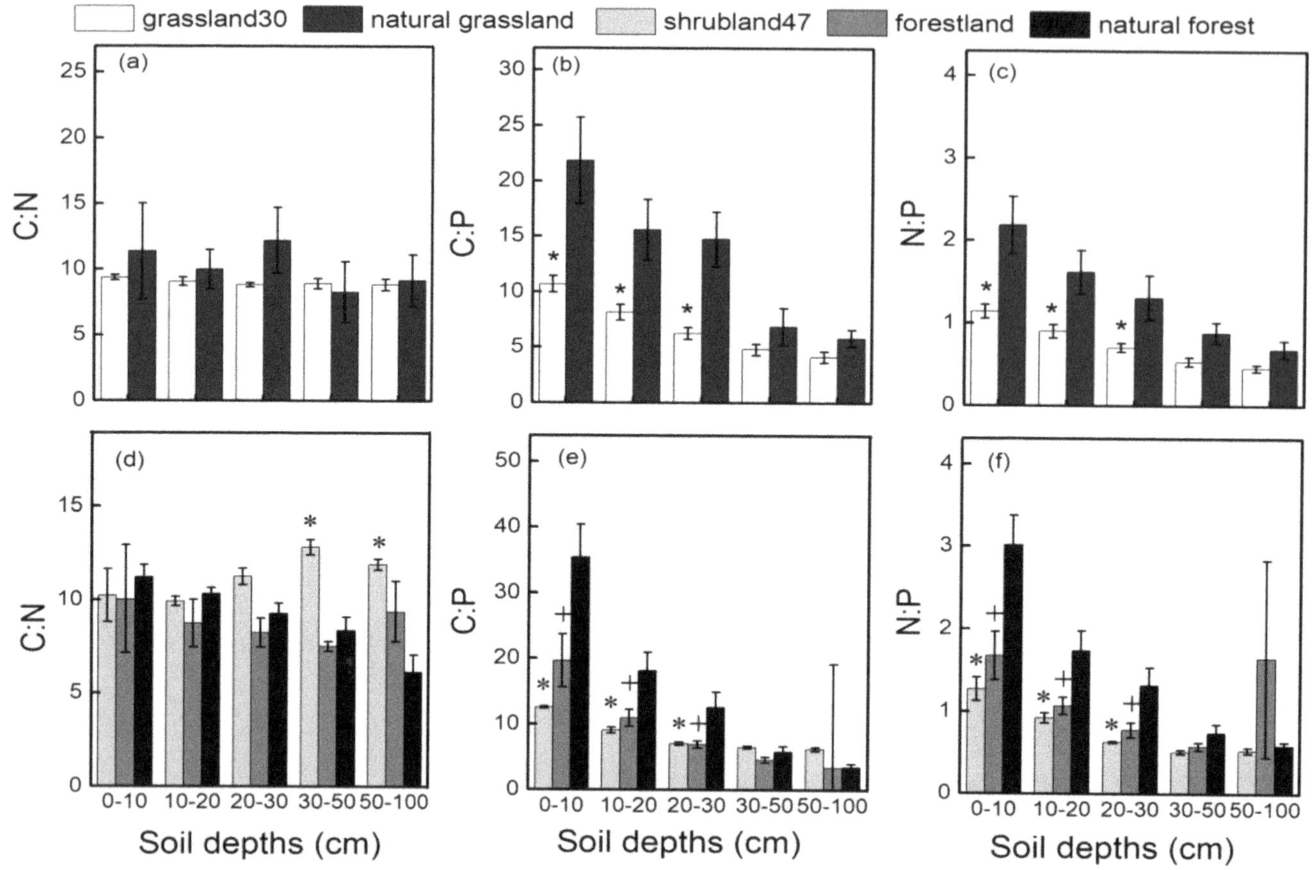

Figure 5. Stoichiometry of soil organic carbon (SOC), soil total nitrogen (TN), and soil total phosphorus (TP) at different soil depths in grassland at 30 years compared to that of natural grassland, and those of shrubland at 47 years and forestland at 56 years compared to that of natural forestland. Note: values are mean ± standard error. (**a**–**c**) means the C:N, C:P and N:P at different soil depths in grassland30 compared to those of natural grassland. (**d**–**f**) means the C:N, C:P and N:P at different soil depths in shrubland47 and forestland56 compared to those of natural forest * denotes significant differences between the natural grassland and grassland at 30 years after vegetation restoration ($p < 0.05$). * denotes significant differences between the natural forest and shrubland at 47 years after vegetation restoration ($p < 0.05$). + denotes significant differences between the natural forest and forestland at 56 years after vegetation restoration ($p < 0.05$).

4. Discussion

4.1. Restoration Ages Altered SOC, Soil TN, and Soil TP Stocks and Stoichiometry

Our results showed that the SOC stocks at 0–30 cm soil depth in orchardland, grassland, and shrubland first increased and then stabilized with restoration age, while the SOC stocks at 0–50 cm in forestland gradually increased with restoration age. SOC is mainly derived from surface litter, root secretions, and animal residues [21,22]. After vegetation restoration, a large input of litter and organic matter can enhance the accumulation of SOC [23], and with an improvement in soil structure, the levels of surface runoff, soil erosion, and soil nutrient loss may be reduced [24]. In addition, soil microbial activity is strengthened as restoration age increases [25], and soil nutrient conversion and storage are further enhanced. However, vegetation restoration involves the coordinated development of the plant community and soil environment, and the plant community structure, soil structure, and microbial diversity reach a stable level as restoration age increases [26–28]. However, in forestland, SOC stocks had not reached a steady level at 57 years, and when these stocks may stabilize needs to be evaluated. In our study, the soil TN stocks at 0–30 cm depth decreased in the later stages of orchardland, shrubland, and forestland restoration. Plants may enter a senescence phase, and the soil nitrogen nutrients absorbed during the plant growth process are then fed back to the soil by litter; thus, the nitrogen absorption rate is lower than the release rate, resulting in the soil TN stock decreasing [29]. We also found that soil TP stocks did not significantly change with restoration age, in contrast to the results of some previous studies [30]. Soil TP is mainly influenced by parent material, land use, and biogeochemical processes in the soil [31,32]. As the parent material and climate were similar for all vegetation types, the variability and migration rate of TP were not as obvious as those of SOC and soil TN, and restoration age had little effect on soil TP. Within the range of restoration ages that we studied, the SOC and soil TN stocks in grassland, shrubland, and forestland at 0–30 cm soil depth failed to reach the stock levels found in natural grassland or natural forest. Severe soil erosion and lack of water are typical characteristics of the Loess Plateau, which may lead to the loss of soil nutrients [24,33]. Additionally, soil tillage can also cause soil nutrient loss by affecting the soil physical structure and microbial activity [34].

The C:N ratio increased with restoration age after vegetation restoration. The main factors affecting the C:N ratio are changes in the SOC and soil TN contents [35]. Both SOC and soil TN contents increased overall with restoration age (Table S3), and the rate of increase of SOC was greater than that of soil TN; consequently, the C:N ratio increased (Table S3). A previous study showed that the C:N ratio was negatively correlated with the rate of decomposition of organic matter [36], so the decomposition rate of organic matter increases with restoration age. The C:P ratio indicates the availability of soil TP in the soil [37], We found that the rate of increase of SOC was also greater than that of soil TP, resulting in an increase in the C:P ratio with restoration age (Table S3). In addition, the rate of increase of soil TN was greater than that of soil TP, resulting in an increase in the N:P ratio with restoration age (Table S3). The N:P ratio was reduced in the later stages of restoration, which may be related to the lower soil TN content (Figure S1). Soil N and soil TP are essential mineral nutrients for plant growth and common limiting elements in ecosystems, and the N:P ratio is a predictor of nutrient limitation [38]. In the grassland, shrubland, and forestland types, the N:P ratio at 0–30 cm soil depth in the later stages of recovery was lower than those of natural grassland and natural forest, which may result from the more alkaline soil and lower soil TN content in the Loess Plateau Region; the soil TP content did not differ significantly between the restoration and control sites. In the grassland, shrubland, and forestland restoration types, the C:P ratio in the 0–30 cm soil layer was lower than those of natural grassland and natural forest at the later stages of recovery, which may be related to the lower SOC content in restoration soils than in natural soils.

4.2. Vertical Distribution of Stocks and Stoichiometry of SOC, Soil TN, and Soil TP

Soil depth is an important factor influencing SOC, soil TN, and soil TP distribution [39]. In our study, the SOC and soil TN stocks decreased with soil depth at all restoration ages in orchardland, grassland, shrubland, and forestland (Figure 6a,b,d,e,h,I,k,l), which is consistent with the results of previous studies [40,41]. Our study also revealed that the overall rates of SOC and soil TN content change decreased with soil depth in orchardland, grassland, shrubland, and forestland (Table S3), which indicated that the SOC and soil TN sequestration rates gradually decreased with soil depth. Meanwhile, SOC, soil TN, and soil TP were most sensitive to change in the surface soil (0–30 cm). The SOC and soil TN content at 0–30 cm represented more than 65% of the total SOC and soil TN stocks from 0–100 cm depth for all restoration types; the soil TP content at 0–30 cm represented more than 60% of the total soil TP stocks from 0–100 cm (Figure S1). Such differences in the SOC, soil TN, and soil TP profiles can be explained partly by root distribution. The surface soil is affected by external environmental factors, soil microorganisms, and the return of nutrients from surface litter, resulting in a concentration of nutrients in the surface soil [42]. With increasing soil depth, the input of organic matter is limited by the permeability of the soil, microbial decomposition activity, and root absorption [21,39]. Moreover, SOC and soil TN stocks are not only affected by soil parent material, but also by the decomposition of litter and absorption and utilization by plants [40], resulting in large spatial variability. While the soil TP content of orchardland and forestland decreased, it showed little variation with soil depth in grassland and shrubland (Figure 6c,f,j,m). Soil TP is mainly affected by the soil parent material, which, in this case, is a sedimentary mineral with low mobility in soil; therefore, there was little vertical variation in soil TP [43].

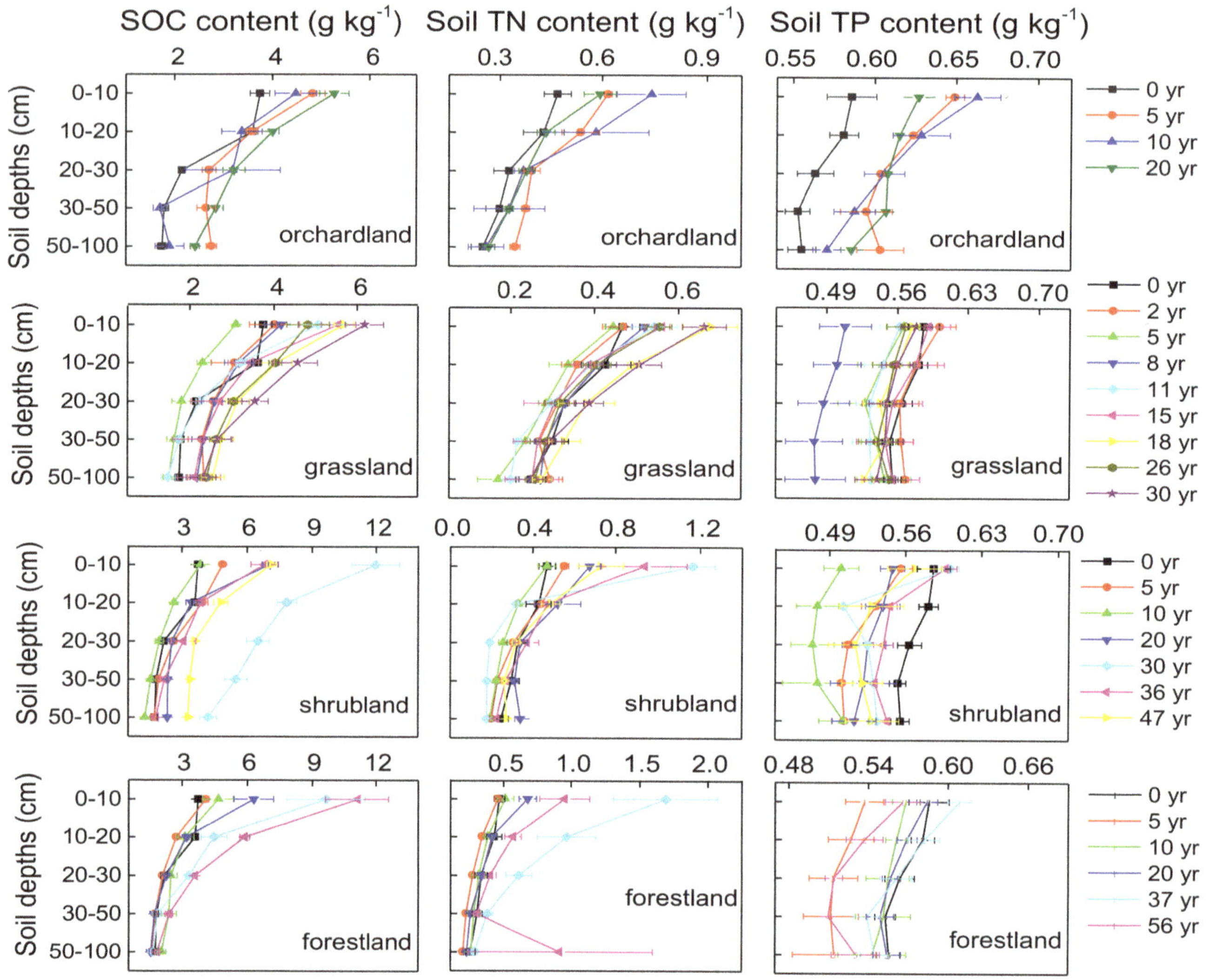

Figure 6. Vertical distributions of soil organic carbon (SOC), soil total nitrogen (TN), and soil total phosphorus (TP) contents for different vegetation types. Note: values are mean ± standard error.

Our study found that the overall C:N ratio gradually decreased with soil depth. It may be that the surface SOC and soil TN contents were higher, but as the soil depth increased, the SOC content change was larger than that of the soil TN content. When the decomposition process occurs, easily decomposed material vanishes, and soil TN is immobilized in decayed products, leaving behind more durable material with slower decomposition rates in the deeper layers [44]. This results in a relatively lower C:N ratio in the deeper soil layers. There was a significant difference in the C:P and N:P ratios at different soil depths. It may be that the soil TP content is relatively stable at different soil depths, and the C:P ratio and N:P ratio are mainly affected by SOC and soil TN content, so they showed greater variation.

4.3. Effect of Restoration Type on SOC, Soil TN, and Soil TP Stocks and Stoichiometry.

Our study demonstrated that the SOC and soil TN stocks in the 0–20 cm soil layer at 5 years were highest in orchardland, which may be related to the use of fertilizer. In other years, SOC, soil TN, and soil TP stocks showed no difference between orchardland, grassland, shrubland, and forestland

(Figure S2). However, the rate of SOC and soil TN change at 0–10 cm soil depth was the highest in forestland, while the rates of SOC and soil TN change at other depths varied among different restoration types (Table S3). The rapid increases in surface SOC and soil TN are closely related to the input of litter [45,46]. Guo et al. [46] showed that forest litter was 19 times greater in mass than that of shrubs in the Loess Plateau Region. At the same time, the presence of the higher amount of organic carbon over a long period of time indicated that forestland likely had developed a higher degree of humification.

The long-term development of soil organic carbon stocks suggested that the afforestation improved the humification process. It has been shown that the process of humification in the soil is critical for the ecosystem, due to its strong contributions to the improvement of fertility and, hence, the storage of both carbon and nitrogen [47,48]. This also explains why the increase in SOC and soil TN in the topsoil of forests was higher than that of shrubs. As soil depth increases, root secretions and soil microorganisms are the main sources of soil nutrition [21,22]. There are significant differences in the effects of different plant roots and litters on the community composition of microorganisms [49], which may explain the large differences in the rates of SOC and soil TN change between different restoration types.

The C:N ratios of orchardland5 and grassland5 at 0–20 cm soil depth were lower than those of shrubland5 and forestland5, and the C:N ratio of orchardland10 was significantly lower than those of grassland10, shrubland10, and forestland10 (Figure S2). The C:N ratio of orchardland was lower than those of grassland, shrubland, and forestland, which may be related to anthropogenic N deposition in orchardland (Figure S2). The lower C:N ratio in grassland5 may occur because grassland retains more organic matter content and greater nutrient absorption takes place through plant roots [8]. There were no significant differences in the C:N ratio at other soil depths of orchardland, grassland, shrubland, and forestland as there were no significant differences in the stocks of SOC and soil TN (Figure S2).

This finding is related to the nutrient conditions of the soil in the study area and the feedback between plant and soil, so the soil stoichiometric changes in different restoration types showed the same characteristics in the same environmental context. In addition, the overall C:P and N:P ratios of orchardland5 and shrubland5 at 0–20 cm soil depth were higher than those of grassland5 and forestland5, which is related to the relatively higher SOC and soil TN content in orchardland5 and shrubland5 (Figure S2). However, there were no significant differences in the C:P and N:P ratios at 30–100 cm soil depth between the four restoration types because there were no significant differences in the SOC, soil TN, and soil TP stocks of orchardland, grassland, shrubland, and forestland (Figure S2).

5. Conclusions

We examined the changes in SOC, soil TN, and soil TP stocks and stoichiometry at depths of 0–100 cm following vegetation restoration in the Loess Hilly Region. Our results revealed that the SOC stocks appeared to increase and reach stable levels. The soil TN stocks first increased and then decreased with restoration age, but they did not reach the levels seen in natural grassland or natural forest in the Loess Hilly Region in the absence of appropriate management. Soil TP stocks failed to improve substantially with restoration age.

The C:N, C:P, and N:P ratios gradually increased with restoration age. At the same time, the SOC, soil TN, and soil TP stocks and the C:N, C:P, and N:P ratios decreased with soil depth. Reforested land had the highest sequestration rate of SOC and TN at 0–10 cm soil depth. The results of this study provide data for the assessment of the long-term SOC, soil TN, and soil TP stocks and stoichiometry after vegetation restoration under different restoration types in the Loess Hilly Region.

Author Contributions: The research design was completed by H.X. and S.X. The manuscript was written by H.X. and Q.Q. The collection and analysis of soil samples were performed by H.X., P.L., Q.Q., Z.G. and E.W.

Acknowledgments: Many thanks to Juan Li, Dou Zhang and Zhao Fang for their technical assistance in the laboratory work. We would like to thank Mancai Guo and Minggang Wang for providing statistics assistance.

References

1. Reed, S.C.; Yang, X.; Thornton, P.E. Incorporating phosphorus cycling into global modeling efforts: A worthwhile, tractable endeavor. *New Phytol.* **2015**, *208*, 324–329. [CrossRef] [PubMed]
2. Lv, J.L.; Yan, M.J.; Song, B.L. Ecological stoichiometry characteristics of soil carbon, nitrogen, and phosphorus in an oak forest and a black locust plantation in the Loess hilly region. *Acta Ecol. Sin.* **2017**, *10*, 3385–3393, (In Chinese with English Abstract). [CrossRef]
3. Rodríguez, E.S.; Lozano, Y.M.; Bardgett, R.D. Influence of soil microbiota in nurse plant systems. *Funct. Ecol.* **2016**, *30*, 30–40. [CrossRef]
4. Rudel, T.K.; Coomes, O.T.; Moran, E.; Achard, F.; Angelsen, A.; Xu, J.; Lambin, E. Forest transitions: Towards a global understanding of land use change. *Glob. Environ.* **2005**, *15*, 23–31. [CrossRef]
5. Winans, K.; Whalen, J.; Cogliastro, A.; Rivest, D.; Ribaudo, L. Soil Carbon Stocks in Two Hybrid Poplar-Hay Crop Systems in Southern Quebec, Canada. *Forests* **2014**, *5*, 1952–1966. [CrossRef]
6. Jiang, P.; Chen, Y.; Cao, Y. C:N:P Stoichiometry and Carbon Storage in a Naturally-Regenerated Secondary Quercus variabilis Forest Age Sequence in the Qinling Mountains, China. *Forests* **2017**, *8*, 281. [CrossRef]
7. Wang, W.; Sardans, J.; Zeng, C.; Zhong, C.; Li, Y.; Peñuelas, J. Responses of soil nutrient concentrations and stoichiometry to different human land uses in a subtropical tidal wetland. *Geoderma* **2014**, *232–234*, 459–470. [CrossRef]
8. Gao, Y.; He, N.; Yu, G.; Chen, W.; Wang, Q. Long-term effects of different land use types on C, N, and P stoichiometry and storage in subtropical ecosystems: A case study in China. *Ecol. Eng.* **2014**, *67*, 171–181. [CrossRef]
9. Zhao, F.; Sun, J.; Ren, C.; Kang, D.; Deng, J.; Han, X.; Yang, G.; Feng, Y.; Ren, G. Land use change influences soil C, N and P stoichiometry under 'Grain-to-Green Program' in China. *Sci. Rep.* **2015**, *5*, 10195. [CrossRef]
10. Fang, J.; Chen, A.; Peng, C.; Zhao, S.; Ci, L. Changes in Forest Biomass Carbon Storage in China Between 1949 and 1998. *Science* **2001**, *292*, 2320–2322. [CrossRef]
11. Post, W.M.; Kwon, K.C. Soil carbon sequestratrion and land-use change: Processes and potential. *Glob. Environ.* **2010**, *6*, 317–327. [CrossRef]
12. Xiao, H.; Li, Z.; Dong, Y.; Chang, X.; Deng, L.; Huang, J.; Nie, X.; Liu, C.; Liu, L.; Wang, D.; et al. Changes in microbial communities and respiration following the revegetation of eroded soil. *Agric. Ecosyst. Environ.* **2017**, *246*, 30–37. [CrossRef]
13. Yang, W.; Cheng, H.; Hao, F.; Ouyang, W.; Liu, S.; Lin, C. The influence of land-use change on the forms of phosphorus in soil profiles from the Sanjiang Plain of China. *Geoderma* **2012**, *189–190*, 207–214. [CrossRef]
14. Liu, X.; Ma, J.; Ma, Z.; Li, L. Soil nutrient contents and stoichiometry as affected by land-use in an agro-pastoral region of northwest China. *Catena* **2017**, *150*, 146–153. [CrossRef]
15. Li, C.; Zhao, L.; Sun, P.; Zhao, F.Z.; Kang, D.; Yang, G.H. Deep Soil C, N, and P Stocks and Stoichiometry in Response to Land Use Patterns in the Loess Hilly Region of China. *PLoS ONE* **2016**, *11*, e0159075. [CrossRef] [PubMed]
16. Fei, W.; Rui, L.I.; Zhong, W. Problems and Proposals on Policy of Converting Cropland Into Forest and Grassland: A Case-based Study. *Nat. Sci. Ed.* **2003**, *3*, 60–65, (In Chinese with English Abstract). [CrossRef]
17. Xin, Z.; Ran, L.; Lu, X.X. Soil Erosion Control and Sediment Load Reduction in the Loess Plateau: Policy Perspectives. *Int. J. Water Resour. Dev.* **2012**, *28*, 325–341. [CrossRef]
18. Nelson, D.W.; Sommers, L.E. Total carbon, organic carbon and organic matter. In Methods of Soil Analysis Part 3—Chemical Methods. *Chem. Microbiol. Prop.* **1982**. [CrossRef]
19. Bremner, J.; Mulvaney, C. Nitrogen—total. *Methods of Soil Analysis Chemical Methods Part.* **1996**, *72*, 532–535. [CrossRef]
20. Schimel, J.P.; Weintraub, M.N. The implications of exoenzyme activity on microbial carbon and nitrogen limitation in soil: A theoretical model. *Soil Biol. Biochem.* **2003**, *35*, 549–563. [CrossRef]

21. Clemmensen, K.E.; Bahr, A.; Ovaskainen, O.; Dahlberg, A.; Ekblad, A.; Wallander, H.; Stenlid, J.; Finlay, R.D.; Wardle, D.A.; Lindahl, B.D. Roots and Associated Fungi Drive Long-Term Carbon Sequestration in Boreal Forest. *Science* **2013**, *339*, 1615–1618. [CrossRef] [PubMed]
22. Fontaine, S.; Barot, S.; Barré, P.; Bdioui, N.; Mary, B.; Rumpel, C. Stability of organic carbon in deep soil layers controlled by fresh carbon supply. *Nature* **2007**, *450*, 277–280. [CrossRef] [PubMed]
23. Yang, Y.; Luo, Y. Carbon:nitrogen stoichiometry in forest ecosystems during stand development. *Glob. Ecol. Biogeogr.* **2011**, *20*, 354–361. [CrossRef]
24. Li, Z.; Liu, C.; Dong, Y.; Chang, X.; Nie, X.; Liu, L.; Xiao, H.; Lu, Y.; Zeng, G. Response of soil organic carbon and nitrogen stocks to soil erosion and land use types in the Loess hilly–gully region of China. *Soil Tillage Res.* **2017**, *166*, 1–9. [CrossRef]
25. Leon, D.G.D.; Moora, M.; Öpik, M.; Neuenkamp, L.; Gerz, M.; Jairus, T.; Vasar, M.; Bueno, C.G.; Davison, J.; Zobel, M. Symbiont dynamics during ecosystem succession: Co-occurring plant and arbuscular mycorrhizal fungal communities. *FEMS Microbiol. Ecol.* **2016**, *92*, 1–9. [CrossRef]
26. Hao, W.F.; Liang, Z.S.; Chen, C.G.; Tang, L. Study of species diversity evolvement process during vegetation restoration of abandoned farmland in the hilly loess plateau. *Pratac. Sci.* **2005**, *22*, 1–8, (In Chinese with English Abstract). [CrossRef]
27. Novara, A.; Mantia, T.L.; Rühl, J.; Badalucco, L.; Kuzyakov, Y.; Gristina, L.; Laudicina, V.A. Dynamics of soil organic carbon pools after agricultural abandonment. *Geoderma* **2014**, *235–236*, 191–198. [CrossRef]
28. O'Brien, S.L.; Jastrow, J.D. Physical and chemical protection in hierarchical soil aggregates regulates soil carbon and nitrogen recovery in restored perennial grasslands. *Soil Biol. Biochem.* **2013**, *61*, 1–13. [CrossRef]
29. Jia, H.E.; Yu-Fu, H.U.; Shu, X.Y.; Wang, Q.; Jia, A.D.; Yan, X.; University, S.A. Effect of Salix cupularis plantations on soil stoichiometry and stocks in the alpine-cold desert of northwestern. *Sin. Acta Agrestia Sin.* **2018**. (In Chinese with English Abstract). [CrossRef]
30. Lane, P.N.J.; Noske, P.J.; Sheridan, G.J. Phosphorus enrichment from point to catchment scale following fire in eucalypt forests. *Catena* **2011**, *87*, 157–162. [CrossRef]
31. Cheng, Y.; Li, P.; Xu, G.; Gao, H. Spatial distribution of soil total phosphorus in Yingwugou watershed of the Dan River, China. *Catena* **2016**, *136*, 175–181. [CrossRef]
32. Kooijman, A.M.; Jongejans, J.; Sevink, J. Parent material effects on Mediterranean woodland ecosystems in NE Spain. *Catena* **2005**, *59*, 55–68. [CrossRef]
33. Comino, J.R.; Sinoga, J.D.R.; González, J.M.S.; Guerra-Merchán, A.; Seeger, M.; Ries, J.B. High variability of soil erosion and hydrological processes in Mediterranean hillslope vineyards (Montes de Málaga, Spain). *Catena* **2016**, *145*, 274–284. [CrossRef]
34. Duan, X.N.; Wang, X.K.; Lu, F.; Ouyang, Z.Y. Soil carbon sequestration and its potential by grassland ecosystems in China. *Acta Ecol. Sin.* **2008**. (In Chinese with English Abstract). [CrossRef]
35. Reich, P.B.; Oleksyn, J. Global patterns of plant leaf N and P in relation to temperature and latitude. *Proc. Natl. Acad. Sci. USA* **2004**, *101*, 11001–11006. [CrossRef] [PubMed]
36. An, H.; Li, G.Q. Effects of grazing on carbon and nitrogen in plants and soils in a semiarid desert grassland, China. *J. Arid Land* **2015**, *7*, 341–349. [CrossRef]
37. Yu, Y.F.; Peng, W.X.; Song, T.Q. Stoichiometric characteristics of plant and soil C, N and P in different forest types in depressions between karst hills, southwest China. *Chin. J. Appl. Ecol.* **2014**, *25*, 947–954, (In Chinese with English Abstract). [CrossRef]
38. Chen, L.; Li, C.L. Research advances in ecological stoichiometry of marine plankton. *Chin. J. Appl. Ecol.* **2014**, *25*, 3047–3055, (In Chinese with English Abstract). [CrossRef]
39. Berger, T.W.; Neubauer, C.; Glatzel, G. Factors controlling soil carbon and nitrogen stores in pure stands of Norway spruce (*Picea abies*) and mixed species stands in Austria. *For. Ecol. Manag.* **2002**, *159*, 3–14. [CrossRef]
40. Liu, X.Z.; Zhou, G.Y.; Zhang, D.Q.; Liu, S.Z.; Chu, G.W.; Yan, J.H. N and P stoichiometry of plant and soil in lower subtropical forest successional series in southern China. *J. Plant Ecol.* **2010**, *34*, 64–71, (In Chinese with English Abstract). [CrossRef]
41. Zeng, Q.; Liu, Y.; Fang, Y.; Ma, R.; Lal, R.; An, S.; Huang, Y. Impact of vegetation restoration on plants and soil C:N:P stoichiometry on the Yunwu Mountain Reserve of China. *Ecol. Eng.* **2017**, *109*, 92–100. [CrossRef]
42. Barreto, P.A.B.; Gama-Rodrigues, E.F.; Fontes, A.G.; Polidoro, J.C.; Moço, M.K.S.; Machado, R.C.R.; Baligar, V.C. Distribution of oxidizable organic C fractions in soils under cacao agroforestry systems in Southern Bahia, Brazil. *Agrofor. Syst.* **2011**, *81*, 213–220. [CrossRef]

43. Zhang, X. Ecological stoichiometry characteristics of robinia pseudoacacia forest soil in different latitudes of loess plateau. *Acta Ecol. Sin.* **2013**, *50*, 818–825. [CrossRef]
44. Han, L.; Li, Z.; Ceng, Y. Carbon, nitrogen, and phosphorous stoichiometry of herbaceous plant leaf and soil in riparian zone of Taihu Lake basin, East China under effects of different land use types. *Chin. J. Ecol.* **2013**, *32*, 3281–3288, (In Chinese with English Abstract). [CrossRef]
45. Vesterdal, L.; Ritter, E.; Gundersen, P. Change in soil organic carbon following afforestation of former arable land. *For. Ecol. Manag.* **2002**, *169*, 137–147. [CrossRef]
46. Guo, S.L.; Ma, Y.H.; Che, S.G.; Sun, W.Y. Effects of artificial and natural vegetations on litter production and soil organic carbon change in loess hilly areas. *Sci. Silvae Sin.* **2009**, *45*, 14–18, (In Chinese with English Abstract). [CrossRef]
47. Pizzeghello, D.; Francioso, O.; Concheri, G.; Muscolo, A.; Nardi, S. Land Use Affects the Soil C Sequestration in Alpine Environment, NE Italy. *Forests* **2017**, *8*, 197. [CrossRef]
48. De Feudis, M.; Cardelli, V.; Massaccesi, L.; Lagomarsino, A.; Fornasier, F.; Westphalen, D.J.; Coccco, S.; Corti, G.; Agnelli, A. Influence of Altitude on Biochemical Properties of European Beech (*Fagus sylvatica* L.) Forest Soils. *Forests* **2017**, *8*, 213. [CrossRef]
49. Cesarz, S.; Fender, A.C.; Beyer, F.; Valtanen, K.; Pfeiffer, B.; Gansert, D.; Hertel, D.; Leuschner, C. Roots from beech (*Fagus sylvatica* L.) and ash (*Fraxinus excelsior* L.) differentially affect soil microorganisms and carbon dynamics. *Soil Biol. Biochem.* **2013**, *61*, 23–32. [CrossRef]

6

Phosphorus in Preferential Flow Pathways of Forest Soils in Germany

Dorit Julich *, Stefan Julich and Karl-Heinz Feger

Institute of Soil Science and Site Ecology, TU Dresden, Pienner Str. 19, 01737 Tharandt, Germany; Stefan.Julich@tu-dresden.de (S.J.); karl-heinz.feger@tu-dresden.de (K.-H.F.)

* Correspondence: dorit.julich@tu-dresden.de

Academic Editors: Scott X. Chang and Xiangyang Sun

Abstract: The transport of nutrients in forest soils predominantly occurs along preferential flow pathways (PFP). This study investigated the composition of phosphorus (P) forms in PFPs and soil matrix in several temperate beech forests with contrasting soil P contents in Germany. The PFPs were visualized using dye tracer experiments. Stained and unstained soil was sampled from three profile cuts per plot and analyzed for P fractions. The results show that labile P concentrations were highest in the O-layer and had the same range of values at all sites (240–320 mg·kg^{-1}), although total P (TP) differed considerably (530–2330 mg·kg^{-1}). The ratio of labile P to TP was significantly lower in the P-rich soil compared to the medium and P-poor soils. By contrast, the ratio of moderately labile P to TP was highest at the P-rich site. The shifts in P fractions with soil depth were generally gradual in the P-rich soil, but more abrupt at the others. The contents of labile and moderately labile P clearly differed in PFPs compared to soil matrix, but not statistically significant. The studied soils are characterized by high stone contents with low potential for P sorption. However, indications were found that labile organically bound P accumulates in PFPs such as biopores.

Keywords: phosphorus fractions; forest soil; preferential flow; PFP

1. Introduction

Phosphorus (P) in forest soils is distributed highly variably, both between and within soil types, with respect to contents, speciation, availability, and sources of P [1]. Soils contain various P forms, generally differentiated in organically bound (P_o) and inorganic (Pi) P. Organic P occurs as orthophosphate monoesters (e.g., inositols, mononucleotides, sugar phosphates), orthophosphate diesters (phospholipids, teichoic acids, nucleic acids, and their degradation products), and phosphonates. Inorganic P is bound in minerals (e.g., apatite) and is present in dissolved form as ortho-, pyro-, and/or polyphosphates [2,3]. The total content and stock of soil P varies from very low (<100 mg·kg^{-1} and <10 g·m^{-2}) to P-rich (>3000 mg·kg^{-1} and >1000 g·m^{-2}), and the mineral P supply and P speciation is strongly dependent on the parent material and pedogenesis [4,5]. Whereas P supply from the mineral phase is the main source of available P in P-rich soils, P availability for plants in P-poor soils is considerably influenced by P speciation [4]. These differences in P supply drive P fluxes in the systems and determine the dominating strategy of ecosystem nutrition [6]. Besides those internal factors, external environmental and anthropogenic impacts (as forest management, N deposition, etc.) substantially influence the distribution of P and nutrient cycling in temperate forest soils [7–10].

The predominant transport pathway for nutrients in forest soils is along preferential flow pathways (PFP), bypassing large parts of the soil matrix [11–13]. An established method to investigate PFPs and their role in water and nutrient transport are tracer experiments [14–17]. Previous tracer studies indicated that PFPs in forest soils can be stable for decades, resulting in differences in biological and chemical parameters between PFPs and the soil matrix, particularly when preferential flow along root channels dominates the flow system [12,18]. Accumulation in stable flow paths has been found in previous studies for radionuclides, SOC, N [15,18], and also for C, Ca, Mg, and Fe [12]. Backnäs et al. [19] investigated P fractions in PFPs as compared to the soil matrix in a Podzol under forest in Southern Lapland, Finland. The results indicated that the leaching of P through PFPs occurred primarily along stone surfaces, where PFPs are larger than PFPs related to coarse grains or roots. Thus, P sorption may be limited due to potentially faster water flow, weaker lateral mass exchange and less sorption interaction than in smaller PFPs [20]. By contrast, P accumulation might occur in small preferential pores with decelerated water flow and longer contact time between the solution and inner pore surfaces [19]. As a result, such PFPs with high potential to accumulate P increase the likelihood for elevated P concentrations in export fluxes [21]. However, there are very few data available regarding P accumulation and transport in forest soils. Therefore, this study investigated the composition of P forms in PFPs and the soil matrix in four temperate forest ecosystems in Germany. Furthermore, the influence of different initial conditions regarding the content of total and available P at the sites on the distribution of P forms in the soil profiles and in PFPs was examined. The correlation between P fractions and chemical parameters was tested to identify the most important factors influencing the occurrence of P fractions in the investigated soils.

2. Materials and Methods

2.1. Site Description

Field experiments and sampling were conducted at four beech dominated forests (*F. sylvatica* L.) across Germany with similar stand age (110–120 years). The four study sites Lüss (LUE), Vessertal (VES), Mitterfels (MIT), and Bad Brückenau (BBR) are core sites of the German DFG Priority Program SPP 1685 which aims at testing the overall hypothesis that the P-depletion of soils drives forest ecosystems from P acquiring systems (efficient mobilization of P from the mineral phase) to P recycling systems (highly efficient cycling of P) [6]. Further, all sites are long-term monitoring sites in Germany and part of the EU-ICP Forests Level II Program (International Co-operative Programme on the Assessment and Monitoring of Air Pollution Effects on Forests). The sites were selected with regard to their considerable different availability of mineral P following the sequence BBR > MIT > VES > LUE. The BBR site was defined as P-rich site, LUE as P-poor site, and MIT and VES as medium P-supplied sites. Three sites (VES, MIT, and BBR) are located in the low mountain ranges, whereas LUE is in the lowlands. The sites differ in parent material, soil type, and climate (Table 1). Soil properties are summarized in Table 2. For better comparisons, the soil horizons were aggregated for each site considering their physical and chemical properties to: O—organic layer, A—topsoil horizon, B—subsoil horizon, C—lower subsoil horizon containing weathered parent material.

Table 1. Overview over the study sites.

	LUE	VES	MIT	BBR
Location	52°50′32″ N 10°16′06″ E	50°36′26″ N 10°46′20″ E	48°58′35″ N 12°52′49″ E	50°21′38″ N 9°55′71″ E
Elevation (m a.s.l.)	115	810	1023	809
MAT (°C)	8.0	5.5	4.5	5.8
MAP (mm)	779	1200	1229	1031
Parent material	Pleistocene sands	Trachyandesite	Gneiss	Basalt
Soil type (WRB *)	Dystric skeletic Cambisol	Hyperdystric skeletic folic Cambisol	Hyperdystric skeletic chromic Cambisol	Hyperdystric folic Cambisol

LUE: Lüss (Lower Saxony); VES, Vessertal (Thuringian Forest); MIT, Mitterfels (Bavarian Forest); BBR, Bad Brückenau (Bavarian Rhön Mountains); MAT, Mean annual temperature; MAP, Mean annual precipitation; * World Reference Base for Soil Resources 2014, World Soil Resources Reports 106, FAO, Rome, Italy.

Table 2. Soil properties at four study sites in beech forests in Germany.

Study Site/ Horizon	Depth (cm)	Texture	pH ($CaCl_2$)	C (%)	N (%)	Al_{ox}	Al_{di}	Fe_{ox}	Fe_{di}	P_{ox}	TP
						($g{\cdot}kg^{-1}$)					
Lüss (LUE)											
O	9 to 0		2.8	36.63	1.59	0.73	0.79	0.87	1.30	0.15	0.49
A	0 to −12	Loamy sand	3.0	1.69	0.07	0.18	0.37	0.29	2.17	<0.04	0.06
B	−12 to −52	Loamy sand	3.7	0.90	0.04	1.04	1.00	1.40	3.21	<0.04	0.08
C	−52 to −100	Sand	4.3	0.12	0.01	0.50	0.41	0.15	1.01	<0.04	0.04
Vessertal (VES)											
O	4 to 0		3.4	32.17	1.72	4.11	1.20	1.28	1.75	0.30	0.86
A	0 to −3	Sandy loam	3.0	7.49	0.45	2.75	1.68	3.14	7.46	0.53	0.70
B	−3 to −65	Sandy loam	4.1	3.03	0.20	9.44	7.40	5.13	9.23	0.83	0.98
C	−65 to −100	Loamy sand	4.1	0.37	0.03	4.64	2.83	0.59	5.55	0.16	0.32
Mitterfels (MIT)											
O	4 to 0		3.5	44.12	2.19	0.95	0.87	1.28	1.45	0.34	0.88
A	0 to −4	Silt loam	3.0	35.50	2.05	3.32	3.83	3.58	4.97	0.47	1.14
B	−4 to −80	Silt loam	3.9	5.45	0.34	6.26	5.62	8.05	12.17	0.53	0.91
C	−80 to −100	Loamy sand	4.3	1.09	0.09	5.73	4.63	3.61	10.89	0.35	0.66
Bad Brückenau (BBR)											
O	3 to 0		4.6	27.33	1.68	5.53	5.51	15.27	21.85	1.29	2.24
A	0 to −7	Silty clay	4.2	17.57	1.28	7.77	8.30	21.15	31.52	1.51	2.45
B	−7 to −30	Silt clay loam	4.3	7.45	0.54	10.58	10.83	24.46	35.41	1.84	2.62
C	−30 to −65	Loam	4.5	4.60	0.34	10.70	9.57	18.87	25.92	1.63	2.39

Horizon: O—organic layer, A—topsoil horizon, B—subsoil horizon, C—lower subsoil horizon containing weathered parent material. Values for C and N are total contents (for C this equals organic C); Al_{ox}, Fe_{ox}, and P_{ox}—ammonium oxalate-extractable contents; Al_{di} and Fe_{di}—dithionite-extractable contents; TP—total P contents.

2.2. Field Experiments

At each study site a test plot of 1 m^2 was established at a minimum distance of 2 m from any trees. On the first day, the plots were irrigated with 30 L of tracer-water solution containing 3 $g{\cdot}L^{-1}$ of Brilliant Blue FCF (within 1 hour) to activate rapid flow paths (PFPs) in the soils. Brilliant Blue dye was used due to its good visibility, retardation behavior in soil, and its non-toxicity to the environment [16,17]. After the dye tracer irrigation, the plots were sprinkled with 2 L of water to wash off the dye residues from the organic layer [19], and then covered overnight with a tarpaulin to exclude disturbances (e.g., rainfall). The irrigated plots were excavated on the following day by successively preparing five horizontal profile cuts per plot (at 20, 35, 50, 65, and 80 cm from the plot border) down to 1 m depth and 1 m width (see Figure 1). Stained (flow region) and unstained (non-flow region) soil was sampled respectively from the three inner cuts at 35, 50, and 65 cm from all morphological horizons. The amounts of soil material per sample varied because the thickness of soil horizons and stained/unstained parts per horizon differed considerably. The samples were sieved (<2 mm), dried at 40 °C, and then stored in plastic bottles until analysis. For the analysis of C, N, and total P the samples were finely grounded.

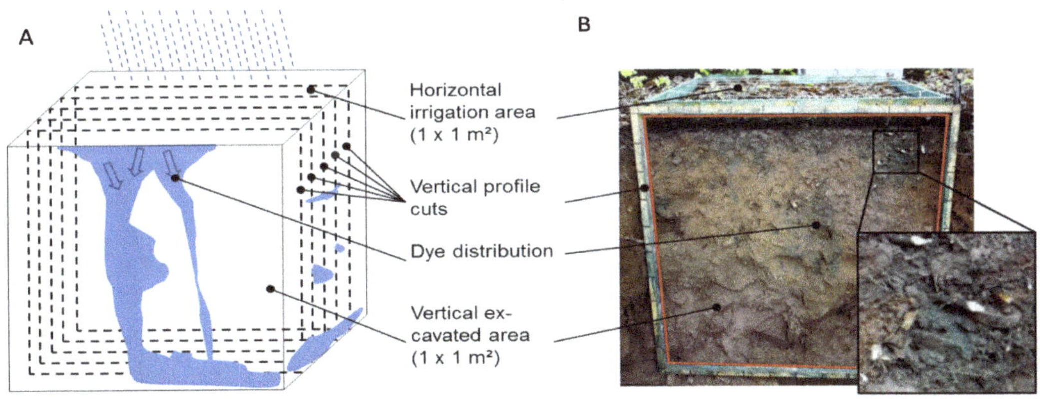

Figure 1. Scheme of tracer experiments (**A**): Dye application at the soil surface (1 m^2), covering overnight, and excavation of five vertical (1 m^2) profile cuts (example on the right (**B**)). The black framed enlargement shows staining of flow pathways along roots and stone surfaces.

2.3. Chemical Analyses

The P fractions in stained and unstained soil samples were analyzed using the Hedley fractionation method [22,23]. This method uses increasingly strong extractants to determine P forms with differing availability for plants. All samples were extracted with a replicate. Additionally, certified reference soil material from the Wageningen Evaluating Programs for Analytical Laboratories, International Soil-analytical Exchange (WEPAL-ISE 865 and 884) and blanks were treated equally to the samples for quality control of the extraction procedure.

In the fractionation procedure, a subsample of 0.5 g soil was first shaken in distilled water containing an anion exchange resin (Dowex 22, 16–50 mesh, Sigma-Aldrich, Taufkirchen, Germany) in netting bags (PP-405/230-41, Bückmann, Mönchengladbach, Germany) [24]. The following extractants applied were: 0.5 M sodium bicarbonate ($NaHCO_3$-P), 0.1 M sodium hydroxide (NaOH-$_i$), 1 M hydrochloric acid (HCl_{dil}-P), hot concentrated 3.5 M hydrochloric acid (HCl_{conc}-P), and finally nitric acid digestion of the residual P fraction (residual P). After each extraction step, the amount of inorganic P in the extract solution was measured using the method of Murphy and Riley [25] and O'Halloran and Cade-Menun [26]. For the determination of total P (TP) in $NaHCO_3$, NaOH, and HCl_{conc}, the extracts were oxidized in an autoclave using ammonium persulfate. Organically bound P (P_o) was calculated as the difference between TP and P_i [23]. The P concentrations in all extracts were measured colorimetrically with a spectrophotometer (UV-mini 1240, Shimadzu Deutschland GmbH, Duisburg, Germany). For a detailed description of the procedure applied here see Tiessen and Moir [23].

According to the concept of Hedley fractionation, the different P fractions represent (1) surface-adsorbed readily available P, (2) strongly-bonded or adsorbed moderate available P, and (3) very strongly-bonded or inaccessible mineral P with very low availability [22,23,27]. Therefore, the analyzed P fractions were classified for further data analyses as labile P (resin-P_i + $NaHCO_3$-$P_{i,o}$), moderately labile P (NaOH-$P_{i,o}$ + HCl_{dil}-P_i), and stable P (HCl_{conc}-$P_{i,o}$ + residual-P) (cf. [28]). The initial TP content of all samples was determined using the same method as the residual P fraction, by digestion with concentrated nitric acid and measurement with ICP-OES (CIROS, SPECTRO Analytical Instruments GmbH, Kleve, Germany). Total C and N contents were measured after dry combustion of grounded samples with a CN analyzer (Vario EL III Elementar GmbH, Hanau, Germany). The soils contained no inorganic C, thus soil organic C (SOC) equals total C. The concentrations of C in PFPs were not corrected for any C added as part of the Brilliant Blue dye. The C content of the dye is 56%, and the amount of total added C in the experiments was 50 g C m^{-2}. Bundt et al. [15] conducted comparable experiments with a dye-derived addition of 75 g C m^{-2} and found a resulting increase of C with a maximum of 2.5 % SOC in in PFPs in 0–9 cm depth, and a maximum of 3.5 % SOC in PFPs in 50–100 cm depth. In the current study, less dye was applied to the plots, and thus the change in C content in PFPs is expected to be very low. Amorphous metal sesquioxides (Al_{ox}, Fe_{ox}, Mn_{ox}) and P_{ox} were determined by extraction with acid 0.2 M ammonium oxalate (100 mL added to a subsample of 2 g). Dithionite-extractable (Fe_{di}, Al_{di}, Mn_{di}) was analyzed with 50 mL Na-citrate-dithionite in a subsample of 0.5 g (modified after Holmgren, 1967). The measurement of element contents in the extracts was conducted with ICP-OES.

2.4. Statistical Analysis

The statistical significance of differences between P contents of different chemical fractions in flow pathways and matrix were tested using the R software package (version R i386 3.1.0, R Foundation for Statistical Computing 2014, Vienna, Austria). Given that most of the data sets exhibited non-normal distributions (tested by Shapiro-Wilk test and q-q-plot), the non-parametric Wilcoxon rank-sum test was used to statistically compare mean values. The statistical tests were conducted pair-wise including all data of the four sites, and at horizon level due to the hierarchical structure of the data (profiles, cuts, horizons). For correlations between chemical parameters and P fractions, the Spearman's rho correlation coefficients were calculated and tested for significance.

3. Results and Discussion

The soil P fractions classified as labile, moderately labile, and stable P for PFP and soil matrix samples are shown in Figure 2. The values represent mean values for the three vertical cuts per profile. For the forest floor (O-layer), it was not possible to distinguish between stained and unstained material, due to the homogeneous distribution of the dye tracer from the irrigation. At the sites VES and MIT, the tracer did not reach the C horizon. Thus, the presented results contain no soil matrix data for the O-layer and no PFP data for the lower subsoil.

The TP values differed at the four study sites in their contents (45–2700 mg·kg^{-1}) and distribution within the soil profiles (Figure 2, TP = total bar length). Whereas TP at the P-poor site (LUE) was clearly highest in the O-layer and strongly reduced in the mineral soil, the P-rich soil (BBR) showed the lowest values in the O-layer and had a slight increase in the subsoil. However, the amount of TP in all horizons at the P-rich site was much higher compared to the other sites, and was more or less equally distributed through the profile. The summarized P fractions were compared to the initial TP contents in order to determine the recovery rate of the analytical method. Considering all samples, the TP recovery was very high, with a correlation coefficient of 0.98.

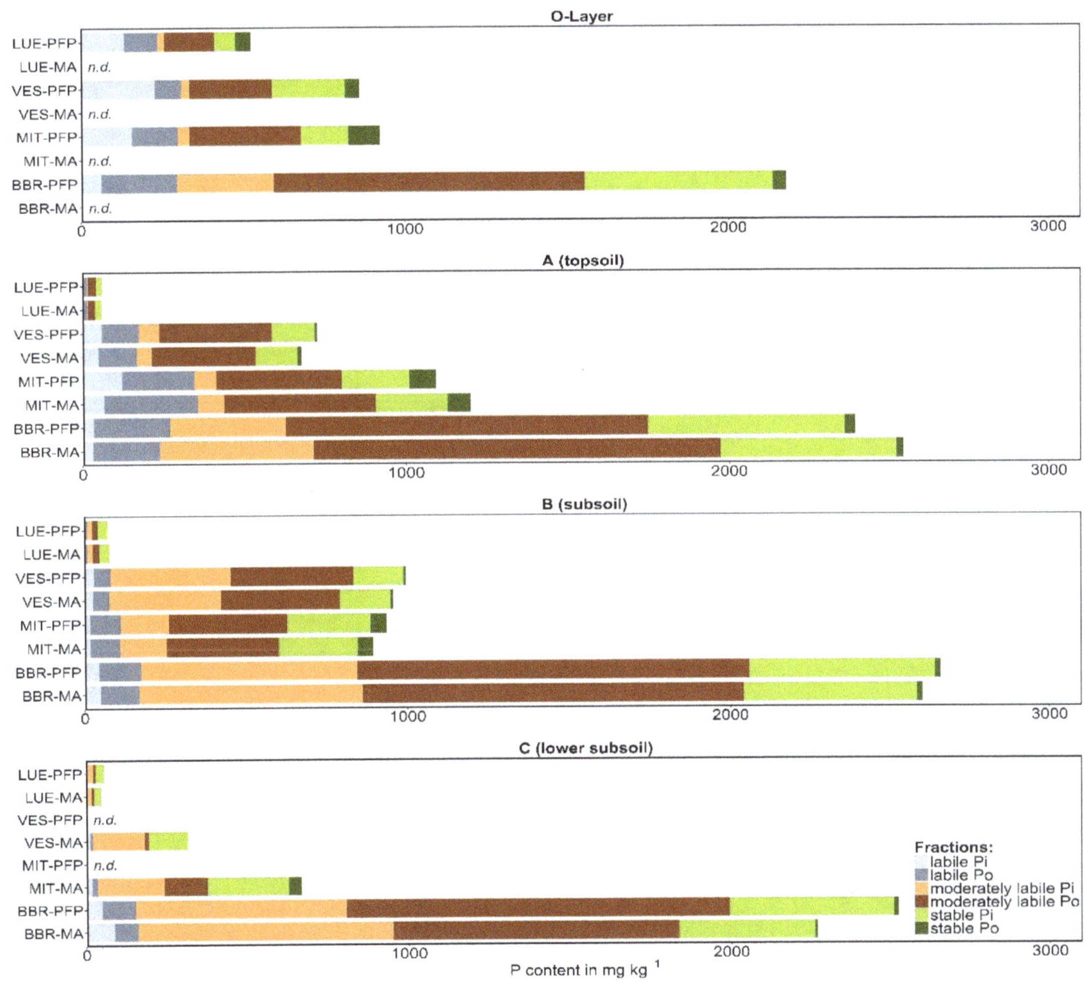

Figure 2. Phosphorus fractions summarized as labile (resin-P_i + $NaHCO_3$-$P_{i,o}$), moderately labile (NaOH-$P_{i,o}$ + HCl_{dil}-P_i), and stable P (HCl_{conc}-$P_{i,o}$ + residual-P) for PFP (preferential flow pathways) and MA (soil matrix) samples; n.d., no data available; study sites in the sequence P-poor to P-rich are: LUE, Lüss, Lower Saxony; VES, Vessertal, Thuringian Forest; MIT, Mitterfels, Bavarian Forest; BBR, Bad Brückenau, Bavarian Rhön Mountains.

3.1. Labile P Fractions

Despite the large difference in TP between the sites, labile P (here defined as resin-P_i + $NaHCO_3$-$P_{i,o}$) was in the same range of values, at least for the medium P-supplied (VES, MIT), and the P-rich (BBR) sites. In the O-layer, the P-poor site (LUE) also had comparable values for labile P as the other sites. The share of labile P to TP in the O-layer was highest at the P-poor site (45 %), slightly lower at the medium sites (36 % and 33 %), and lowest in the P-rich site (14 %) (Figure 3a.1). There was a steep decrease in the ratio labile P to TP with soil depth at the P-poor to medium sites; and a more gradual decrease with depth at the P-rich site.

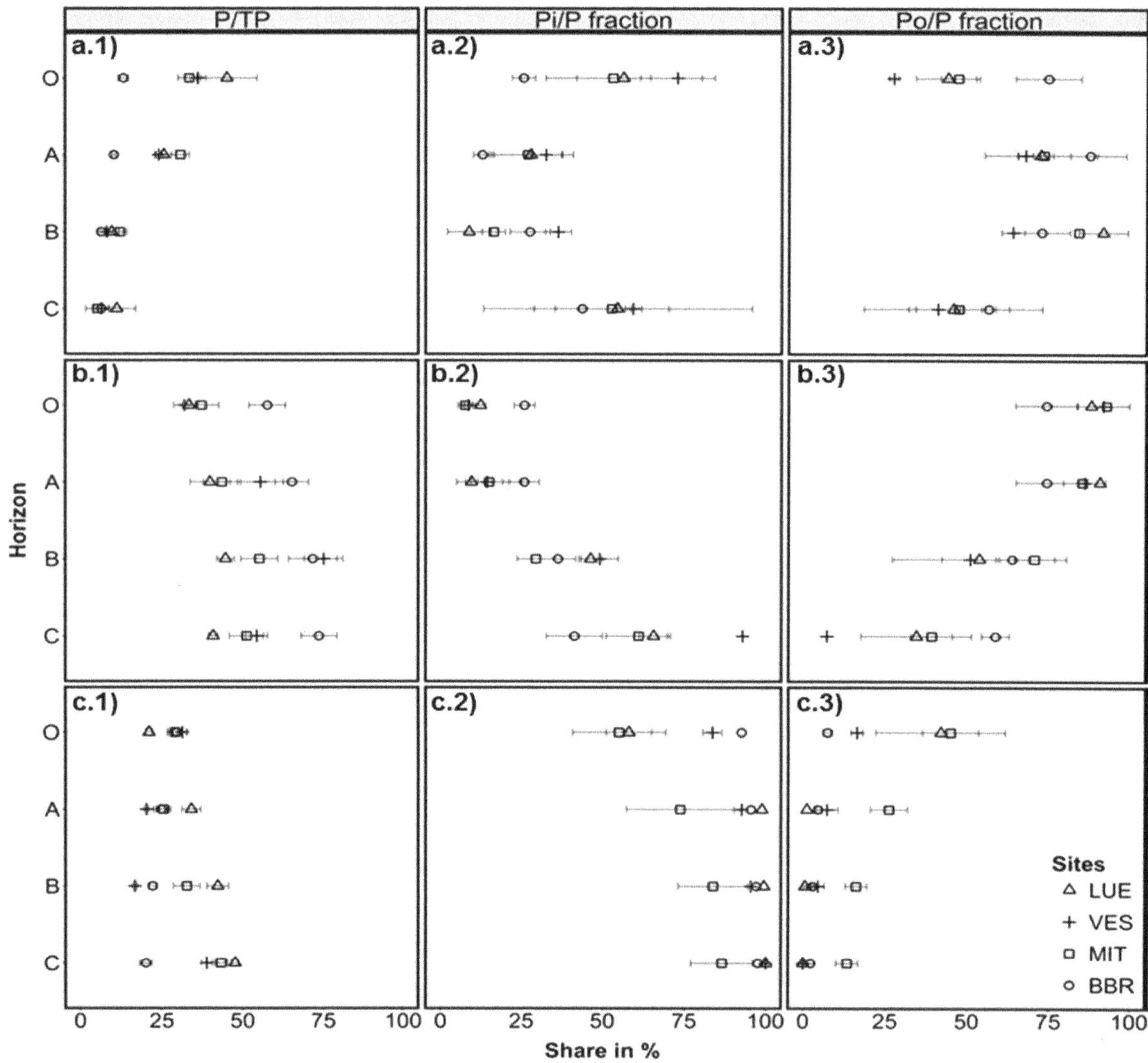

Figure 3. Share of **a.1**) the labile (resin-P_i + $NaHCO_3$-$P_{i,o}$), **b.1**) moderately labile (NaOH-$P_{i,o}$ + HCl_{dil}-P_i), and **c.1**) stable (HCl_{conc}-$P_{i,o}$ + residual-P) fractions on total P (TP) in organic layer (O) and soil (A, B, C horizons). Share of inorganic P (**a.2**: labile P_i, **b.2**: moderately labile P_i, **c.2**: stable P_i) and organically bound P (**a.3**: labile P_o, **b.3**: moderately labile P_o, **c.3**: stable P_o) on the particular pools. Error bars: standard deviation of three samples (profile cuts). Study sites in the sequence P-poor to P-rich are: LUE, Lüss, Lower Saxony; VES, Vessertal, Thuringian Forest; MIT, Mitterfels, Bavarian Forest; and BBR, Bad Brückenau, Bavarian Rhön Mountains.

Organically bound P dominated the labile P pool in the A and B horizon for all sites, and also in the O-layer for BBR. At VES, 27% of labile P in the O-layer consisted of P_o, while MIT and LUE had approximately equal amounts of P_i and P_o (Figure 3a.2,a.3). Thus, most of the easily plant-available P_i

was located in the O-layers at the medium and P-poor sites, whereas labile P was clearly dominated by P_o at the P-richest site. This finding was unexpected, as very low P supply from the mineral phase is assumed to enhance tight P recycling in forest ecosystems [6] (and therefore the fast uptake of P_i) which would result in a higher share of P_o at the P-poor site. In a recent study conducted at the same P-rich and P-poor study sites, Bünemann et al. [29] found that biological/biochemical processes were the dominant control of P_i availability in the low-P-sorbing sandy soil of the P-poor site. By contrast organic P mineralization was negligible when compared to the high P_i availability at the P-rich study site, which has a high-P-sorbing soil and very high TP content. Further, the high contents of amorphous and crystalline sesquioxides at the P-rich site (Table 2) may act as a sink for labile P_i, when it is adsorbed onto those surfaces and precipitated (depending on pH) (cf. [30,31]).

The differences in labile P between PFP and soil matrix were marginal considering the sum of P_i and P_o of this pool (Figure 2), but the ratio of P_i to P_o differed. Thus, the medium P-supplied site showed clearly higher contents of labile P_i and less labile P_o in PFPs compared to soil matrix. By contrast, labile P_o was significantly higher in PFPs of the P-rich site. However, the variation of results for labile P_i and P_o was quite high between the three profile cuts within the soil plot (see error bars in Figure 3), which indicates that conclusions from the results have to be drawn carefully.

3.2. Moderately Labile P Fractions

The sum of moderately labile P forms (NaOH-$P_{i,o}$ + HCl_{dil}-P_i) was the dominant fraction in mineral soil at the P-rich sites (Figures 2 and 3). At the P-poor site, this pool was equal or slightly lower than the stable P pool. At all sites, moderately labile P was predominantly organically bound in the O-layer and topsoil. In the subsoil, the ratio of P_i to P_o shifted towards P_i (Figure 3b.2,b.3). The shifts with depth were relatively gradual in the P-rich soil, but were more abrupt at the medium and P-poor sites.

The NaOH-extracted P represents the moderately available P forms related to Al, Fe, and humic substances. In acidic forest soils on silicate parent material, NaOH-P was observed to be involved in P_i adsorption (mainly to Al-oxyhydroxides) as well as P_o adsorption (to Fe-oxyhydroxides or Al-saturated soil organic matter) [4]. The share of HCl_{dil}-extractable P (apatite-bound) was very small for most samples, which indicates highly weathered soils [32]. In the P-rich soil, HCl_{dil}-P was found to be in the range of 150–300 $mg \cdot kg^{-1}$, which confirms that the supply with mineral P is still high at this site. These results are in line with findings of Prietzel et al. [4] who identified apatite-P in this P-rich soil using P *K*-edge X-ray absorption near-edge structure (XANES) spectroscopy.

The sum of moderately labile P showed a variable distribution between PFP and soil matrix (Figure 2). In the topsoils, the P-poor to medium sites (LUE and VES) had slightly higher contents of this pool in PFPs, whereas clearly higher contents in soil matrix were observed at the medium to P-rich sites (MIT and BBR). Here, both higher P_i as well as higher P_o values contributed to the increased pool contents. For the subsoils, slightly higher contents of moderately labile P in PFPs as compared to soil matrix were detected at all sites which was mainly caused by higher P_o contents, especially at the P-rich site. For the P-poor site, the differences are hardly interpretable because the chemical analyses of the very low P concentrations led to high data uncertainty.

3.3. Stable P Fractions

Stable P in the soils (HCl_{conc}-$P_{i,o}$ + residual-P) followed the sequence of decreasing initial TP in the order of: BBR > MIT > VES > LUE, and was characterized by low contents of P_o compared to P_i in all soils (Figure 2). Elevated P_o portions in the O-layer were observed at the P-poor site and at the medium site MIT (Figure 3c.3), which indicates the presence of P associated with very stable organic compounds. However, the ratio of stable P to TP in the O-layer was comparable at all sites with around 30 % of TP at the P-rich to medium sites, and slightly lower at the P-poor site with 21 %. The depth distribution of this relation showed a slight decrease with depth in the P-rich soil, but an increase in the medium P-supplied soils and particularly increased in the P-poor soil. Here, almost 50 % of the

P in the subsoil is stable bound and not available for plants, which is in line with the assumption of decreasing availability of mineral-bound P from the P-rich site to the P-poor site.

The contents of stable P in PFPs and soil matrix were on the same level. Only for the P-rich soil, slightly higher values in PFPs were observed.

The distribution of P fractions revealed for the P-rich site comparable labile P concentrations as for the other sites, but a clearly lower ratio of labile P to TP, and much more P_o than P_i in this pool. Considering the assumption of negligible P_o mineralization [29], it can be assumed that the P-rich soil has enough labile P available from mineral sources and has no need for increased P_o mineralization. P_o therefore remains stable, but P_i is consumed and reduced thus relative to the amount of labile P_o. This may also explain the higher ratio of moderately labile P to TP at the P-rich site compared to the other sites. In this pool, P is moderately bound to inorganic and organic compound, and can be acquired if necessary.

3.4. Factors Affecting the Distribution of P Fractions in the Soil Profiles

The statistical test of correlations between parameters included in the first step all data from the four study sites and covered a wide range of values of chemical soil properties. The results confirmed a close relationship between labile P and the contents of soil organic C and total N, which is in accordance with previous studies [4,30,33,34] (for the correlation matrix and scatter plots see Table S1 of the supplementary material). This correlation is stronger for labile P_o (N: R^2 = 0.98, C: R^2 = 0.97) than for labile P_i (N: R^2 = 0.94, C: R^2 = 0.93). Also in agreement with previous studies [30,34], the easily available resin-P was strongly correlated with $NaHCO_3$-P_i (R^2 = 0.79) and $NaHCO_3$-P_o (R^2 = 0.75) indicating the contribution of inorganic as well as organic labile P fractions to the resin-P_i pool.

Moderately labile P was strongly related to sesquioxides (Al, Fe, Mn). In this fraction, P_i showed close correlations to Al_{ox}, Al_{di}, Mn_{ox}, Mn_{di}, and had slightly lower correlations with Fe_{ox} and Fe_{di}. Moderately labile P_i was also positively correlated with pH, but negatively correlated to the C/N ratio. By contrast, the strongest correlation of P_o in this fraction was found with Fe-oxyhydroxides, and to a lesser extent with Al- and Mn-oxyhydroxides. This strong contribution of Fe compounds to P_o adsorption, but the adsorption of mainly P_i to Al-oxyhydroxides, is in line with the study of Prietzel et al. [4] for acidic forest soils.

For stable P, a close correlation of P_i with Fe, Al, and Mn compounds was identified, whereas P_o was strongly related to C and N. This shows that P_o is occluded in stable organic compounds or particulate organic matter.

To indicate the dominant processes influencing the distribution of P_i and P_o in soils, the ratio C to P_o was calculated (Table 3). Low C/P_o values (<200) indicate the net mineralization of P, whereas high ratios (>300) are indicative for immobilization as result of microbial P_i utilization during decomposition of organic matter [35,36]. When comparing the test sites with considerable differences in soil P contents, the C to P_o ratio was highest at the P-poor site and lowest at the P-rich site. The strong reduction of C/P_o with depth at all sites can be attributed to the sharp decrease of organic C compared to P. In the O-layers, C/P_o was > 300 at the P-poor and medium sites, which indicates P immobilization (incorporation into biomass). C/P_o-values between 200 and 300, which occurred in the O-layer of the P-rich site, indicate a stable rate of cycling between P_i and P_o [35]. These findings support the hypothesis of intense P recycling at P-poor sites [6] leading to P depletion and high C to P_o ratios. The C/P_o ratio also differed considerably between PFP and soil matrix samples at all site with mostly clearly higher values in PFPs (Table 3). For the P-rich site, the difference between PFP and matrix was more pronounced in the topsoil and diminished with soil depth. In the P-poor soil, the contrasting behavior was observed: the difference increased with depth indicating stronger P_o depletion in the subsoil PFPs.

Table 3. Ratios of organic C to sum of organic P (Po) at four beech forest study sites in Germany for PFP (preferential flow pathways), Ma (soil matrix), and RC (root channel-PFP) samples.

C/P_o	LUE		VES		MIT		BBR	
Horizon	**PFP**	**Ma**	**PFP**	**Ma**	**PFP**	**Ma**	**PFP**	**Ma**
O	1143		840		910		223	
A	512	494	155	174	593	371	139	97
B	356	274	89	66	101	93	58	51
C	204	88		170		59	42	43

Horizon: O—organic layer; A—topsoil horizon; B—subsoil horizon; C—lower subsoil horizon containing weathered parent material. Study sites in the sequence P-poor to P-rich are: LUE—Lüss, Lower Saxony; VES—Vessertal, Thuringian Forest; MIT—Mitterfels, Bavarian Forest; and BBR—Bad Brückenau, Bavarian Rhön Mountains.

For all chemical parameters, the differences between PFP and the soil matrix were tested statistically (for test results see Table S2 of the supplementary material). The most statistical differences were found for the P-rich site. Significantly higher values in PFPs occurred here for C, labile P_o, and C to TP ratio with $p < 0.01$, and for N, stable P_o and stable P_i with $p < 0.05$. For the other sites, almost no statistically significant differences of P fractions and other parameters were identified. At the P-poor site, the ratio C to TP and the C content were statistically higher in the PFPs ($p < 0.05$). At the medium sites, the content of metals differed substantially, with high contents of Fe_{ox}, Mn_{ox}, Al_{di} and Mn_{di} in PFPs, but lower values of Fe_{di} compared to matrix samples. However, these statistical tests included all data per site and have to be considered carefully due to the hierarchical structure of the data. For the same tests at the horizon level per site, the data sets were much smaller (three samples for PFP and matrix respectively) and revealed no significant differences between PFP and matrix for all parameters (Table S3). In previous studies, Hagedorn and Bundt [18] suggested that PFPs in forest soils can be stable for decades, resulting in biological and chemical gradients between PFPs and soil matrix [15]. Bogner et al. [12] confirmed this hypothesis for PFPs along root channels and found an enrichment of C, N, Ca, Mg, and Fe in PFPs compared to the soil matrix. Such an accumulation in PFPs was also documented for different P fractions in flow pathways along stones by Backnäs et al. [19]. Our study sites are characterized by very high stone contents, resulting in very heterogeneous conditions for water and solute flow. PFPs were therefore located mostly along stone surfaces with low potential for P sorption [19], and preferential flow occurred less in root channels or other biopores where P accumulation would be expected to be more prevalent. Single biopores could be identified and sampled separately, which contained decayed root material. Indications were found in these biopores of an accumulation of labile organically bound P (P_o), but could not be statistically confirmed. With respect to the distribution of subsurface flow pathways at hillslopes (midrange mountain sites), it has been shown that a predominant surface-near lateral flow does not support chemical gradients between PFPs and soil matrix. Thus, non-significant results in the statistical test are a result of specific site conditions, spatial soil heterogeneity, and analytical uncertainties with regard to very low P concentrations at the P-poor site.

Nevertheless, several factors affecting the distribution of P within the tested soils became apparent. The results showed that the labile $P_{i,o}$ pool and stable P_o were strongly related to C and N, whereas moderately labile $P_{i,o}$ as well as stable P_i correlated with metal-oxides and -oxyhydroxides (Al, Fe, Mn). At the medium to P-rich sites, where higher contents of sesquioxide in topsoil matrix compared to PFPs were observed, moderately labile P_i and P_o also was clearly higher in soil matrix. The higher P_o contents of this moderately labile pool in subsoil PFPs could be related to Fe_{ox} accumulated in the flow pathways.

4. Conclusions

As expected, considerable differences were found for the four study sites in both the P contents and the distribution of P fractions within the soil profiles. Despite these large differences, the labile P contents in the O-layers were in the same range of values at all sites. At the site with very low supply of P from the mineral phase, the predominant portion of labile P was available in the O-layer, but was greatly reduced in the mineral soil. This indicates that for this site intense P recycling happens in the O-layer, which is also supported by a high C to P_o ratio in this layer. By contrast, labile P was available throughout the whole profile at the P-rich site. Here, the supply of plant-available P seems to be sufficient enough that there is no need for tight P recycling, and may therefore explain the very low share of labile P to TP, when compared to the P-poor site. The contents of labile and moderately labile P clearly differed in PFPs compared to soil matrix, and indicate that labile organically bound P (P_o) accumulates in PFPs with a high content of organic matter. However, these results could not be proved statistically at the tested sites. Therefore, more sites particularly for soils with a higher proportion of preferential flow through biopores should be studied to verify the indications.

Acknowledgments: We thank Gisela Ciesielski and Manuela Unger for technical support in the laboratory. Furthermore, our thanks go to Maximilian Kirsten, Jianhong Liang, Peer Appelfelder, and Nils Heide for their assistance in the field and laboratory works. Finally, we thank Dan Hawtree for helpful comments and linguistic help. This work was funded within the framework of Priority Program SPP 1685 "Ecosystem Nutrition: Forest Strategies for Limited Phosphorus Resources" by the German Research Foundation (DFG), grant No. JU 2940/1-1. We further acknowledge support by the Open Access Publication Funds of the TU Dresden.

Author Contributions: D.J. designed and performed the field experiments, and did the data analyses; all authors wrote the manuscript.

References

1. Bol, R.; Julich, D.; Brödlin, D.; Siemens, J.; Kaiser, K.; Dippold, M.A.; Spielvogel, S.; Zilla, T.; Mewes, D.; von Blanckenburg, F.; et al. Dissolved and colloidal phosphorus fluxes in forest ecosystems—An almost blind spot in ecosystem research. *J. Plant Nutr. Soil Sci.* **2016**, *179*, 425–438. [CrossRef]
2. Newman, R.H.; Tate, K.R. Soil phosphorus characterisation by ^{31}p nuclear magnetic resonance. *Commun. Soil Sci. Plant Anal.* **1980**, *11*, 835–842. [CrossRef]
3. Cade-Menun, B.; Liu, C.W. Solution phosphorus-31 nuclear magnetic resonance spectroscopy of soils from 2005 to 2013: A review of sample preparation and experimental parameters. *Soil Sci. Soc. Am. J.* **2014**, *78*, 19. [CrossRef]
4. Prietzel, J.; Klysubun, W.; Werner, F. Speciation of phosphorus in temperate zone forest soils as assessed by combined wet-chemical fractionation and XANES spectroscopy. *J. Plant Nutr. Soil Sci.* **2016**, *179*, 168–185. [CrossRef]
5. Walker, T.W.; Syers, J.K. The fate of phosphorus during pedogenesis. *Geoderma* **1976**, *15*, 1–19. [CrossRef]
6. Lang, F.; Bauhus, J.; Frossard, E.; George, E.; Kaiser, K.; Kaupenjohann, M.; Krüger, J.; Matzner, E.; Polle, A.; Prietzel, J.; et al. Phosphorus in forest ecosystems: New insights from an ecosystem nutrition perspective. *J. Plant Nutr. Soil Sci.* **2016**, *179*, 129–135. [CrossRef]
7. Prietzel, J.; Kaiser, K. De-eutrophication of a nitrogen-saturated Scots pine forest by prescribed litter-raking. *J. Plant Nutr. Soil Sci.* **2005**, *168*, 461–471. [CrossRef]
8. Neary, D.G.; Ice, G.G.; Jackson, C.R. Linkages between forest soils and water quality and quantity. *For. Ecol. Manag.* **2009**, *258*, 2269–2281. [CrossRef]
9. Van Miegrot, H.; Johnson, D.W. Feedbacks and synergism among biogeochemistry, basic ecology, and forest soil science. *Forest Ecol. Manag.* **2009**, *258*, 2214–2223. [CrossRef]
10. Santos, R.M.B.; Sanches Fernandes, L.F.; Pereira, M.G.; Cortes, R.M.V.; Pacheco, F.A.L. A framework model for investigating the export of phosphorus to surface waters in forested watersheds: Implications to management. *Sci. Total Environ.* **2015**, *536*, 295–305. [CrossRef] [PubMed]
11. Uchida, T.; Tromp-van Meerveld, I.; McDonnell, J.J. The role of lateral pipe flow in hillslope runoff response: An intercomparison of non-linear hillslope response. *J. Hydrol.* **2005**, *311*, 117–133. [CrossRef]
12. Bogner, C.; Borken, W.; Huwe, B. Impact of preferential flow on soil chemistry of a podzol. *Geoderma* **2012**, *175*, 37–46. [CrossRef]
13. Van der Heijden, G.; Legout, A.; Pollier, B.; Bréchet, C.; Ranger, J.; Dambrine, E. Tracing and modeling

preferential flow in a forest soil—Potential impact on nutrient leaching. *Geoderma* **2013**, *195–196*, 12–22. [CrossRef]
14. Van Verseveld, W.J.; McDonnell, J.J.; Lajtha, K. The role of hillslope hydrology in controlling nutrient loss. *J. Hydrol.* **2009**, *367*, 177–187. [CrossRef]
15. Bundt, M.; Jäggi, M.; Blaser, P.; Siegwolf, R.; Hagedorn, F. Carbon and nitrogen dynamics in preferential flow paths and matrix of a forest soil. *Soil Sci. Soc. Am. J.* **2001**, *65*, 1529–1538. [CrossRef]
16. Flury, M.; Flühler, H. Brilliant Blue FCF as a dye tracer for solute transport studies—A toxicological overview. *J. Environ. Qual.* **1994**, *23*, 1108–1112. [CrossRef]
17. Weiler, M.; Flühler, H. Inferring flow types from dye patterns in macroporous soils. *Geoderma* **2004**, *120*, 137–153. [CrossRef]
18. Hagedorn, F.; Bundt, M. The age of preferential flow paths. *Geoderma* **2002**, *108*, 119–132. [CrossRef]
19. Backnäs, S.; Laine-Kaulio, H.; Kløve, B. Phosphorus forms and related soil chemistry in preferential flowpaths and the soil matrix of a forested podzolic till soil profile. *Geoderma* **2012**, *189*, 50–64. [CrossRef]
20. Jarvis, N.J. A review of non-equilibrium water flow and solute transport in soil macropores: principles, controlling factors and consequences for water quality. *Eur. J. Soil Sci.* **2007**, *58*, 523–546. [CrossRef]
21. Jensen, M.B. Subsurface transport of phosphorus in relation to its mobilization and immobilization in structured soil (steady state flow experiments). *Acta Agric. Scand. Sect. B-Soil Plant Sci.* **1998**, *48*, 11–17.
22. Hedley, M.J.; Stewart, J.W.B.; Chauhan, B.S. Changes in inorganic and organic soil phosphorus fractions induced by cultivation practices and by laboratory incubations. *Soil Sci. Soc. Am. J.* **1982**, *46*, 970. [CrossRef]
23. Tiessen, H.; Moir, J.O. Characterization of Available P by Sequential Extraction. In *Soil Sampling and Methods of Analysis*; Carter, M.R., Gregorich, E.G., Eds.; Canadian Society of Soil Science: Boca Raton, FL, USA, 2008; Chapter 25; pp. 293–306.
24. Sibbesen, E. A simple ion-exchange resin procedure for extracting plant-available elements from soil. *Plant Soil* **1977**, *46*, 665–669. [CrossRef]
25. Murphy, J.; Riley, J.P. A modified single solution method for the determination of phosphate in natural waters. *Anal. Chim. Acta* **1962**, *27*, 31–36. [CrossRef]
26. O'Halloran, I.P.; Cade-Menun, B.J. Total and Organic Phosphorus. In *Soil Sampling and Methods of Analysis*; Carter, M.R., Gregorich, E.G., Eds.; Canadian Society of Soil Science: Boca Raton, FL, USA, 2008; pp. 265–291.
27. Crews, T.E.; Brookes, P.C. Changes in soil phosphorus forms through time in perennial versus annual agroecosystems. *Agric. Ecosyst. Environ.* **2014**, *184*, 168–181. [CrossRef]
28. Negassa, W.; Leinweber, P. How does the Hedley sequential phosphorus fractionation reflect impacts of land use and management on soil phosphorus: A review. *J. Plant Nutr. Soil Sci.* **2009**, *172*, 305–325. [CrossRef]
29. Bünemann, E.K.; Augstburger, S.; Frossard, E. Dominance of either physicochemical or biological phosphorus cycling processes in temperate forest soils of contrasting phosphate availability. *Soil Biol. Biochem.* **2016**, *101*, 85–95. [CrossRef]
30. Tiessen, H.; Stewart, J.W.B.; Cole, C.V. Pathways of phosphorus transformations in soils of differing pedogenesis. *Soil Sci. Soc. Am. J.* **1984**, *48*, 853. [CrossRef]
31. Renneson, M.; Dufey, J.; Legrain, X.; Genot, V.; Bock, L.; Colinet, G. Relationships between the P status of surface and deep horizons of agricultural soils under various cropping systems and for different soil types: A case study in Belgium. *Soil Use Manag.* **2013**, *29*, 103–113. [CrossRef]
32. Hashimoto, Y.; Watanabe, Y. Combined applications of chemical fractionation, solution ^{31}P-NMR and P K-edge XANES to determine phosphorus speciation in soils formed on serpentine landscapes. *Geoderma* **2014**, *230*, 143–150. [CrossRef]
33. Frizano, J.; Johnson, A.H.; Vann, D.R.; Scatena, F.N. Soil phosphorus fractionation during forest development on landslide scars in the Luquillo Mountains, Puerto Rico. *Biotropica* **2002**, *34*, 17–26. [CrossRef]
34. Thomas, S.M.; Johnson, A.H.; Frizano, J.; Vann, D.R.; Zarin, D.J.; Joshi, A. Phosphorus fractions in montane forest soils of the Cordillera de Piuchué, Chile: Biogeochemical implications. *Plant Soil* **1999**, *211*, 139–148. [CrossRef]
35. McDowell, R.W.; Stewart, I. The phosphorus composition of contrasting soils in pastoral, native and forest management in Otago, New Zealand: Sequential extraction and ^{31}P NMR. *Geoderma* **2006**, *130*, 176–189. [CrossRef]
36. Stutter, M.I.; Shand, C.A.; George, T.S.; Blackwell, M.S.A.; Dixon, L.; Bol, R.; MacKay, R.L.; Richardson, A.E.; Condron, L.M.; Haygarth, P.M. Land use and soil factors affecting accumulation of phosphorus species in temperate soils. *Geoderma* **2015**, *257*, 29–39. [CrossRef]

7

Immediate Changes in Organic Matter and Plant Available Nutrients of Haplic Luvisol Soils Following Different Experimental Burning Intensities in Damak Forest, Hungary

Jack M. Bridges [1,*], George P. Petropoulos [2] and Nicola Clerici [3]

[1] Geography and Earth Sciences, University of Aberystwyth, Wales SY23 3DB, UK
[2] Department of Soil & Water Resources, Hellenic Agricultural Organization (HAO) "Demeter" (former NAGREF), 1 Theofrastou St., 41335 Larissa, Greece; petropoulos.george@gmail.com
[3] Biology Program, Faculty of Natural Sciences and Mathematics, Universidad del Rosario, Carrera 26 # 63B-48, Bogotá 111221, Colombia; nicola.clerici@urosario.edu.co
* Correspondence: bridgesj874@gmail.com

Abstract: One of the major pedological changes produced by wildfires is the drastic modification of forest soil systems properties. To our knowledge, large research gaps are currently present concerning the effect of such fires on forest Haplic Luvisols soils in Central Europe. In this study, the effects of experimental fires on soil organic matter and chemical properties at different burning intensities in a Central European forest were examined. The study was conducted at Damak Forest, in Hungary, ecosystem dominated by deciduous broadleaf trees, including the rare Hungarian oak *Quercus frainetto* Ten. The experimental fires were carried out in nine different plots on Haplic Luvisol soils transferred from Damak Forest to the burning site. Three types of fuel load were collected from the forest: litter layer, understorey and overstorey. Groups of three plots were burned at low (litter layer), medium intensity (litter and understorey) and high intensity (litter, understorey and overstorey). Pre-fire and post-fire soil samples were taken from each plot, analysed in the laboratory and statistically compared. Key plant nutrients of organic matter, carbon, potassium, calcium, magnesium and phosphorus were analysed from each sample. No significant differences in soil organic matter and carbon between pre- and post-fire samples were observed, but high intensity fires did increase soil pH significantly. Calcium, magnesium and phosphorus availability increased significantly at all fire intensity levels. Soil potassium levels significantly decreased (ca. 50%) for all intensity treatments, in contrast to most literature. Potassium is a key nutrient for ion transport in plants, and any loss of this nutrient from the soil could have significant effects on local agricultural production. Overall, our findings provide evidence that support the maintaining of the current Hungarian fire prevention policy.

Keywords: soil properties; experimental fires; nutrients; UV-spectroscopy analysis; thermal infrared thermometer

---

1. Introduction

Wildfires affect agricultural and forest land covers more than any other land cover type [1]. Prescribed fires involve basically controlled burning of forests as a mean to reduce fire fuel levels. This practice minimises the extent, severity and danger of potential wildfires [2]. In Europe, prescribed fire use has disappeared from many countries due to the intensification of agriculture and socio-economical changes [3]. In most European countries prescribed burning is prohibited. Where permitted (e.g., in UK, Germany, France, Spain, Slovenia), there are strict controls on the timing and location of the burning extension. European fire management policies are still aimed at fire suppression instead of using prescribed fires as a preventative measure. At present, only 10,000 ha $year^{-1}$ of the Mediterranean basin is prescribed burnt; for comparison, this is about 3% of the wildfire extent in Spain, France and Portugal alone [4,5].

The use of prescribed fires in Europe is mostly an 'unnatural' event for both biotic and abiotic systems. Ecosystem alterations can arise as a consequence of fires, such as niche alteration for native species [6], invasion of alien species [7], and drastic changes to forest soil systems properties [2]. Climate change is an additional threat to wildfires frequency, due to projected higher mean temperatures, increase in the number of summer days and the decline of average precipitation. Climate change models suggest that by 2100, temperatures in central Europe will increase up to about 3 °C [8]. This increase in temperatures will cause soils to become drier for longer periods and therefore would extend the wildfire season globally. Southern Europe has already experienced record summer temperatures in 2017. The Intergovernmental Panel on Climate Change's (IPCC) Fifth Report (AR5) on forest fire risk shows as a general trend that each region, below the 55th parallel north, will either increase in forest fire risk or stay at the same level [9]. Temperate forests are particularly sensitive to the effects of climate change due to their long lifespan preventing rapid adaptation. The large amount of C in forest soils means that any deviation of the level will have a significant effect on global C balance and climate change [10]. The future IPCC projections support that the effects of wildfires will be an increasing concern for both the scientific community and the governmental agencies.

A major factor influencing the effects of fires on soil properties is fire severity, a parameter primarily controlled by combustion and site factors. Combustion factors include quantity, moisture and type of fuel, whereas site factors include topography, wind direction, air temperature and humidity [2]. There are two key components of fire behaviour that influence fire severity: intensity and duration. Intensity is defined as the rate of thermal production in the fire, whereas duration is the length of time soil is exposed to fire. The most varied change to forest soils during burning is the loss of soil organic matter (OM) and plant available phosphorous and nitrogen in mineral soils. Fire severity plays a significant role in the effect on this important carbon pool. Substantial consumption of OM occurs between 200 and 250 °C and is complete at ca. 460 °C [11]. Fire does not maintain a consistent temperature in the soil. Some parts of the burn area might experience over 460 °C, whereas other areas may only experience less than 200 °C. This means that the OM may only experience minor volatilisation at low temperatures compared to complete oxidation at high temperatures. There is consensus about combustion causing reduction or total removal of the forest floor [12]. The reduction of OM is not uniform with depth: for example, some studies have found that concentrations of OM decreases significantly by the heating effect at 1 cm depth, but not at 2 and 3 cm depth in a pine forest ecosystem, North-East Spain [13].

Studies have found differences between prescribed burning and wildfire effects on soil 10 years following the fire event: i.e., lower carbon content following prescribed fire and higher soil C content following wildfire, due to the accumulation of charcoal and the encroachment of post-fire N-fixing vegetation [14]. Studies support these findings, describing, e.g., OM returning to normal after one year [13]. They also showed that fire has a fertilising effect, due to the dissolution of ashes and the mineralisation of charcoal enriching the soil.

The key nutrients within forest soils, mainly contained in its organic matter component (OM) are: nitrogen (N), available phosphorus (P), exchangeable potassium (K), exchangeable calcium (Ca) and exchangeable magnesium (Mg). Each nutrient reacts differently to fire, depending on its individual volatilisation threshold [15]. The combustion of nutrients bound in vegetation and soil organic matter add inorganic forms of K, Ca, Mg, P and N to the soil [16]. Each nutrient has its own response to burning, e.g., early studies have found that concentrations of K, Ca and Mg ions can increase, whereas N and S often decrease [17]. Fire intensity is directly linked to the temperature an object experiences during fire, which in turn impacts the type of nutrient and the amount volatilised. Fire acts as a rapid mineralising agent [18]. This mineralisation has been studied extensively for N and P, because they are the two key macronutrients for plants. Soil nutrients that have low volatilisation thresholds are the first to be mineralised. There is an immediate reduction in soil organic N due to volatilisation [19]. Substantial proportions of soil organic N survive low intensity fires, however, moderate to high intensity fires convert most soil organic nitrogen to inorganic forms [2]. The effects of fire on properties of forest soils have been extensively reviewed and found that generally, losses of P through volatilisation are low [2]. However, the combustion of vegetation and litter cause major modifications on the P cycle. Micronutrients are also affected immediately after burning; e.g., total content and reducible forms of Mn increase significantly following fire due to the ash produced [20].

Soil pH has been also found to increase post fire. This increase has been attributed to the addition of base cations in ash, and to organic acid denaturing due to the heating of the soil [2]. However, significant increases only occur at temperatures >450 °C [21]; this temperature links with the total combustion of fuel as the ash produced has the capacity to neutralise soil acidity. Through analysing mixed forest soils, topsoil pH could increase as much as three units immediately after burning due to the production of K and Na oxides [22]. This significant increase can only be found in non-calcareous soils; calcareous soils are already alkaline, and in them the pH increases induced by fire are often negligible.

Currently, very little is known about the effect fires have on forest soils in Central Europe. Most research in Europe comes from Mediterranean forest studies, while there are few investigations about the areas that, according to the IPCC climate change scenarios, will have in the future climate similar to that of the Mediterranean region [9]. Soil in Mediterranean forests have low moisture and available nutrients, and plants are adapted by co-existing to prevent competition in resource scare environments [23]. Field research on fires, to our knowledge, is limited in Hungary in particular. Primary research is generally on the effects to biodiversity, mainly in grasslands and based on questionnaires to fire departments [24,25]. In particular, little research has been done on the effects of fire on Haplic Luvisols in Central Europe. Luvisols occupy ca. 5% of the total continental land area on Earth, and are found mainly in west-central Russia, the United States, Central Europe, the Mediterranean and southern Australia [26].

In purview of the above, the present study objectives were to: (1) characterise the effect of different fire intensities in a burn experiment on the organic matter, pH, potassium, phosphorus, calcium and magnesium of forest Haplic Luvisol through laboratory analysis; (2) examine the effectiveness of the experimental burning design at measuring changes in soil properties, and, (3) provide recommendations for fire authorities in Hungary about potential risks to the dominant soil Haplic Luvisol from both prescribed fires and wildfires. For this purpose, a field experiment of three fire treatments on forest Haplic Luvisol in Damak forest, NE Hungary was implemented.

2. Materials and Methods

2.1. Study Area

The study was carried out in Damak forest (48°31′ N, 20°82′ E), in the county of Borsod-Abaúj-Zemplén, Hungary (Figure 1). The area is characterised by a temperate continental climate. The mean annual temperature of the area is 9.5 °C; warmest period ranges from May to September (20.0 °C in July). Maximum precipitation within this warm period is 82 mm (July); mean annual precipitation is 567 mm. Forest typology is classified as mixed broadleaf and coniferous woodland (Figure 2). Dominant species are Hungarian oak, common nettle *Urtica dioica* L., black locust *Robinia pseudoacacia* L., Scots pine *Pinus sylvestris* L. and hornbeam *Carpinus betulus* L.

Soils in the studied area are of the Haplic Luvisol type [27,28]. This is a non-calcareous soil, with neutral pH and high in organic matter. In stark contrast to the Hungarian Plains (80 km South-West, higher precipitation and lower temperatures in Damak forest has led to the development of this soil type. As this is a forest soil, there is an organic horizon, comprised primarily of leafy material that is separated from the mineral horizon [27]. For this study, we identified a Haplic Luvisol that to our knowledge was negligibly affected by human disturbance, at about 250 m from the experimental site.

In Damak Forest, where soil samples were collected, there have been no recent significant disturbances that could affect this study, in particular prescribed or wild fires.

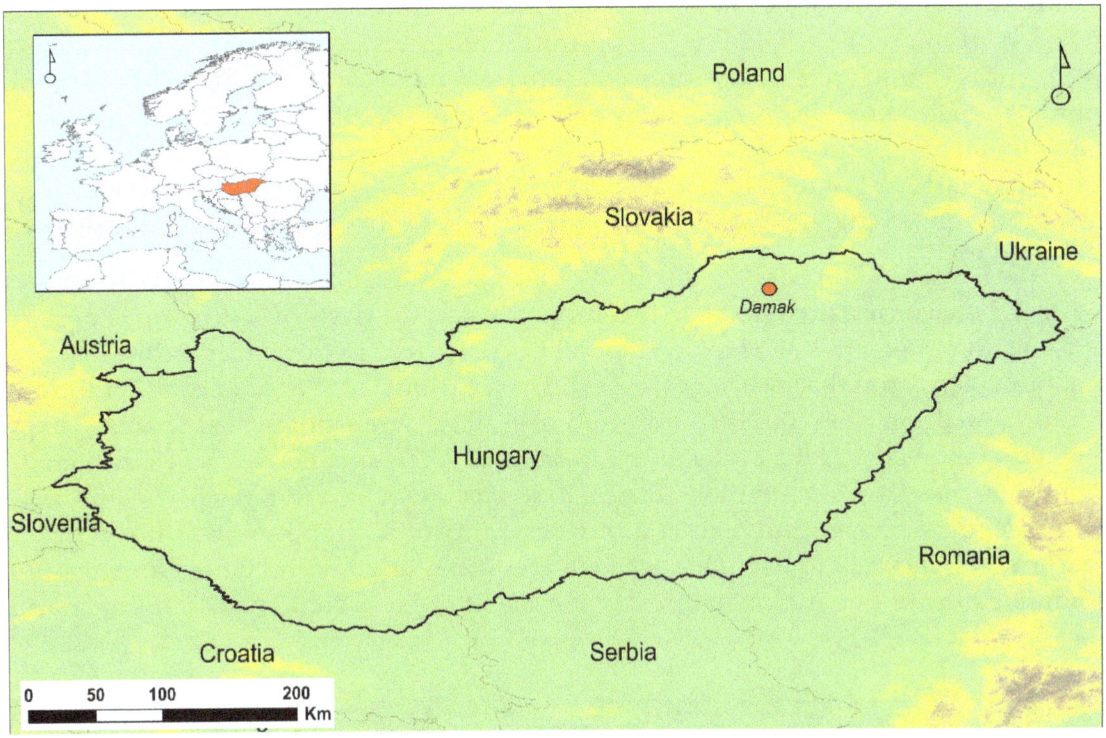

Figure 1. Study area: Damak Forest (red point; geographical coordinates 48°31′ N, 20°82′ E) in Borsod-Abaúj-Zemplén county, Hungary. Inset map: Hungary (red) in Europe.

Figure 2. Characteristic vegetation and strata in Damak forest: (**a**) soil to litter layer interface, (**b**) litter layer, (**c**) understorey and (**d**) overstorey. Photo: Jack M. Bridges.

2.2. Experimental Design

Experimental burning is used to simulate fire conditions in a safe area, where the effects of fire can be accurately observed. Studies using experimental burning are usually carried out between late spring and early autumn [29,30]. This is the optimum period where soil is dry and fuel is burnable. Experimental burns should recreate the same conditions as a real forest fires. One of the problems with the literature of experimental burning is that there are no universally established fuel load amounts for low, medium and high intensity fires. Studies have recommended for high and medium intensity fires, to use 40 and 20 t ha^{-1} of fuel, respectively [31]. For smaller scale experimental burning plots this equates to a minimum of 4 kg m^{-2} of fuel for high intensity and 2 kg m^{-2} for medium intensity.

The experimental approach is built along the lines of previously published studies [29,30,32] measuring fire effects on soil properties on the basis of setting up experimental fires of different burning intensities. The experimental burning area in our study is located in a 1500 m^2 paddock, about 200 m from the source of forest soil, providing a safe area to conduct the experiment. A total of nine 0.5 × 0.5 m plots and 0.225 m^3 of forest soil was transported to the area. The litter layer from the soil was removed and the upper 10 cm of mineral soil was collected. Soil was then distributed evenly to form the nine 0.1 m depth plots (Figure 3) to recreate the conditions in the forest. To prevent factors such as slope, drainage and texture affecting our results, all the soil was taken from the exact same location within Damak forest. This area was representative of the whole forest, and no single species dominated the under and overstorey, while the litter layer contained similar material to most of the forest floor.

Figure 3. Characteristics of the experiment burn plots: (**a**) 0.1 m depth, (**b**) 0.5 m width, (**c**) plots (n = 9), and (**d**) three plots per fuel load. Photo: Jack M. Bridges.

Three sets of three plots were burned close to the soil samples collection site with three distinct fire intensity treatments; low (F1), medium (F2) and high (F3). Three different fuel layers were used to achieve these fire intensities (Table 1), to represent degrees of heterogeneity in forest structure. The plots were ignited by a 30 s exposure from a propane torch in a circular motion around the plot. Pre-fire samples (F0) were taken from each plot for comparison. The vegetation used as the fuel load came from the same area as the soil, creating realistic conditions of a wildfire/prescribed fire. The first treatment (F1), involved the addition of ca. 4 kg m^{-2} of litter layer to three plots. The second treatment (F2), ca. 12 kg m^{-2} of litter layer and understory fuel was added to another three plots. In the third treatment (F3), ca. 40 kg m^{-2} of litter layer understorey and overstorey fuel was added to three plots (Table 2). Allocation of fire treatment to the plots was randomised to prevent pseudo-replication in a Latin square design [33].

Table 1. One-way ANOVA mean difference between low (F1), medium (F2) and high (F3) fire intensity treatments, and pre-fire (F0) for pH (soil), soil organic matter (OM), carbon (C), calcium (Ca), potassium (K), magnesium (Mg) and phosphorus (P). $n = 9$ samples for F0 and $n = 3$ samples for F1, 2 and 3 treatments. ** Significant difference at $p < 0.01$. * Significant difference at $p < 0.05$.

Treatment	F0						
	pH (soil)	OM (%)	C (mg/kg)	Ca (mg/kg)	K (mg/kg)	Mg (mg/kg)	P (mg/kg)
F1	0.03	0.53	0.50	242.00 **	190.00 **	6.00 **	0.25
F2	0.44 *	0.14	0.70	1272.00 **	190.33 **	22.00 **	33.25 **
F3	0.45 *	0.82	4.10	1452.00 **	187.33 **	13.74 **	39.7 **

Table 2. Weight of fuel for each plot.

Subplot	Weight of Litter Layer (kg)	Weight of Understorey Fuel (kg)	Weight of Overstorey Fuel (kg)	Total Fuel Weight (kg)
1A	1	0	0	1
1B	1	0	0	1
1C	1	0	0	1
2A	1	1.6	0	2.6
2B	1	1.7	0	2.7
2C	1	1.4	0	2.4
3A	1	1.2	7.2	9.4
3B	1	1.5	6.1	8.6
3C	1	1.2	6.9	9.1

Subplot 1 (A, B, C) = Low intensity fire. Subplot 2 (A, B, C) = Medium intensity fire. Subplot 3 (A, B, C) = High intensity fire.

During fire, temperatures on the surface of the soil were measured by a thermal infrared thermometer. Plot temperatures were recorded at 2 min intervals until the maximum temperature fell below 100 °C. This value was chosen as it has been shown to be the threshold temperature for the most relevant changes in soil [29,31,34].

2.3. *Sample Collection and Laboratory Analysis*

The remaining litter was discarded and the whole ash layer was collected and weighed from each plot; sample mean ash weight from F1 plots was 9.00 g, 36.00 g from F2 plots and 200.33 g from F3 plots. Prior to soil sampling, the litter layer was removed. An effective and representative method of soil sampling was employed. This involved taking 10 cm cores in a zigzag path across the plot and mixing to form an average sample; this was repeated for each plot [35]. The samples were taken using a trowel, and this was cleaned after each sample was taken to prevent cross-contamination. The samples were stored in a Kraft wet-strength paper bag labelled with the sample subplot and whether it was pre-fire or post-fire.

Samples were then air dried and sieved to isolate and keep the <2 mm diameter soil particles, which were then used for the analysis of chemical properties. One plot had three laboratory samples made up to improve the accuracy of the instrumentation (sample 3A).

Chemical properties analysis was then carried out in a laboratory. Organic matter content (%OM) was determined by Loss-on-Ignition (LOI). Samples were properly heated to ensure the complete loss of moisture from the soil [36,37]. F0, F1, F2 and F3 samples were then weighed (Figure A2, Appendix A). The samples were then heated to 430 °C in a muffle furnace for 16 h and weighed. This removed most of the carbon and OM [38]. By comparing the two weights before and after ignition, LOI can be calculated, which is the value for OM. C makes up ca. 50% of the OM value [39,40], therefore C was estimated by multiplying LOI by 0.5.

An acetic acid extraction (Figure A1, Appendix A) provided an estimate of the plant available content of the macro and micro nutrients contained in the samples. Using atomic absorption spectrophotometry (AAS) [41], elements Ca, Mg and K were measured (Figure A3, Appendix A). UV-visible Spectroscopy [42] was used to determine the value of P (Figure A4, Appendix A). The pH of the ash layer, F0, F1, F2 and F3 samples were measured using a Whatman pHA wet-bulb pH meter, pH obtained with water and a proportion of 1:2.5 soil to liquid (Figure A5, Appendix A).

An in-house standard reference material (ABS3) validated the precision of the sample measurements. This standard is soil based and was stored in a sealed container in cool, dark storage prior to use. All dry samples were weighted with the same mass balance to ensure consistency. Sample 3A was repeated three times and a control (blank) was used to ensure the reliability of all methods. Analysis of variance (ANOVA), at $\alpha = 0.05$, between F0 and the three post-fire samples (F1, F2 and F3) was performed to detect any significant statistical differences among results.

3. Results

3.1. Burn Characteristics

Temperatures at the soil surface during the low intensity fires (F1) averaged 228 °C. The highest temperature recorded was 400 °C. Mean burn time >100 °C was 0.79 min (Figure 4). Weight of ash averaged 11 g ($n = 3$ for all values). Due to the infrared thermometer's maximum temperature threshold of 420 °C, a true average of the soil surface temperature could not be calculated for the medium and high intensity fires. For this reason, we measured the percentage of time >420 °C. This resulted an average of 43% of the total burn time for medium intensity fires (F2). Highest temperature recorded was ≥420 °C. Mean burn time >100 °C was 15.98 min (Figure 4). Weight of ash averaged 36 g ($n = 3$ for all values). Temperatures at the soil surface during the high intensity fires (F3) averaged >420 °C for 85% of the total burn time. Highest temperature recorded was ≥420 °C. Mean burn time >100 °C was 190.67 min (Figure 4). Weight of ash averaged 200.6 g ($n = 3$ for all values).

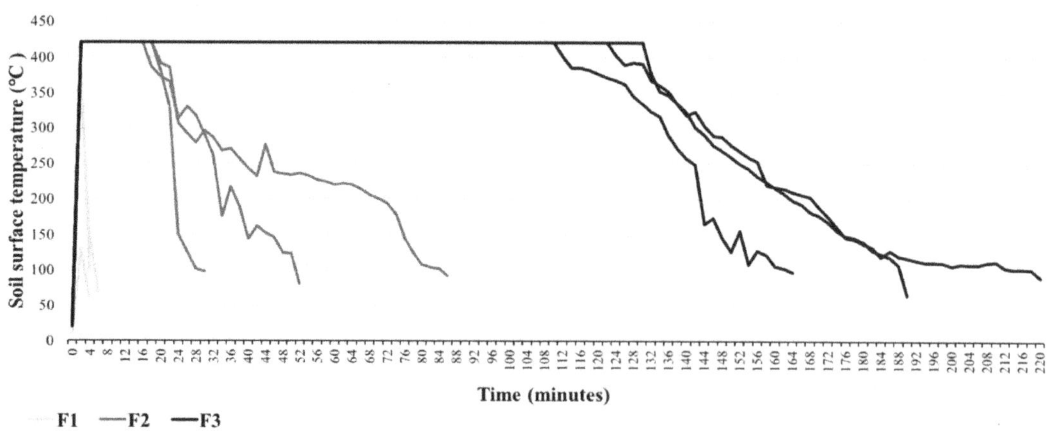

Figure 4. The burn time and soil surface temperatures of each subplot for F1, F2 and F3 treatments.

3.2. Fire Intensity Treatment Effects on Haplic Luvisol Properties

3.2.1. Soil and Ash pH

Results obtained for the pH of soil showed a slight decrease for F1 (low intensity fire treatment; Figure 5). This was followed by an increase in pH after F2 and F3 (medium and high fire treatment, respectively). F3 was significantly increased at $p > 0.05$ compared with the F0 measured pH (Table 1), F1 and F2 were not significantly different. This shows that immediately after the burn, medium and high intensity fires can change the Haplic Luvisol pH from slightly acidic to neutral. The pH of ash increased by ca. 1.0 for each treatment: F1 8.42, F2 9.79, F3 10.33. This is a dramatic change as the pH scale is logarithmic (Table 3). The coefficient of variation for F1 and F2 was 5%, whereas it was 14% for F3.

Table 3. Mean values of pH (soil and ash), soil organic matter (OM), carbon (C), calcium (Ca), potassium (K), magnesium (Mg) and phosphorus (P) in low (F1), medium (F2) and high (F3) fire intensity treatments, and pre-fire (F0). $n = 9$ samples for F0 and $n = 3$ samples for F1, F2 and F3.

Variable	F0	F1	F2	F3	Reference Material Accuracy
pH (soil)	6.39 ± 0.34	6.38 ± 0.12	7.17 ± 0.63	7.01 ± 0.36	N/A
pH (ash)	N/A	8.42 ± 0.55	9.79 ± 0.43	10.33 ± 1.54	N/A
OM (%)	6.60 ± 1.70	6.07 ± 1.88	6.74 ± 1.45	5.78 ± 1.15	N/A
C (mg/kg)	33.00 ± 8.50	33.50 ± 9.40	33.70 ± 7.25	28.90 ± 5.75	N/A
Ca (mg/kg)	2048.00 ± 39.00	2290.00 ± 47.00	3320.00 ± 78.00	3500.00 ± 12.00	97.00
K (mg/kg)	402.48 ± 0.44	212.48 ± 0.03	212.15 ± 0.03	215.15 ± 0.05	85.00
Mg (mg/kg)	25.67 ± 0.02	31.67 ± 0.05	47.67 ± 0.05	39.41 ± 0.02	59.00
P (mg/kg)	13.30 ± 2.08	13.55 ± 1.96	46.55 ± 4.75	53.00 ± 4.00	80.00

3.2.2. Organic Matter and Carbon

Results obtained for organic matter (OM) content varied greatly among the plots (Table 3). F1 and F3 lowered the OM% by a small amount (0.53% and 0.82% respectively), though not significantly (Table 1), whereas, F2 increased the amount of OM by only 0.14%. The coefficient of variation for all treatments was around 20% (Table 3). These results show that organic matter in Haplic Luvisol is not significantly affected by fire.

The carbon content of the soil samples (Table 3) remained at similar levels for F1 and F2 treatments compared to F0, 33 mg/kg. For F3, the amount of carbon decreased by 13%, from 33.00 to 29 mg/kg (Table 3). Carbon levels in the soil samples did not significantly change after any of the treatments (Table 1). Coefficient of variation was 30% for all treatments. Carbon levels in Haplic Luvisol after a fire event of any intensity are not affected.

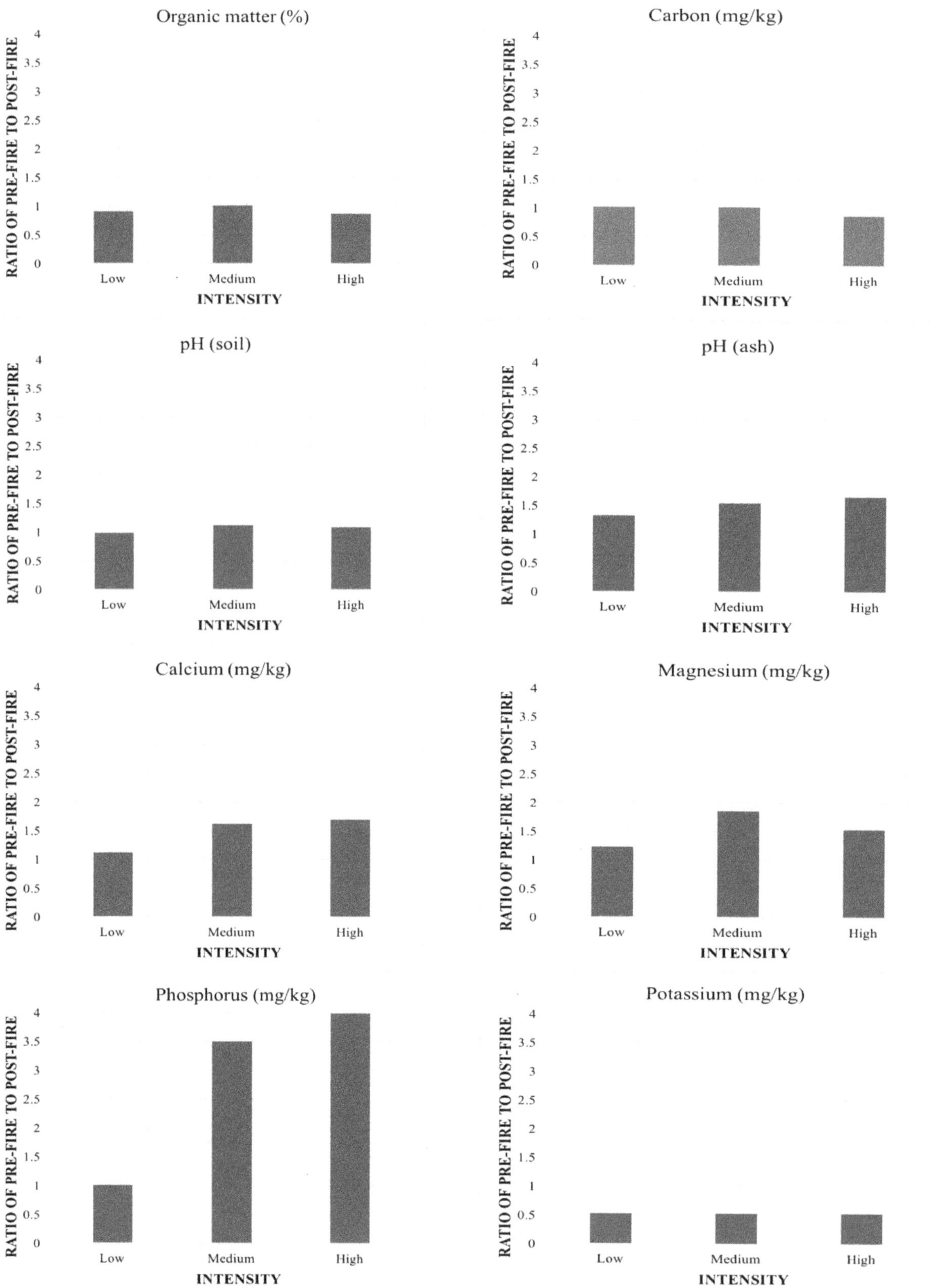

Figure 5. Ratio of pre-fire to post-fire change for pH (soil and ash), organic matter, carbon, calcium, magnesium, potassium and phosphorus. >1 = positive change, 1 = no change, <1 = negative change. Note the different scales on each graph.

3.2.3. Available calcium

Available calcium in soil increases with fire intensity (Figure 5), higher intensity fires result in more available calcium present. There was a noticeable increase after F2 and F3, over 1000 mg/kg more calcium measured than after F1 (Table 3). Each fire treatment showed a significant increase at $p < 0.01$ in comparison to F0 (Table 1). The coefficient of variation for calcium was 2% for F0, F1 and F2; while F3 was 0.3%. The reference material accuracy was 97% for calcium. This was the highest out of all the parameters measured (Table 3), meaning the accuracy of the Atomic Absorption Spectroscopy was very high. Fire therefore, has a significant effect on the amount of calcium present in Haplic Luvisol immediately after a fire event of any intensity.

3.2.4. Available Potassium

The potassium levels in the soil decreased significantly for F1, F2 and F3 at a significance of $p < 0.01$ (Table 1). The negative correlation has shown that higher intensity fires decrease the amount of potassium present in the soil immediately after the burn. The decreasing ratio is constant for all three treatments (Figure 5), dropping from 402.48 mg/kg (F0) to 212.48, 212.15 and 215.15 mg/kg (F1, F2 and F3, respectively). The coefficient of variation was; 0.11% for F0 and 0.01% for F1, F2, F3 (Table 3). Reference material accuracy was calculated at 85%, showing a high accuracy of the AAS instrumentation. Therefore, the level of potassium in Haplic Luvisol can be significantly affected by fire of any intensity immediately after the fire event.

3.2.5. Available Magnesium

Magnesium levels varied across the three treatments (Figure 5); F2 had the highest level at 47.67 mg/kg followed by F3 (39.41 mg/kg) then F1 (31.67 mg/kg) in comparison with the pre-fire level of 25.67 mg/kg. All treatments showed a significant increase in plant available magnesium at $p > 0.01$ (Table 1). The coefficient of variation was; 0.08% for F0, 0.16% for F1, 0.10% for F2 and 0.05% for F3 (Table 3). The reference material accuracy was 59%, which shows a medium accuracy of the AAS instrumentation for magnesium parameter.

3.2.6. Available Phosphorous

The higher the fire intensity, the higher the phosphorus availability. (Figure 5). The amount of P stays constant between F0 and F1, at 13 mg/kg, rising to 46.55 and 53.00 mg/kg after F2 and F3, respectively. These increases in P are significant at $p > 0.01$ (Table 1). The coefficient of variation was; 15.6% for F0, 14.5% for F1, 10.2 % for F2 and 7.5% for F3. Reference material accuracy was calculated at 80% (Table 3), this is a high accuracy for the UV-visible spectroscopy.

4. Discussion

4.1. Burn Characteristics

Recreating conditions of a natural burn, whether prescribed fire or wildfire, is one of the biggest challenges of experimental burning. The amount of biomass and type of species can dictate the intensity of the fire; the challenge is that no forest composition is the same. There are other dominating factors such as weather and topography that influence fire behaviour; weather conditions determine whether the fuel load is 'available' to burn. Weather and topography cannot be controlled in this study so therefore the focus was the effect of different fuel load and type of fuel.

For this experiment, extensive research was carried out to produce the best experimental burn design as detailed in the Materials and Methods section. Some studies [43,44] that used the same recommendations of fuel load for medium and high intensity burns, 20 and 40 t ha^{-1} respectively, found that medium intensity fires burned for ca. 17 min and high intensity fires burned for ca. 37 min [45]. The Damak experiment followed the same fuel load recommendations; however, the mean burn time >100 °C for F2 and F3 were 55.67 and 190.67 min, respectively. An explanation for this is could be attributed to the size of burn area; the area for one study was 105 m^2 [44] and another was 80 m^2 [43]. A larger burn area would create a more severe fire and therefore, burn at a quicker rate. These differences are possibly due to different instrumentation, being that these studies used thermocouples that measured temperature directly within the soil. Our experiment used an infrared thermometer that measured the soil surface temperature. The burning material on the soil surface would make both the burn time and temperature a higher reading compared to studies that use thermocouples due to the soils insulating properties.

4.2. Effect of Fire Intensity on Soil Properties

The pH of noncalcareous soils (such as Haplic Luvisol) should increase after burning, due to the effect of organic acid denaturisation [2] and the release of alkaline cations. Below the threshold of medium fire intensity, burning does not affect soil pH. This is likely to be due to some of the plots for F1 being at a lower pH originally compared to the average pH of all the plots pre-fire. This would mean that the F1 pH would appear to decrease compared to the pH of the plots before burning (F0). pH is not consistent in soil. Substances such as decomposing leaf litter and plant materials can have a higher pH than the underlying soil. Therefore, when F1 was applied to the soil, some of the leaf material is likely to have mixed with the soil. This would produce a higher pH for F1 plots compared to the F0. The fact that the soil sample bags were left for two months before laboratory analysis, might have been enough time for the plant material to change the original soil pH. Studies have found a decrease in pH after low intensity followed by an increase at higher intensity burn sites [32]. It is important to note that the results for F1 pH were not significantly different to F0.

The pH of F2 was higher than F0 but not significantly. However, F3 pH was significantly higher than F0. This shows that the pH after high intensity fires is significantly affected, supported by other studies [2]. Studies have found that significant increases in soil pH only occurred at temperatures >450–500 °C [21], this supports our F3 pH data. However, this shows that the maximum temperature for F2 cannot have exceeded 450 °C. Many authors report increased pH values after a prescribed fire [46].

The pH of ash above the soil for F1, F2 and F3 were significantly higher than both the original soil pH and each other. This was to be expected as the carbonates and oxides that are produced from combustion are highly alkaline. This is supported by studies that found that topsoil pH (including the ash layer), under a mixed conifer forest, increased three units immediately after the burning [22]. Our high intensity burn plots experienced an increase of six pH units within the ash layer. It is likely that ash leaching, chemicals draining from the ash into the underlying soil by the action of percolating liquid (usually rainwater), could turn the soil from slightly acidic to more alkaline. This would possibly have an effect on tree species, especially in forests like Damak. Common pine *Pinus sylvestris* and nettle would likely cope with these changes to the soil, because the latter prefers high levels of phosphate. However, species like hornbeam can only grow in neutral soils, so in the case the pH becomes too alkaline it will not be suitable for its growth. There would possibly be a profound effect on the native Hungarian oak, as they are specially adapted to heavy acidic soils. Additionally, following a forest fire in the region, invasive species such as black locust would probably pioneer alkaline rich soil: this species has a tolerable pH range of 4.6–8.2, but prefers dry alkaline soils.

The varied results for the OM and C post-fire in our experiment are supported by other studies [47]. The effect of fire on soil OM is highly dependent on a number of factors that are seen in our study including; fire intensity, varied distribution of organic matter in the soil, nature of burned vegetation. Effects have been highly variable in many studies [15,48] with effects ranging from complete combustion of organic matter to increases of up to 30%. This is a prominent discussion in soil research as there is no generalised trend between organic matter and burning, it is variable controlled. For example, soil heated under laboratory conditions have frequently reported significant losses in OM and C of up to 100% [49], whereas after a wildfire there have been records of an increase [10]. Another study suggests this could be due to the addition of necromass in forests (dead wood and organisms) after a fire event which is not as prevalent in a laboratory experiment [50]. One study produced a comprehensive dataset of soils effected by wildfire or prescribed fire. Their study concluded that lower C content was found after prescribed fires while higher C content was found after wildfires [14]. This was attributed to the accumulation of charcoal and encroachment of post-fire N-fixing vegetation. We do, however, acknowledge the presence of some limitations due to the application of the LOI method to determine OM and OC. LOI is based on two assumptions: (1) LOI equals OM and (2) the ratio OM/OC is the same for all treatments. LOI incorporated more than just OM, for example hydrated covers of clays, which are expected to be partially lost when a thermal shock occurs, therefore LOI was not exactly equal to OM in this case. In addition, as different temperatures were reached among fire treatments, the OM/OC ratio would not be the same in all of them, likely: OM/OC ratio decreases with the thermal treatment. For these reasons, what we found in our results related to OM and OC can be mainly attributed to using a broad non-specific technique of measuring OM and OC, with its inherent limitations. Further research should imply a more precise method to determine OM and OC, like a modern dichromate oxidation method. This would ensure a precise quantification of OM and OC.

Calcium results suggest that all fire intensities significantly affect calcium levels in Haplic Luvisol. This finding is supported by other studies examining the effect of fire on Ca levels in soil [2]. As the volatilisation threshold temperature of Ca occurs at 1484 °C [15], it is relatively insensitive to fire (temperature of burning woody fuel ca. 1100 °C). Therefore, it was not expected in this study to observe any reductions in the amount of Ca. The reason why Ca increases significantly after fire is thought to have come from the mineralisation of the organic matter [32]. Studies suggest that the exchange capacity of surface horizons is depleted after fire which increases the number of cations that are not volatised (Ca and Mg) [51].

The soil samples measured after the burnings would contain Ca released from the overlying biomass. This could be evidence as to why there was a major increase in Ca between F1 and F2/F3. The increased biomass (ca. 2 and 8 kg more for F2 and F3, respectively) would contribute much more Ca cations than 1 kg of litter (F1). Another possibility is that the temperatures in F2 and F3 reached the volatilisation threshold of K and P, ca. 750 °C [52].

The potassium (K) level after all fire treatments was significantly lower than the pre-fire levels, ca. 50% reduction. This is not supported by much of the literature measuring K levels post-fire events [2,53]. As the volatilisation of K occurs at ca. 750 °C, this reduction would be expected in very high fire intensity events, however, we have observed the reduction in F1 (low intensity) so the reduction in K cannot be attributed to volatilisation. The instrumentation accuracy was high for K (87%), so the reduction cannot be an instrumentation error. There is a possibility that being left in the soil sample bags for 2 months the K cations bound to inorganic substances; however, similar reductions would be observed in F0 if this was the case. A probable reason is that the available potassium was immobilised within mineral structures driven by thermal treatment.

There are very limited studies exploring the effect of fire on K levels in Haplic Luvisol. The studies that have been carried out support our findings that K levels are reduced, after a fire event, in Haplic Luvisol [54]. One study measured significantly lower K levels in the 0–5 cm soil layer post-fire for Umbric leptosols, attributing the loss to erosion and leaching [55]. Studies suggest available K should increase after low intensity prescribed fires [46]. However, there were no erosional or leaching processes occurring in our experiment therefore, the reason why there is a reduction of K in Haplic Luvisol is likely to be due to the composition of the soil. Another explanation could be the moisture level of the soil. It has been found that soil moisture significantly affects the availability of K [56]. The Damak soil was dry (observed in fieldwork) which could have reduced the level of K recorded. However, dry soil would be present in F0 measurements so the K values should be similar to F1, F2 and F3 K values. The results from this study show that fire was the cause of the reduction in K. The effect fire has on K levels in Haplic Luvisol is a therefore a key finding.

Mg levels post-fire were all significantly increased. This increase has been found in many post-fire soil analysis studies [2]. The volatilisation of Mg occurs at ca. 1107 °C [15] therefore there should be no reduction in its availability post-fire. As with Ca, the increase in available Mg is likely to come from the mineralisation of the organic matter in the soil.

Following a prescribed fire, we were expecting increased available phosphorus [46]. The fact that available phosphorous increases with fire intensity can be attributed to the conversion of organic P to orthophosphate [57]. This occurs during the combustion of organic matter and is dependent on the type of soil, vegetation species and leaching [58]. We observed an increase in P for Haplic Luvisol underlying a temperate deciduous forest with no leaching (as samples were taken immediately post-fire). This increase confirms other findings in this area of soil science [59]. The general trend is that P increases significantly dependant mainly on the fire intensity [2]. Small amounts of P can be lost through volatilisation. The threshold temperature for P is ca. 774 °C [60], therefore, losses would be expected in F2 and F3 as the burn time and temperatures could reach >774 °C. However, it is important to note these losses would be very low in comparison to the converted P. In this experiment, there is a clear positive relationship between available P and fire intensity. This result has been found in many levels [2].

In relation to the changes observed in soil OM and C, were somehow inconclusive. This is likely because soil was taken at a depth of 10 cm. Therefore, there would be a dilution effect on the impact of fire on the upper soil (organic matter is mainly affected in the upper 2 cm layer). It could also depend on the amount of OM in the soil pre-fire and soil type. This is supported by other studies [10]. However, in a forest setting, organic matter has shown to return to pre-fire levels after one year [13]. Therefore, any effect of fire on the OM and C would be recoverable in the long term.

One limitation of this study was the instrumentation used to measure soil temperature. This can be observed in the burn time (Figure 4), being much longer than studies that used thermocouples [43,44]. It is likely that the infrared thermometer measured flames from the burning vegetation instead of the soil surface, thus giving the higher temperature (flame). An accurate measurement of the soil surface temperature could not be measured. Therefore, assumptions had to be made in relation to whether certain cations reached their volatilisation threshold. Ideally, a thermocouple would be used that could be inserted into the soil surface and measure the temperature. In addition to that, this study was limited to 0.5 m^2 plots through health and safety restrictions. With accessible funding and Government approval, larger plots could be used, as is the case in Spain, where researchers have had access to 80 m^2 plots [29,31]. A larger scale plot would be a better way to recreate the conditions of a prescribed fire/wildfire. Furthermore, the time between soil sample collection and analysis would ideally be as short as possible, however for our experiment, the gap between collection and analysis was two months. Samples were stored in dark and dry conditions; nevertheless, several processes could still take place within the soil. For example, the exchangeable cations released by the fire treatments could bind to inorganic substances and affect the results for that cation. The soil samples contained root systems and small amounts of vegetation. Their decomposition over two months could lower the pH through the release of carbon dioxide. However, in our study, the processes described above would have had negligible effect on our samples; pH increased in all samples and the change, for all cations measured, was significantly increased. Lastly, the same prescribed fire experiment could have also been implemented at other locations of Hungary with the same Haplic Luvisol soil type, to provide more conclusive results on the burning effect on forest Haplic Luvisol.

The immediate changes to forest soils following burning have been discussed. This study contributes to the concept that burning temperate forests in Central Europe can have a direct impact on soil fertility and forest management, whether that is from prescribed burning or wildfires. The use of small scale experimental fires to measure this effect has mirrored larger experimental burning and wildfire studies.

5. Conclusions

The main purpose of this study was to measure and analyse the immediate effect of fire on forest Haplic Luvisol. This was carried out using representative small scale fire plots that enables fire effect on soil to be measured without damaging large areas of land. These experimental fires carried out on soil matter in Damak Forest, North East Hungary, produced significant changes to the Haplic Luvisol organic matter and chemical properties. Increased changes were most obvious in exchangeable Ca, followed by available P and exchangeable Mg, with pH only significant in the high intensity fire treatment. Significant decrease was observed in all the fire intensities for exchangeable K. This is a key finding of this study, as it is one of the few studies that have found losses in exchangeable K post-fire. This contrasts with the general behaviour of exchangeable K in soils [2,46]. As expected, a higher intensity fire resulted in a greater change in nutrient levels, with the exception of Mg.

Another relevant finding of this study is the increase of pH, which was significant in the high intensity fires. Through leaching, the Haplic Luvisol will likely have a much higher pH, which can directly affect pH sensitive vegetation (e.g., native rare species such as *Quercus frainetto* Ten.).

This study indicates that there are potential negative effects on soil fertility based on the decrease of potassium in the experiment. Potassium is a key nutrient for the ion transport within vegetation. Any loss of this from the soil would have major influences on the agricultural sector of the Hungarian economy. With an increased forest fire risk across Europe, from the 2041–2070 A1B emission scenario [61], nutrient rich soil, such as Haplic Luvisol, will be adversely affected by losses in K. As such, this study provides a reference point for further investigation into the effect of fire intensity on Haplic Luvisol and for any soil type that could be subject to burning. To enable stronger conclusions to be drawn in future investigations, we recommend conducting a larger scale of experimental burning, to follow burned soils over a recovery period and to evaluate impacts on productivity. Nevertheless, we believe our findings provide useful evidence for Haplic Luvisol changes under fire disturbance, while also supporting the maintaining of the current Hungarian fire prevention policy.

Author Contributions: Conceptualisation, J.M.B. and G.P.P.; methodology, J.M.B. and G.P.P.; formal analysis, J.M.B.; investigation, J.M.B., G.P.P. and N.C.; writing—original draft preparation, J.M.B., G.P.P. and N.C.; writing—review and editing, J.M.B., G.P.P., and N.C.; visualisation, J.M.B.; supervision, G.P.P. and N.C.; project administration and funding acquisition J.M.B.

Acknowledgments: We thank János and Katalin Tóth, who granted permission for the experiment to be carried out on their land. We are grateful to Roxána Filetóth for facilitating the arrangement and in particular for her help with the fieldwork. Ian Saunders in the DGES Laboratory at Aberystwyth University who helped with the analysis of the soil samples. Andrew Thomas is acknowledged for insightful discussions on the manuscript.

Appendix A

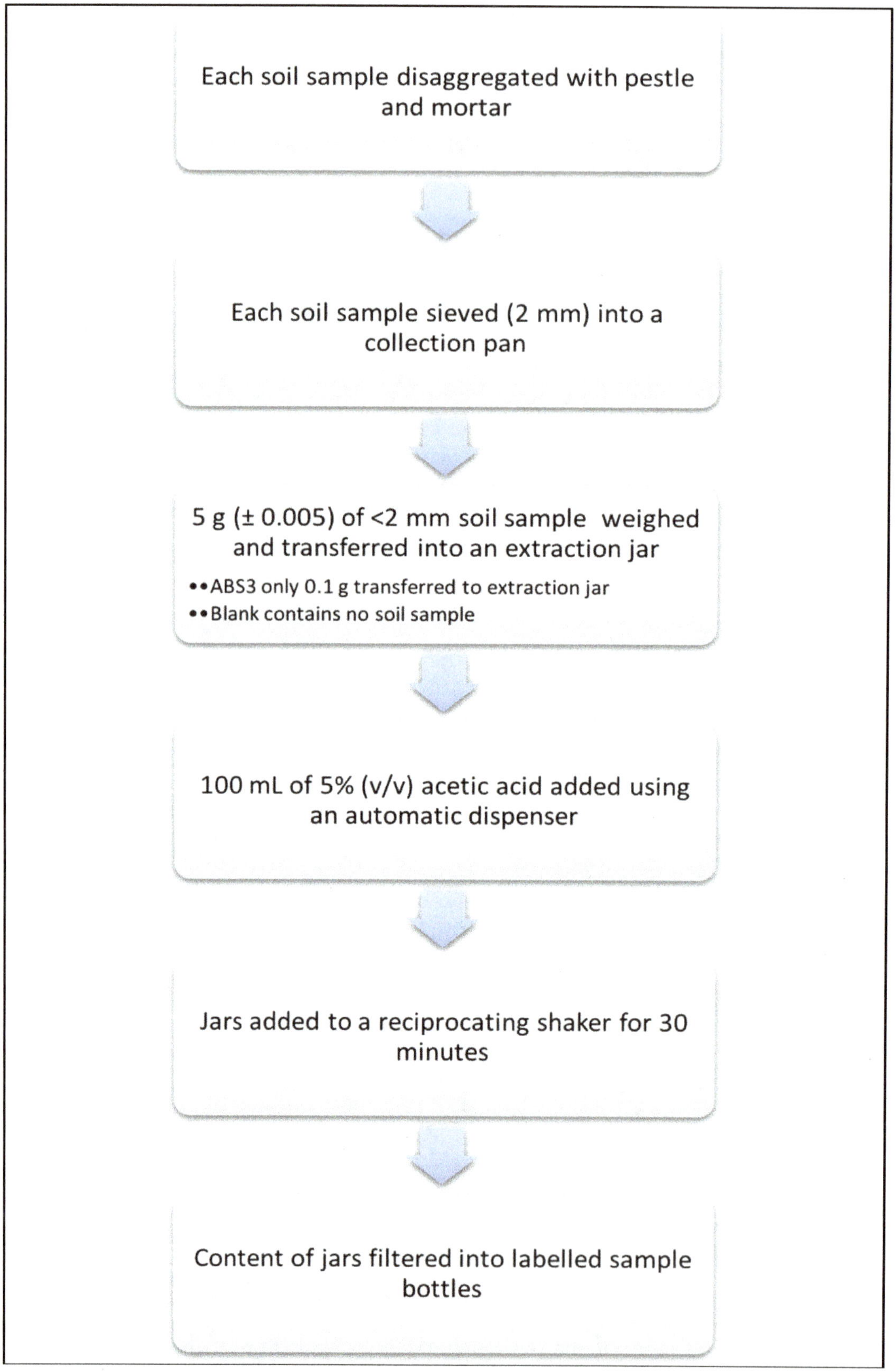

Figure A1. Method used to prepare soil samples for AAS and UV-spectroscopy analysis for plant available concentration.

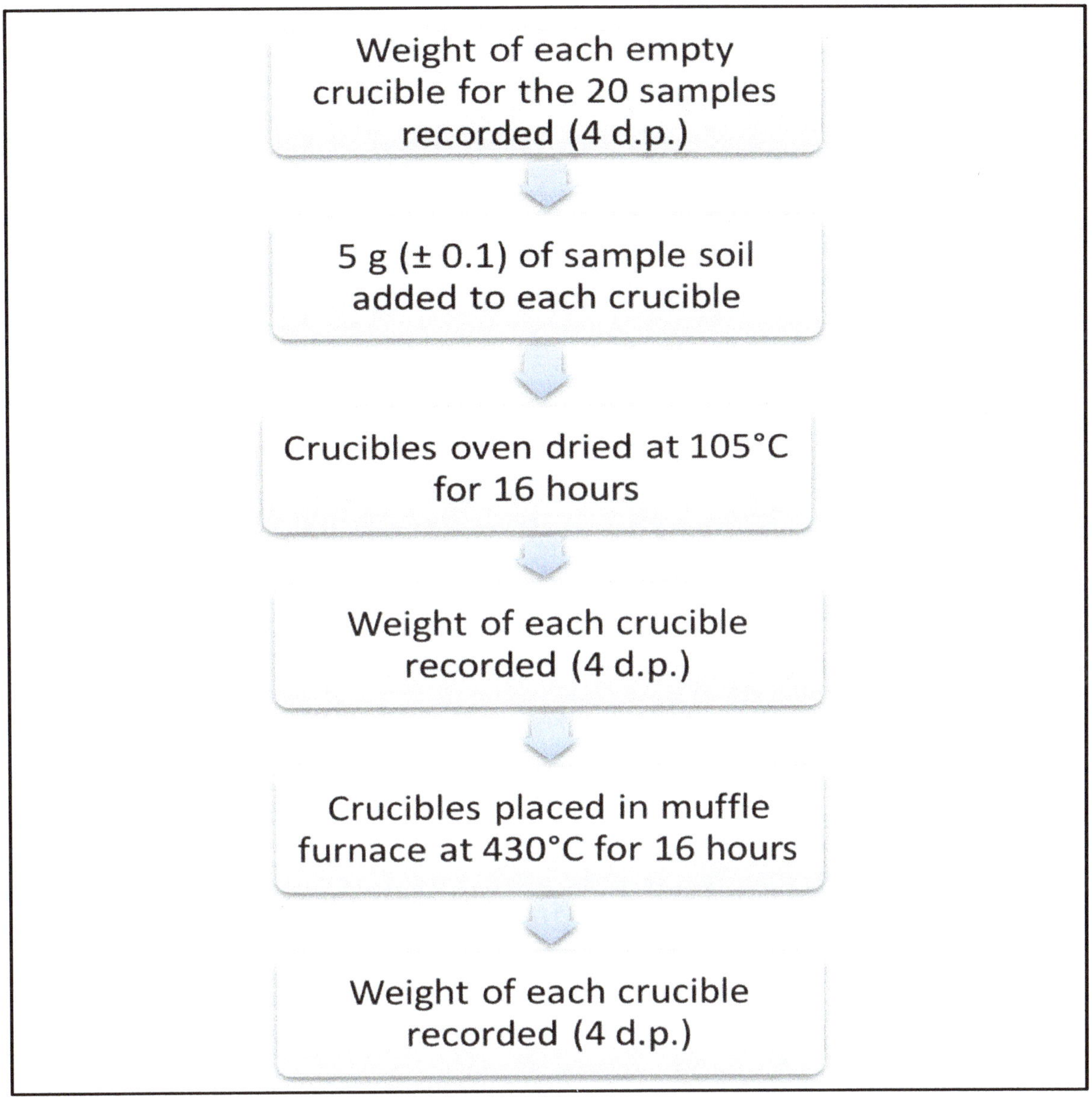

Figure A2. Method used for organic matter (OM) and C analysis of all samples.

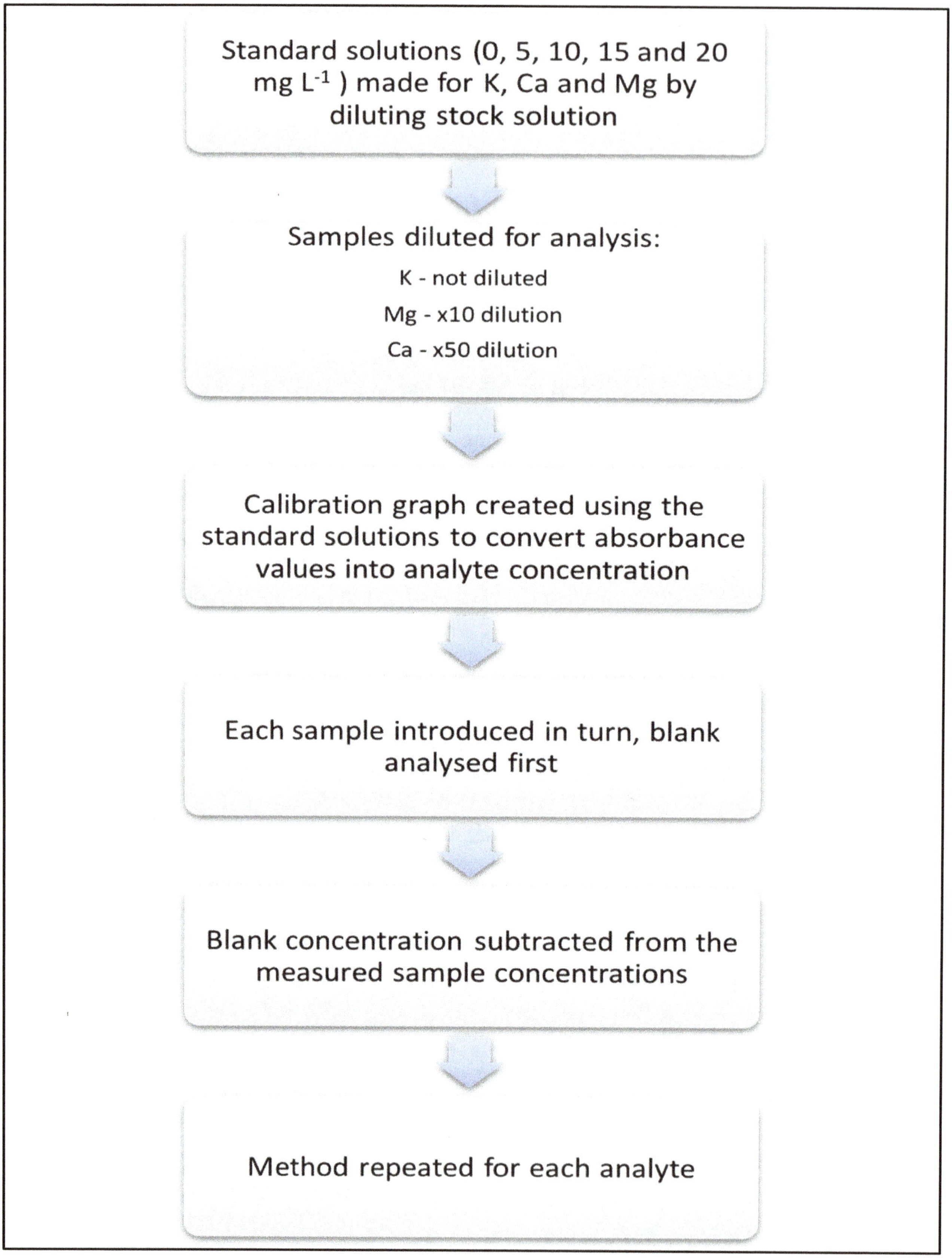

Figure A3. Method used for Ca, K and Mg analysis of all samples.

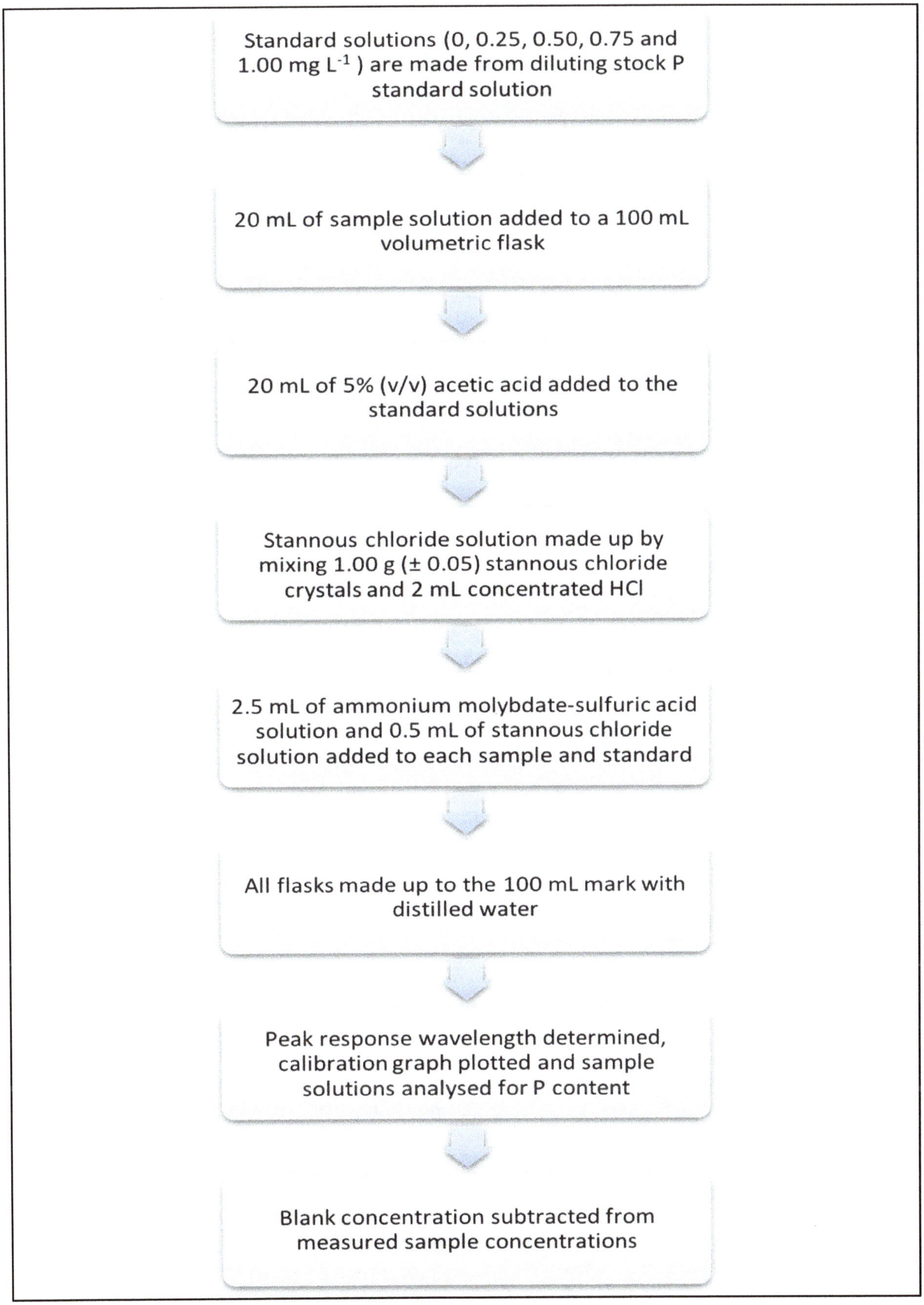

Figure A4. Methods used for P analysis of all samples.

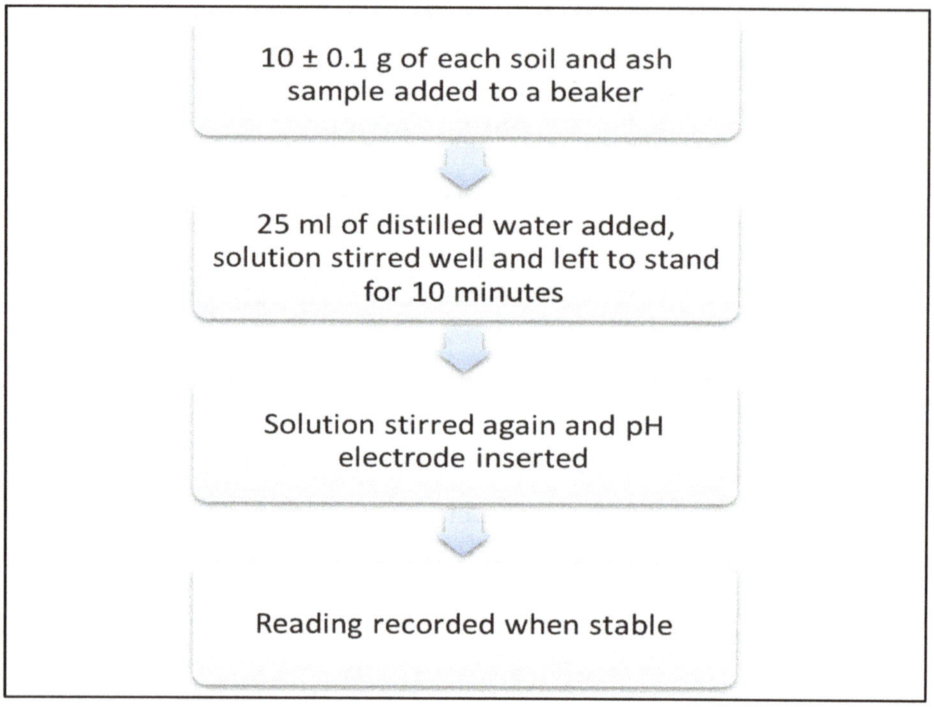

Figure A5. Method used for soil and ash pH analysis.

References

1. Barros, A.M.G.; Pereira, J.M.C. Wildfire Selectivity for Land Cover Type: Does Size Matter? *PLoS ONE* **2014**, *9*, e84760. [CrossRef] [PubMed]
2. Certini, G. Effects of fire on properties of forest soils: A review. *Oecologia* **2005**, *143*, 1–10. [CrossRef] [PubMed]
3. Castellnou, M.; Kraus, D.; Miralles, M. Prescribed Burning and Suppression Fire Techniques. In *Best Practices of Fire Use—Prescribed Burning and Suppression Fire Programmes in Selected Case-Study Regions in Europe*; Montiel, C., Kraus, D., Eds.; European Forest Institute: Joensuu, Finland, 2010; pp. 3–17.
4. Ascoli, D.; Bovio, G. Prescribed burning in Italy: Issues, advances and challenges. *iForest—Biogeosci. For.* **2013**, *6*, 79–89. [CrossRef]
5. Fernandes, P.; Davies, G.; Ascoli, D.; Fernández, C.; Moreira, F.; Rigolot, E.; Stoof, C.; Vega, J.; Molina, D. Prescribed burning in southern Europe: Developing fire management in a dynamic landscape. *Front. Ecol. Environ.* **2013**, *11*, 4–14. [CrossRef]
6. Tilman, D.; Lehman, C. Human-caused environmental change: Impacts on plant diversity and evolution. *Proc. Natl. Acad. Sci. USA* **2001**, *98*, 5433–5440. [CrossRef]
7. Moritz, M.A.; Batllori, E.; Bradstock, R.A.; Gill, A.M.; Handmer, J.; Hessburg, P.F.; Leonard, J.; McCaffrey, S.; Odion, D.C.; Schoennagel, T.; et al. Learning to coexist with wildfire. *Nature* **2014**, *515*, 58–66. [CrossRef]
8. Christensen, J.H.; Hewitson, B.; Busuioc, A.; Chen, A.; Gao, X.; Held, R.; Jones, R.; Kolli, R.K.; Kwon, W.K.; Laprise, R.; et al. Regional climate projections. In *Climate Change 2007: The Physical Science Basis. Contribution of Working Group I to the Fourth Assessment Report of the Intergovernmental Panel on Climate Change*, 1st ed.; Solomon, S., Qin, D., Manning, M., Chen, Z., Marquis, M., Averyt, K.B., Tignor, M., Miller, H.L., Eds.; Cambridge University Press: Cambridge, UK, 2007; pp. 847–940.
9. IPCC. *Fifth Assessment Report*; Cambridge University Press: Cambridge, UK, 2014.
10. González-Pérez, J.; González-Vila, F.; Almendros, G.; Knicker, H. The effect of fire on soil organic matter—A review. *Environ. Int.* **2004**, *30*, 855–870. [CrossRef]
11. Giovannini, G.; Lucchesi, S.; Giachetti, M. Effect of heating on some physical and chemical parameters related to soil aggregation and erodibility. *Soil Sci.* **1988**, *146*, 255–261. [CrossRef]
12. Simard, D.G.; Fyles, J.W.; Paré, D.; Nguyen, T. Impacts of clearcut harvesting and wildfire on soil nutrient status in the Quebec boreal forest. Can. *J. Soil Sci.* **2001**, *81*, 229–237.

13. Badía, D.; Martí, C.; Aguirre, A.; Aznar, J.; González-Pérez, J.; De la Rosa, J.; León, J.; Ibarra, P.; Echeverría, T. Wildfire effects on nutrients and organic carbon of a Rendzic Phaeozem in NE Spain: Changes at cm-scale topsoil. *Catena* **2014**, *113*, 267–275. [CrossRef]
14. Johnson, D.W.; Curtis, P.S. Effects of forest management on soil C and N storage: Meta analysis. *For. Ecol. Manag.* **2001**, *140*, 227–238. [CrossRef]
15. DeBano, L.F. The role of fire and soil heating on water repellency in wildland environments: A review. *J. Hydrol.* **2000**, *231*, 195–206. [CrossRef]
16. Schlesinger, W.H.; Dietze, M.C.; Jackson, R.B.; Phillips, R.P.; Rhoades, C.C.; Rustad, L.E.; Vose, J.M. Forest biogeochemistry in response to drought. *Glob. Chang. Biol.* **2016**, *22*, 2318–2328. [CrossRef] [PubMed]
17. Hough, D. *Long Corner Creek Hydrologic Project: Aspects of the Geology, Physiography and Soils*; Soil Conservation Authority: Victoria, Australia, 1981.
18. John, T.V.S.; Rundel, P.W. The role of fire as a mineralizing agent in a Sierran coniferous forest. *Oecologia* **1976**, *25*, 35–45. [CrossRef]
19. Binkley, D.; Fisher, R. *Ecology and Management of Forest Soils*, 4th ed.; Wiley-Blackwell: Chichester, UK, 2013.
20. Parra, J.; Rivero, V.; Lopez, T. Forms of Mn in soils affected by a forest fire. *Sci. Total Environ.* **1996**, *181*, 231–236. [CrossRef]
21. Arocena, J.M.; Opio, C. Prescribed fire-induced changes in properties of sub-boreal forest soils. *Geoderma* **2003**, *113*, 1–16. [CrossRef]
22. Ulery, A.L.; Graham, R.C.; Amrhein, C. Wood-ash composition and soil pH following intense burning. *Soil Sci.* **1993**, *156*, 358–364. [CrossRef]
23. Sardans, J.; Peñuelas, J. Plant-soil interactions in Mediterranean forest and shrublands: Impacts of climatic change. *Plant Soil* **2013**, *365*, 1–33. [CrossRef]
24. Deák, B.; Valkó, O.; Török, P.; Végvári, Z.S.; Hartel, T.; Schmotzer, A.; Kapocsi, I.; Tóthmérész, B. Grassland fires in Hungary—Experiences of nature conservationists on the effect of fire on biodiversity. *Appl. Ecol. Environ. Res.* **2014**, *12*, 267–283. [CrossRef]
25. Végvári, Z.; Valkó, O.; Deák, B.; Török, P.; Konyhás, S.; Tóthmérész, B. Effects of land use and wildfires on the habitat selection of Great Bustard (Otis tarda L.)—implications for species conservation. *Land Degrad. Dev.* **2016**, *27*, 910–918. [CrossRef]
26. FAO—UNESCO. *Soil Map World*, 5th ed.; UNESCO: Paris, France, 1981.
27. FAO. *World Reference Base for Soil Resources 2006: A Framework for International Classification, Correlation and Communication*; FAO: Rome, Italy, 2006.
28. European Soil Bureau Network and the Hungarian Soil Science Society. The Soils of Hungary. 2011. Available online: http://enfo.agt.bme.hu/drupal/sites/default/files/Soils%20of%20Hungary.pdf (accessed on 23 September 2017).
29. Campo, J.; Gimeno-García, E.; Andreu, V.; González-Pelayo, O.; Rubio, J. Aggregation of under canopy and bare soils in a Mediterranean environment affected by different fire intensities. *Catena* **2008**, *74*, 212–218. [CrossRef]
30. Rubio, J.L.; Forteza, J.; Andreu, V.; Cerni, R. Soil profile characteristics influencing runoff and soil erosion after forest fire: A case study (Valencia, Spain). *Soil Technol.* **1997**, *11*, 67–78. [CrossRef]
31. Gimeno-Garcia, E.; Andreu, V.; Rubio, J. Changes in organic matter, nitrogen, phosphorus and cations in soil as a result of fire and water erosion in a Mediterranean landscape. *Eur. J. Soil Sci.* **2000**, *51*, 201–210. [CrossRef]
32. Giovannini, G.; Lucchesi, S. Modifications induced in soil physico-chemical parameters by experimental fires at different intensities. *Soil Sci.* **1997**, *162*, 479–486. [CrossRef]
33. Hurlbert, S.H. Pseudoreplication and the design of ecological field experiments. *Ecol. Monogr.* **1984**, *54*, 187–211. [CrossRef]
34. Whelan, R.J. *The Ecology of Fire*, 2nd ed.; Cambridge University Press: Cambridge, UK, 1995.
35. Jackson, M.L. *Soil Chemical Analysis*; Prentice-Hall: Englewood Cliffs, NJ, USA, 1958.
36. Schulte, E.E.; Hopkins, B.G. Estimation of soil organic matter by weight loss-on-ignition. In *Soil Organic Matter: Analysis and Interpretation, SSSA Special Publication*, 46th ed.; Magdoff, F., Tabatabai, M.A., Hanlon, E.A., Eds.; Soil Society of America: Madison, WI, USA, 1996; pp. 21–31.

37. Combs, S.M.; Nathan, M.V. Soil Organic Matter. In *Recommended Chemical Soil Test. Procedures*, 1st ed.; Warncke, D., Brown, J., Eds.; Missouri Agricultural Experiment Station SB 1001: Columbia, MO, USA, 1998; pp. 53–59.
38. Zhang, H.; Wang, J.J. Loss on Ignition Method. In *Soil Test Methods from the Southeastern United States*; Sikora, F.J., Moore, K.P., Huluka, G., Kissel, D.E., Miller, R., Mylavarapu, R., Oldham, J.L., Eds.; Southern Cooperative Series Bulletin: Stillwater, OK, USA, 2014; pp. 155–158.
39. Pribyl, D.W. A critical review of the conventional SOC to SOM conversion factor. *Geoderma* **2010**, *156*, 75–83. [CrossRef]
40. Ball, D.F. Loss-on-ignition as an estimate of organic matter and organic carbon in non-calcareous soil. *Eur. J. Soil Sci.* **1964**, *15*, 84–92. [CrossRef]
41. Welz, B.; Sperling, M. *Atomic Absorption Spectrometry*, 3rd ed.; John Wiley & Sons: New York, NY, USA, 2008.
42. Rossel, R.V.; Walvoort, D.J.J.; McBratney, A.B.; Janik, L.J.; Skjemstad, J.O. Visible, near infrared, mid infrared or combined diffuse reflectance spectroscopy for simultaneous assessment of various soil properties. *Geoderma* **2006**, *131*, 59–75. [CrossRef]
43. González-Pelayo, O.V.; Andreu, J.; Campo, E.; Gimeno-García, E.; Rubio, J.L. Hydrological properties of a Mediterranean soil burned with different fire intensities. *Catena* **2006**, *68*, 186–193. [CrossRef]
44. Granged, A.; Jordán, A.; Zavala, L.; Muñoz-Rojas, M.; Mataix-Solera, J. Short-term effects of experimental fire for a soil under eucalyptus forest (SE Australia). *Geoderma* **2011**, *167*, 125–134. [CrossRef]
45. Etienne, M.; Legrand, C. A non-destructive method to estimate shrubland biomass and combustibility. In Proceedings of the 2nd International Conference on Forest Fire Research, Comissão de Coordenação da Região Centro Coimbra, Coimbra, Portugal, 1994; pp. 425–434.
46. Alcañiz, M.; Outeiro, L.; Francos, M.; Úbeda, X. Effects of prescribed fires on soil properties: A review. *Sci. Total Environ.* **2018**, *613*, 944–957. [CrossRef] [PubMed]
47. González-Pérez, J.A.; González-Vila, F.J.; Polvillo, O.; Almendros, G.; Knicker, H.; Salas, F.; Costa, J.C. *Wildfire and Black Carbon in Andalusian Mediterranean Forest*; Viegas, D.X., Ed.; Millpress: Rotterdam, The Netherlands, 2000; pp. 1–7.
48. Chandler, C.; Cheney, P.; Thomas, P.; Trabaud, L.; Williams, D. *Fire in Forestry Volume 1: Forest Fire Behavior and Effects*; Wiley: New York, NY, USA, 1983.
49. Fernández, I.; Cabaneiro, A.; Carballas, T. Organic matter changes immediately after a wildfire in an Atlantic forest soil and comparison with laboratory soil heating. *Soil Biol. Biochem.* **1997**, *29*, 1–11. [CrossRef]
50. Rashid, G.H. Effects of fire on soil carbon and nitrogen in a Mediterranean oak forest of Algeria. *Plant Soil* **1987**, *103*, 89–93. [CrossRef]
51. Smith, D.W. Concentrations of soil nutrients before and after fire. *Can. J. Soil Sci.* **1970**, *50*, 17–29. [CrossRef]
52. Neary, D.G.; Ryan, K.C.; DeBano, L.F. Wildland Fire in Ecosystems: Effects of Fire on Foils and Water. General Technical Report. 2005. Available online: https://www.fs.fed.us/rm/pubs/rmrs_gtr042_4.pdf (accessed on 7 October 2017).
53. DeBano, L.F.; Ffolliott, P.; Neary, D. *Fire's Effects on Ecosystems*; John Wiley & Sons: New York, NY, USA, 1998.
54. Ene, C.; Rogobete, G.; Adam, I.; Ciobanu, V.D. Influence of fires on soils of Berzasca and Moldova Noua forestries (wouth-west of Romania). In Proceedings of the 12th WSEAS international conference on Mathematics and computers in biology, business and acoustics, Braşov, Romania, 11–13 April 2011; pp. 114–121.
55. Fonseca, F.; De Figueiredo, T.; Nogueira, C.; Queirós, A. Effect of prescribed fire on soil properties and soil erosion in a Mediterranean mountain area. *Geoderma* **2017**, *307*, 172–180. [CrossRef]
56. Kuchenbuch, R.; Claassen, N.; Jungk, A. Potassium availability in relation to soil moisture. *Plant Soil* **1986**, *95*, 233–243. [CrossRef]
57. Cade-Menun, B.; Berch, S.; Preston, C.; Lavkulich, L. Phosphorus forms and related soil chemistry of Podzolic soils on northern Vancouver Island. I. A comparison of two forest types. *Can. J. For. Res.* **2000**, *30*, 1714–1725. [CrossRef]
58. Kutiel, P.; Shaviv, A. Effects of soil type, plant composition and leaching on soil nutrients following a simulated forest fire. *For. Ecol. Manag.* **1992**, *53*, 329–343. [CrossRef]
59. Schaller, J.; Tischer, A.; Struyf, E.; Bremer, M.; Belmonte, D.U.; Potthast, K. Fire enhances phosphorus availability in topsoils depending on binding properties. *Ecology* **2015**, *96*, 1598–1606. [CrossRef]

60. Raison, R.J.; Khanna, P.K.; Woods, P.V. Mechanisms of element transfer to the atmosphere during vegetation fires. *Can. J. For. Res.* **1985**, *15*, 132–140. [CrossRef]
61. Lung, T.; Lavalle, C.; Hiederer, R.; Dosio, A.; Bouwer, L.M. A multi-hazard regional level impact assessment for Europe combining indicators of climatic and non-climatic change. *Glob. Environ. Chang.* **2013**, *23*, 522–536. [CrossRef]

8

Spatial Distribution of Soil Nitrogen, Phosphorus and Potassium Stocks in Moso Bamboo Forests in Subtropical China

Xiaolu Tang, Mingpeng Xia, Fengying Guan * and Shaohui Fan *

Key laboratory of Bamboo and Rattan, International Centre for Bamboo and Rattan, Beijing 100102, China; lxtt2010@163.cm (X.T.); mp_xia@yahoo.com (M.X.)

* Correspondence: guanfy@icbr.ac.cn (F.G.); fansh@icbr.ac.cn (S.F.)

Academic Editors: Scott X. Chang and Xiangyang Sun

Abstract: Moso bamboo is famous for fast growth and biomass accumulation, as well as high annual output for timber and bamboo shoots. These high outputs require high nutrient inputs to maintain and improve stand productivity. Soil nitrogen (N), phosphorus (P), and potassium (K) are important macronutrients for plant growth and productivity. Due to high variability of soils, analysing spatial patterns of soil N, P, and K stocks is necessary for scientific nutrient management of Moso bamboo forests. In this study, soils were sampled from 138 locations across Yong'an City and ordinary kriging was applied for spatial interpolation of soil N, P, and K stocks within 0–60 cm. The nugget-to-sill ratio suggested a strong spatial dependence for soil N stock and a moderate spatial dependence for soil P and K stocks, indicating that soil N stock was mainly controlled by intrinsic factors while soil P and K stocks were controlled by both intrinsic and extrinsic factors. Different spatial patterns were observed for soil N, P, and K stocks across the study area, indicating that fertilizations with different ratios of N:P:K should be applied for different sites to maintain and improve stand productivity. The total soil N, P, and K stocks within 0–60 cm were 0.624, 0.020, and 0.583 Tg, respectively, indicating soils were important pools for N, P, and K.

Keywords: ordinary kriging; geostatistical analysis; spatial variability; Moso bamboo

---

1. Introduction

Soil nitrogen (N), phosphorus (P), and potassium (K) are important macronutrients which can limit or co-limit plant growth [1,2]. Human activities, such as fertilization, reclamation, and weeding, have greatly affected the biogeochemical cycling of N, P, and K, thereby altering the pattern, magnitude, and extent of nutrient limitation on land [3]. Effective, efficient, and site-specific management and estimation of soil N, P, and K have attracted great interests for scientists looking to improve nutrient input efficiency, thus increasing stand productivity and reducing environmental risks [4]. Due to various climatic conditions [5], parent materials [6], topography [7], vegetation types [8], soil texture [9], and land use [10], soils are characterized by a highly spatial and temporal variability. This has made the accurate estimation of spatial nutrient content difficult. However, rational soil management requires a deep understanding of the spatiotemporal variability of soil N, P, K, and their maps. For example, areas of particular concern, such as nutrient deficiency, can be identified. Additionally, soil N and P levels are closely related to soil organic carbon cycling [11], which may lead to dynamic effects on greenhouse gas emissions that potentially result in feedback to global climate change. Therefore, it is necessary to improve our understanding of the spatial distribution of soil N, P, and K stocks when evaluating current and potential soil productivity, identifying potential environmental protections, and assessing climate change [12].

Fast developments of geostatistical techniques, such as kriging [11], have significantly advanced the estimation of spatial variability of soil N, P, and K stocks. The most common interpolation techniques calculate the unmeasured property at a given place using measured neighbours data with a weighted mean [13]. The core of geostatistics is that experimental semivariograms are calculated to analyse the spatial autocorrelation of the edaphic variables and determine the range of spatial dependence [13]. The spatial data structure, choice of variogram models, search radius, and number of closest neighbouring points can determine the performance of Kriging approaches. Compared to other interpolation approaches (e.g., Inverse Distance Weighting and splines methods [14,15]), ordinary kriging (OK) can provide more accurate estimates of the spatial distribution of soil nutrients because: (1) OK can provide the best linear unbiased estimates and information on the distribution of the estimation error [15]; (2) OK has relatively strong statistical advantages [16]. Additionally, OK is easy to conduct with high accuracy compared to other kriging approaches, such as cokriging, universal kriging, and factorial kriging. For example, cokriging requires an additional correlated covariant, which leads to the substantial increase of field and lab work and does not increase the statistical power [15]. Thus, OK has been widely used in spatial interpolation for different soil properties. For example, Liu et al. [11] predicted the spatial patterns of soil N and P stocks for the whole Loess Plateau using OK, and total N and P stocks amounted to 0.217 and 0.205 Pg (1 Pg = 10^{15} g), respectively. Martín et al. [17] predicted the spatial distribution of soil carbon stocks within 0–30 cm with OK across Spain.

Bamboo is an important forest type in Southern China, representing an area of 6.16 million ha, more than 70% of which are Moso bamboo (*Phyllostachys heterocycla* (Carr.) Mitford *cv. Pubescens*) forests. Moso bamboo forests are famous for their rapid growth and fast biomass accumulation. However, annual timber harvest and bamboo shoots remove a large amount of nutrients, which means that Moso bamboo forests require more nutrient input compared to other forest types in order to maintain high stand productivity. Consequently, intensive management with fertilization, reclamation, and regular understory removal is becoming increasingly popular in Southern China, especially in the main bamboo producing provinces, such as Zhejiang and Fujian Provinces [18–20]. These activities have significantly changed the nutrient levels by affecting the microbial processes, soil structures, and chemical compositions [21,22]. Although many studies focus on the spatial patterns of soil N, P, and K stocks in different ecosystems [23,24], site specific maps of soil N, P, and K stocks for scientific management of Moso bamboo forests are still lacking in our study area. In contrast, a better understanding of the spatial distribution of soil N, P, and K stocks is strongly required for land management and for maintaining and improving stand productivity of Moso bamboo forests. Therefore, the objectives of this study were to: (1) predict the spatial variability of soil N, P, and K stocks of Moso bamboo forests in Yong'an City, China and (2) estimate total soil N, P, and K stocks within 0–60 cm.

2. Materials and Methods

2.1. Study Area

The study was conducted in the Moso bamboo forests (Figure 1, polygons) across Yong'an City, Fujian Province, Southern China (117°56′–117°47′ E, 25°33′–26°12′ N). It has a subtropical southeast monsoon climate with a mean annual temperature of 19.3 °C, while the lowest temperature is −11°C and the highest temperature is 40 °C. The average annual precipitation is about 1600 mm [25,26]. The elevation of the study area ranged from 580 m to 1605 m above sea level [25]. The accumulated temperature of ≥10 °C is 4520–5800 °C, lasting for 225–250 days, and the relative humidity is about 80% [25]. The main soil type is Oxisol. Yong'an City has a forest cover of 82%, including an area of 5.85 × 10^4 ha of Moso bamboo forests [26]. Moso bamboo forests are mainly distributed below 800 m, most of which are pure stands and sometimes mixed with *Keteleeria cyclolepis* Flous., *Cunninghamia lanceolata* (Lamb.) Hook., *Myrica rubra* (Lour.) S. et Zucc., *Choerospondias axillaris* (Roxb.)

Burtt et Hill., *Liriodendron chinense* (Roxb.) Burtt et Hill., and *Schima Superba* Gardn. et Champ., etc. To improve the stand productivity and increase the income, fertilization treatment has been widely applied in most of the Moso bamboo forests. The fertilization treatment was conducted across the whole study area using the same protocol established by the local Forest Bureau. Similar fertilization types—organic fertilizers with N, P, and K addition, were offered every year by the Forest Bureau because the residents can get an economic subsidy if they follow the protocol and use the recommended fertilizers.

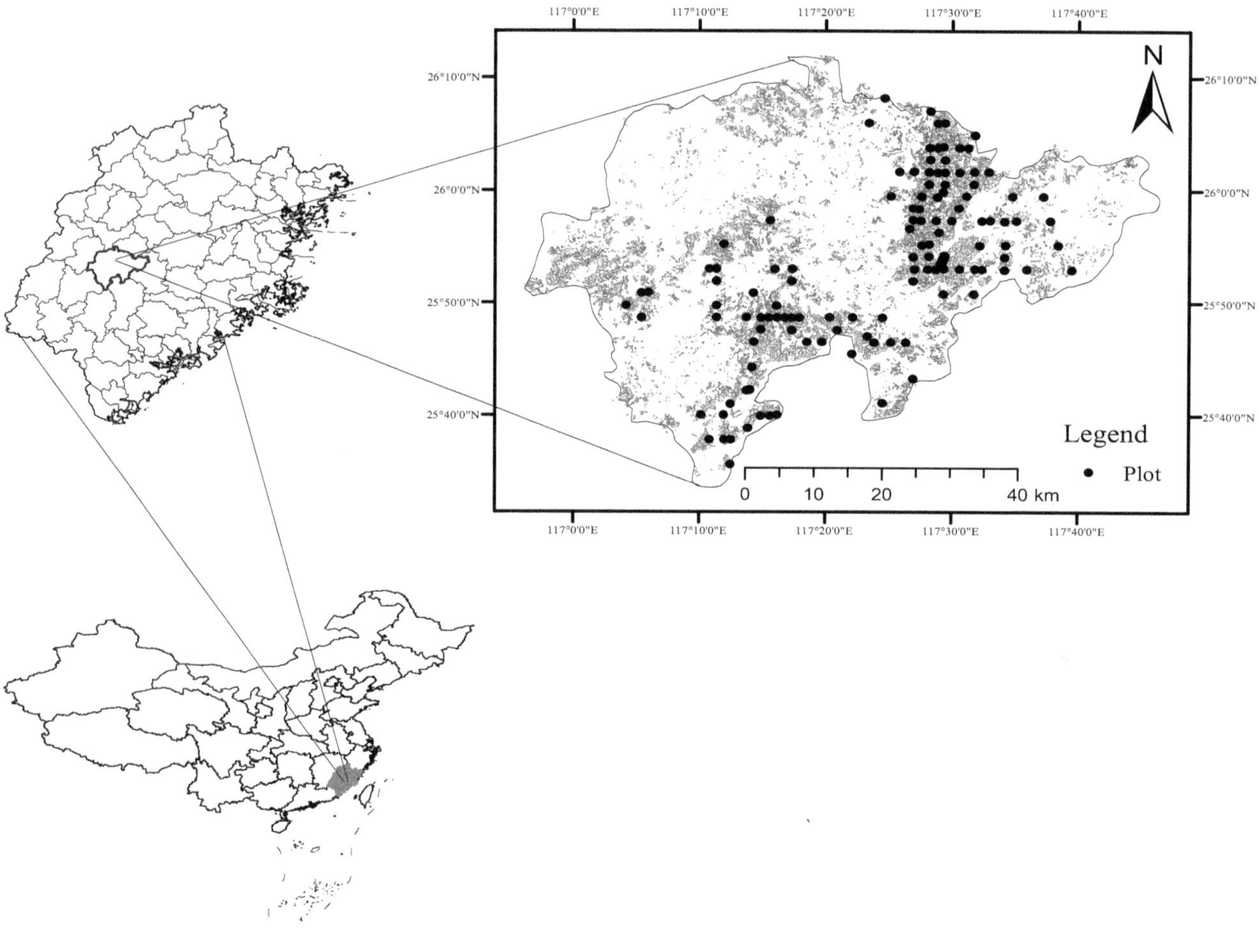

Figure 1. A map of the study area showing the distribution of bamboo forests (polygons) and sampling locations (solid circle).

2.2. Soil Sampling

The representative soils were sampled in the sub-compartments of forest resource management sites in Fujian province, China. These sample sites are established by the local Forest Bureau for soil mapping according to the protocol of forest management inventory [27] (Figure 1). In the targeted sub-compartments, a total of 138 sites were selected and a cluster of three circular plots with a size of 33.3 m^2 were established for each site. Due to the unfavourable access of sampling plots, most of the samples were collected in the East and Southern portions of the study area, which covered 78% of the total Moso bamboo forests. Thus, it is acceptable with these plots to study the spatial distribution of soil N, P, and K stock across the whole study area using the unbiased interpolation approach of OK (see below).

In each plot centre, soils were sampled down to 60 cm by the consideration of three layers: 0–20 cm, 20–40 cm, and 40–60 cm. In the field, soil samples in the sample layer were mixed for one cluster of the three circular plots. In the laboratory, soil samples were air-dried at room temperature and prepared for sieving through 2-mm and 0.15-mm for N, P, and K content analysis. Identifiable plant residues and root materials were removed during sieving. Because the majority of bamboo roots were distributed within 0–40 cm [28], soil samples down to 60 cm were able to meet the research purpose to study spatial patterns of soil N, P, and K stocks. Soil bulk density was determined by a cutting ring approach for each soil layer [29]. During the implementation of fertilization, stones and rocks were removed in Moso bamboo forests. As a result, few stones and rocks were observed in core samples, thus, we did not correct for gravel content. Additionally, site information about elevation, coordinate, soil depth, soil type, soil organic matter content, and bamboo diameters were recorded and determined.

Soil N content (g kg^{-1}) was analysed using the Kjeldahl digestion procedure (5 mL concentrated H_2SO_4, 1 g of soil passed through a 0.15-mm sieve, heated for 90 min, titrated with 0.2 mol L^{-1} standardized $Na_2B_4O_7$). Soil total phosphorus concentration (g kg^{-1}) was determined by alkaline digestion (NaOH 0.2 g, <0.25 mm soil 0.25 g, 300 °C for 15 min and then 750 °C for 15 min) followed by molybdate colorimetric measurement. Soil K content (g kg^{-1}) was determined by sodium hydroxide flame photometer (NaOH 0.2 g, <0.25 mm soil 0.25 g, 300 °C for 15 min and then 750 °C for 15 min, 80 °C water 10 mL). All the analyses followed the national standard protocol of China [29].

Thus, soil N, P, and K stocks were calculated as [30–32]:

$$SON\ (\text{or P or K})\ \text{stock} = SON\ (\text{P,K}) \times BD \times D/10 \quad (1)$$

where *SON* (or P or K) are the soil N, P, and K contents (g kg^{-1}); *BD* is bulk density (g cm^{-3}); *D* is the depth of soil layer (cm). Total soil N, P, and K stocks within 0–60 cm were summed for all soil layers.

2.3. Statistical and Geostatistical Analyses

A standard statistical analysis (mean, standard deviation, first quartile, third quartile, etc.) was conducted in *R* to illustrate trends of the soil N, P, and K stocks [33]. The coefficient of variation (CV) was used to describe the degree of general variation. To meet the assumption of normality for geostatistical analysis, the raw data was log-transformed and transformed back by weighted mean in GS + 10.0 [34]. OK was performed in GS + 10.0 and the distribution maps of soil N, P, and K stocks were produced in ArcGIS 10.2 [35].

2.4. OK

In the last several decades, many spatial interpolation approaches have been developed, such as kriging; OK normally performed better and is easier to apply compared to other approaches, thereby it has been widely used in the spatial interpolation of soil properties [11,17,23]. In the OK process, semivariograms are used to describe the spatial autocorrelation and provide parameters for optimal spatial interpolation based on the theory of regionalized variables [36]. Well-known theoretical models (e.g., exponential, Gaussian, and spherical models) are commonly used to fit the experimental semivariograms [37]. The experimental semivariograms are expressed as a function of the distance between the sampled points and calculate the integrity of spatial continuity in one or multiple directions using the following expression [38]:

$$\gamma(h) = \frac{1}{2N(h)} \sum_{i=1}^{N(h)} \left[z(x_i) - z(x_{i+h})\right]^2 \quad (2)$$

where $z(x_i)$, and $z(x_{i+h})$ are values of z at locations x_i and x_{i+h}, respectively; h is the lag, and $N(h)$ is the number of pairs of sample points separated by h. In this study, spherical, exponential, linear and Gaussian models were used to attempt to describe the semivariograms of soil N, P, and K stocks within 0–60 cm, then the best models determined by the smallest residuals and the largest determination coefficients were selected for spatial interpolation.

In the semevariograms, three major parameters were derived—the nugget (C_0), the sill ($C + C_0$), and the range (A_0)—in order to identify the spatial structure of soil N, P, and K stocks at a given scale. The sill ($C + C_0$) represents the total variation and the ratio of nugget and sill is considered as a criterion to classify the spatial dependence [17]. The range (A_0) represents the separation distance when semivariogram is stabilized, beyond which the measured data are spatially independent [39]. More details about semivariograms and the kriging approach can be found in Goovaerts [37]. In our study, OK was applied for the spatial patterns of soil N, P, and K stocks. The most likely value which could be expected in a particular grid cell when using m neighbouring observations was defined as [17]:

$$R(x) = \sum_{j=1}^{m} \delta_j z(x_i) \tag{3}$$

where δ_j means the optimal weight under the condition of $\sum \delta_j = 1$, m = 16 in this study.

2.5. Model Validation

To evaluate the prediction accuracy of soil N, P, and K stocks, leave-one-out cross-validation was used [11,17,40]. In the validation process, one datum was omitted and this value was estimated by the remaining data. Thereafter, the estimated value was compared with the real value of the omitted point [17]. This process was repeated for all the observations. Four commonly used indices (i.e., absolute mean error (AME), mean error (ME), root mean square error (RMSE), and determination coefficient (R^2)) were used to compare the interpolation accuracy. These indices were calculated as follows [11,15]:

$$\text{AME} = \frac{1}{n} \sum_{i=1}^{n} |(P_i - M_i)| \tag{4}$$

$$\text{ME} = \frac{1}{n} \sum_{i=1}^{n} (P_i - M_i) \tag{5}$$

$$\text{RMSE} = \sqrt{\frac{1}{n} \sum_{i=1}^{n} (P_i - M_i)^2} \tag{6}$$

$$R^2 = 1 - \frac{\sum_{i=1}^{n} (P_i - M_i)^2}{\sum_{i=1}^{n} (M_i - \overline{M})^2} \tag{7}$$

where P_i, M_i, and $\overline{M}$ are predicted values, measured values, and the mean values of the measured data, respectively.

3. Results and Discussion

3.1. Descriptive Statistics

The summary of statistics of soil N, P, and K stocks within 0–60 cm is shown in Table 1. N stock in Moso bamboo forest increased from a minimum of 5.30 Mg ha^{-1} to a maximum of 20.20 Mg ha^{-1} with an average of 11.57 Mg ha^{-1}. These values fell within the range of N stocks in different forest types across China (1.0–49.9 Mg ha^{-1}) [41]. However, the mean N stock was even higher than the average N stock of China's forests within one meter (8.4 Mg ha^{-1}) [41] and the Moso bamboo in Jiangxi Province (2.82 Mg ha^{-1} within 60 cm) [42]. These differences were mainly attributed to the fertilization application in Moso bamboo forests in the study area, whereas no stand management occurs in most of China's forests.

Soil P stock ranged from 0.18 to 0.68 Mg ha^{-1} and the mean value was 0.38 Mg ha^{-1}. Although this range lied within the reported values of main Chinese soil types, the mean value was much lower than the average across China (6 Mg ha^{-1} within 60 cm) [43]. Similarly, average soil K was 10.80 Mg ha^{-1} and the lowest soil K stock was 5.73 Mg ha^{-1} and highest soil K stock was 18.72 Mg ha^{-1}, which were lower than soil K stocks in subtropical forests in China [44]. These low values of soil P and K stocks may be related to the high output of P and K due to the high output of timber and bamboo shoots every year. In Moso bamboo forests, P is the most limited nutrient for stand productivity [4]. Although fertilization was applied in Moso bamboo forests, P and K input could not compensate for the P and K output from the stand. This result further indicated that more P and K fertilization should be applied in Moso bamboo forest in order to maintain and improve stand productivity.

Table 1. Summary of statistics of soil N, P, and K stocks within 0–60 cm (Mg ha^{-1}).

Nutrient	Mean	Minimum	Maximum	Median	SD	CV (%)	1st *Qu*	3rd *Qu*	Skewness	Kurtosis	*p* of S-W
N Stock	11.57	5.30	20.20	10.98	3.23	27.92	8.81	13.66	0.42	−0.33	0.053
P Stock	0.38	0.18	0.68	0.36	0.11	29.10	0.30	0.45	0.60	−0.14	0.010
K Stock	10.80	5.73	18.72	10.81	3.15	29.17	8.11	13.03	0.32	−0.72	0.014

SD = standard deviation; CV = coefficient of variance; 1st *Qu* = 25% quartile; 3rd = 75% quartile; S-W test = Shapiro-Wilk test.

A CV of 10% indicates a low variability and 10%–90% indicates a moderate variability, and CV >90% indicates higher variability [45]. In this study, the CVs of soil N, P, and K stocks were 28%–29%, indicating a moderate variability of soil N, P, and K stocks within 0–60 cm. *p* values of Shapiro-Wilk test were 0.053 for N stock, 0.010 for P stock, and 0.014 for K stock, indicating a normal distribution of N stock and a non-normal distribution of P and K stocks at a significance level of 0.05. However, in order to make the comparison of spatial interpolation of soil N, P, and K stocks consistent, a natural log-transformation was conducted for the three stocks in order to meet the assumption of normal distribution.

3.2. *Geostatistical Analysis of Soil N, P, and K Stocks*

The semevariograms are presented in Figure 2 and their parameters are shown in Table 2. The exponential model performed best for soil N and P stocks while the spherical model was best for soil K stock based on the smallest residuals and the highest determination coefficient. These models produced high determination coefficients, ranging from 0.64 to 0.74.

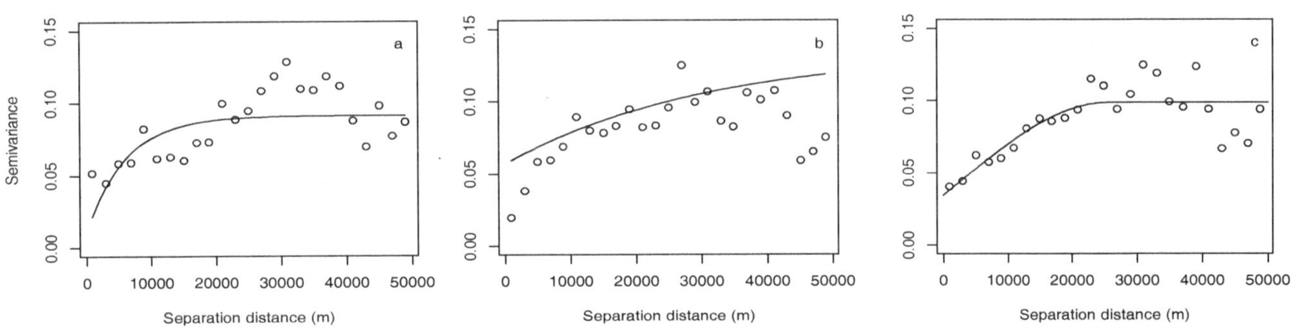

Figure 2. Experimental semivariograms of soil N (**a**), P (**b**), and K (**c**) stocks.

Table 2. Models and their parameters fitted for the semivariograms of soil N, P, and K stocks within 0–60 cm.

Nutrient	Models	Nugget (C_0)	Sill ($C_0 + C$)	Nugget/Sill	Range (A_0, *m*)	Determination Coefficient	Residuals
N stock	Exponential	0.0111	0.0912	12.17	6100	0.642	0.00458
P stock	Exponential	0.0570	0.1340	42.54	30,570	0.741	0.00110
K stock	Spherical	0.0343	0.0986	34.79	25,700	0.643	0.00469

Nugget values present undetectable experimental errors, field variation within the minimum sampling space, and inherent variability [23]. In this study, nugget values were lowest for soil N stock (0.0111), and highest for soil P stock (0.0570). These positive values suggested a positive nugget effect, a sampling error, or random and inherent variability of soil N, P, and K stocks [23]. Sill values represent total spatial variation [11]. The sill values ranged from 0.0912 for N stock to 0.1340 for P stock.

The ratio of nugget-to-sill represents a spatial dependence. If the ratio is lower than 25%, it indicates a strong spatial dependence. If the ratio is higher than 75%, it indicates a weak spatial dependence, while a ratio between 25% and 75% indicates a moderate spatial dependence [46]. A strong spatial dependence is attributed to soil intrinsic properties, such as soil parent material, soil texture, topography, and vegetation [23,47]. A weak spatial dependence indicates that the spatial variability is mainly regulated by extrinsic variations, such as soil fertilization and cultivation practices [23,46]. Therefore, a moderate spatial dependence is controlled by both intrinsic and extrinsic factors. In this study, the nugget-to-sill ratio was 12% for soil N stock, indicating that soil N stock in the study area was mainly controlled by intrinsic factors and that external factors exerted little effects on soil N stock. This may be related to N fertilization application in Moso bamboo forests in that the N input from fertilization can meet the requirement of N output from timber and bamboo shoot harvest. This conclusion was also consistent with the above result (Table 1) that N stock in this study area was higher than other forest types. The nugget-to-sill ratios were 43% for soil P stock and 35% for soil K stock, indicating a moderate spatial dependence that was controlled by both intrinsic and extrinsic factors. This was evidenced by local stand management and complex topography. Intensive managements, such as fertilization and weeding, were widely conducted in Moso bamboo forests in order to improve timber and bamboo shoot output in the study area [19,25]. Meanwhile, the study area was characterized by a high variability of elevations varying from 580 to 1605 m above sea level [25], which could directly affect the soil processes and nutrient contents [11]. In the study area, intensive fertilization treatment was conducted every year, which was expected to be the dominant factor that led to the change of the spatial variability of soil N, P, and K stocks. However, our results (nugget-to-sill ratio) indicated that intrinsic factors played a more important role in controlling spatial patterns of soil N, P, and K stocks.

The ranges (A_0) indicate different influence zones of environmental factors at different scales [23]. Within the range, soil properties are not spatially independent, while beyond the range, soil properties were spatially independent [39]. In this study, the smallest range was found for soil N stock (6100 m), and increased to 25,700 m for soil K stock and 30,570 m for soil P stock. The results indicated that soil N stock was more heterogeneous compared to soil P and K stocks, which was associated with soil processes because soil N stock was mainly controlled by intrinsic factors (see above). Generally, the spatial range was larger than the sampling intervals (Figure 1), which suggested that the sampling strategy in this study was appropriate for studying spatial patterns of soil N, P, and K stocks. However, only a small number of sampling plots were located in the northwest of the study area due to a low cover of bamboo forests. In contrast, the majority of the sampling sites were located in the southern and eastern part of the study area, where most of the bamboo forests were distributed, thus the sampling strategy could meet our research purpose and could predict the spatial distributions of soil N, P, and K stocks accurately.

3.3. Cross-Validation of OK

The predicted values of OK were plotted against the measured values to evaluate the interpolation performance (Figure 3). The linear model and 1:1 dashed line intersected for soil N, P, and K stocks. Before the intersection, the linear model overestimated soil N, P, and K stocks, and vice versa. This conclusion supported the findings of many previous studies because of the nature of the algorithms of OK, which aimed to achieve unbiased predictions of mean values [11,15]. AME, ME, and RMSE were calculated and presented in Table 3. The closer that the AME, ME, and RMSE values are to zero, the better the model performed. ME of soil N, P, and K stocks ranged from −0.0022 to

−0.2115, which were close to zero, indicating that OK produced relatively unbiased values for spatial interpolation. The ME values were negative, illustrating that OK generally underestimated soil N, P, and K stocks.

Table 3. Cross-validation indices for ordinary kriging (OK).

Nutrient	AME	ME	RMSE
N stock	2.1506	−0.2115	2.5557
P stock	0.0626	−0.0022	0.0797
K stock	2.0250	−0.1634	2.3942

AME = absolute mean error; ME = mean error; RMSE = root mean square error.

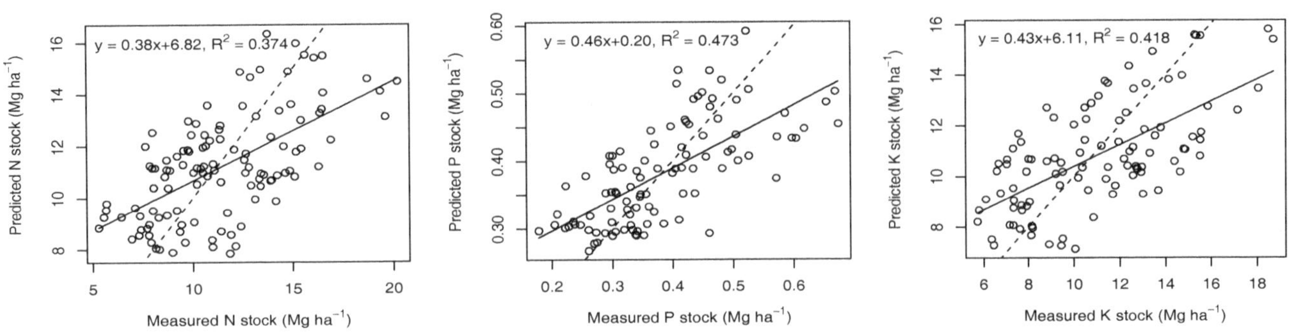

Figure 3. Cross-validation of OK interpolation for soil N, P, and K stocks.

Determination coefficients ranged from 0.37 to 0.47 (Figure 3), which may appear relatively low, but they were similar to many previous studies regarding the spatial interpolation of soil properties [47,48]. This problem is associated with the dataset, which was not sampled with a probabilistic design [48]. For example, the majority of sampling plots were allocated in the southern and eastern parts of Yong'an City in the current study (Figure 1). Therefore, a dataset with a probabilistic design is recommended for spatial interpolation of soil properties. Another possible explanation might be the strong local variation of soil N, P, and K stocks caused by the variability in environmental conditions and fertilization practices. On the other hand, the low correlation between the predicted and measured data indicates that a better methodology, such as one using Artificial Neural Network and Random Forest, should be developed to improve the accuracy of spatial interpolation of soil N, P, and K stocks, including consideration of both intrinsic and extrinsic factors.

3.4. Spatial Prediction of Soil N, ,P and K Stocks

The spatial patterns of soil N, P, and K stocks predicted by OK are shown in Figure 4. Predicted soil N, P, and K stocks ranged from 7.64 Mg ha^{-1} to 16.02 Mg ha^{-1}, from 0.19 Mg ha^{-1} to 0.63 Mg ha^{-1}, and from 6.85 Mg ha^{-1} to 17.13 Mg ha^{-1}, respectively. The lowest soil N, P, and K stocks were found in the middle or southern part of study area, where the city centre was located with a high population, indicating that human activities led to a significant decrease of soil N, P, and K stocks. The highest soil N stock was observed in the east of the study area, the highest soil P stock was found in the southwest and north, and the highest soil K stock was in the north. However, plant growth was constrained by the most limited nutrient. These different spatial distribution patterns suggested that fertilizers with different N:P:K ratios should be applied to maintain and improve the stand productivity. For instance, a relatively lower ratio of P but higher K ratio fertilizer should be applied in the south of the study area (Figure 4b,c), while a relatively lower ratio of N fertilization should be implemented in the east (Figure 4a). Further studies are strongly encouraged to study the N:P:K ratio for the whole study area in order to manage Moso bamboo forests scientifically. The total soil N, P, and K stocks of the whole study area were 0.624, 0.020, and 0.583 Tg (1 Tg = 1 $\times$ 10^{12} g), respectively, indicating soils were important pools of soil N, P, and K.

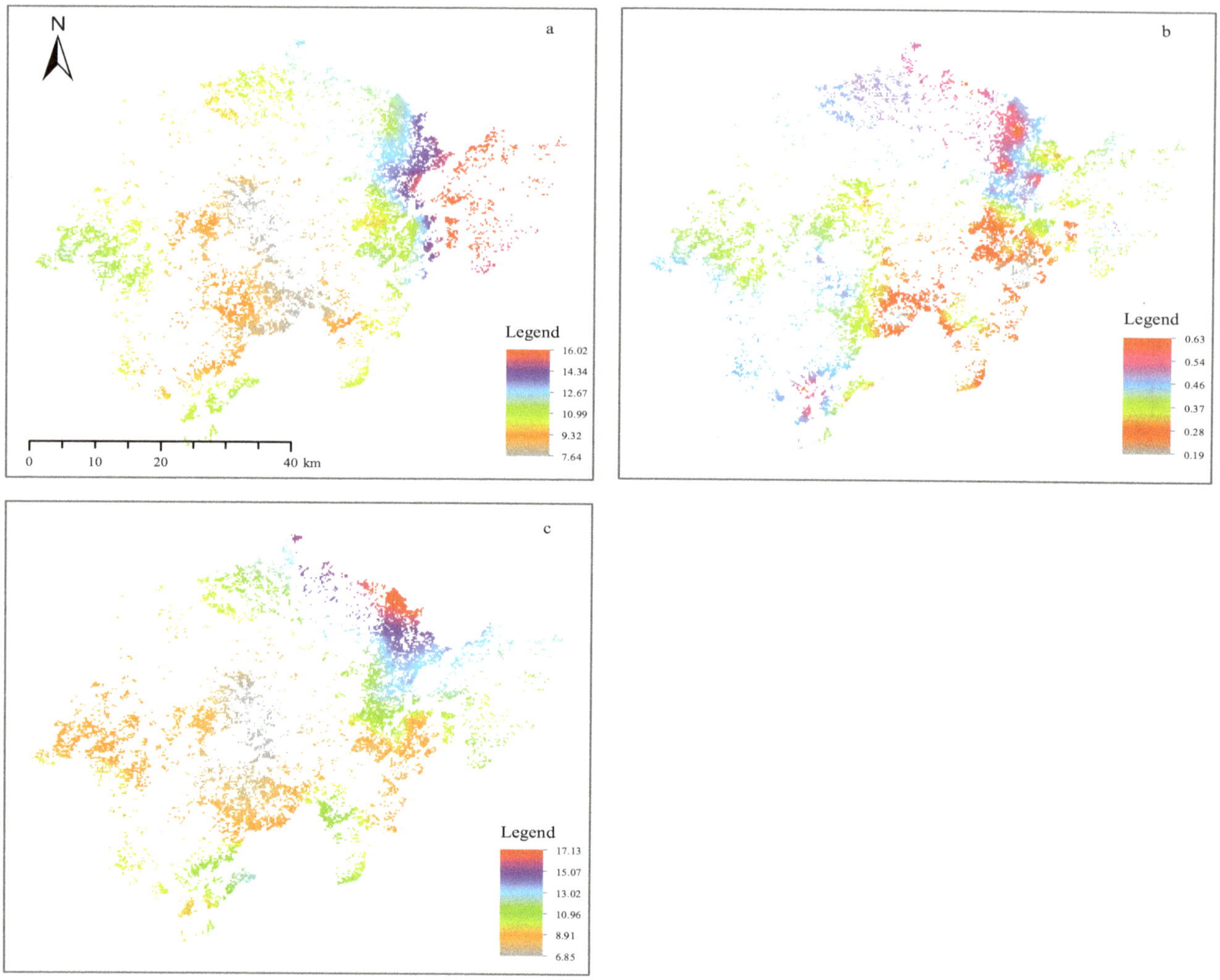

Figure 4. Spatial distribution of soil N (**a**), P (**b**), and K (**c**) stocks.

4. Conclusions

In this study, the current status and spatial interpolation of soil N, P, and K stocks within 0–60 cm were analysed using the measured data from 138 locations in Moso bamboo forests in Yong'an City, China. OK was applied for spatial interpolation of soil N, P, and K stocks across the whole study area. Exponential and spherical models performed well in describing the spatial distribution of soil N, P, and K stocks with determination coefficient from 0.64 to 0.74, and cross-validation demonstrated that OK performed well for the spatial interpolation of soil N, P, and K stocks. Soil N stocks showed a strong spatial dependence, indicating soil N stocks were mainly controlled by intrinsic factors. Soil P and K stocks showed a moderate spatial dependence, suggesting that soil P and K stocks were controlled by both intrinsic (e.g., soil parent material, soil texture, topography) and extrinsic (e.g., soil fertilization and cultivation practices) factors. Soil N, P, and K stocks showed different spatial patterns across the whole study area, indicating that fertilizers with different ratios of N:P:K should be applied for different sites.

Acknowledgments: This study was supported by special research fund of International Centre for Bamboo and Rattan (1632013010) and "Twelfth Five-year" National Technology Support Program (2012BAD23B04). We thank Yanfang Gao for support with fieldwork and Ping He for the technical support.

Author Contributions: Xiaolu Tang, Fengying Guan, and Shaohui Fan conceived and designed the experiments; Xiaolu Tang, Mingpeng Xia, and Shaohui Fan analysed the data; Xiaolu Tang, Mingpeng Xia, Fengying Guan, and Shaohui Fan wrote the paper. All authors contributed to the review of the manuscript.

Abbreviations

The following abbreviations are used in this manuscript:

AME	Absolute Mean Error
BD	Bulk Density
D	Depth
K	Potassium
ME	Mean Error
N	Nitrogen
OK	Ordinary Kriging
P	Phosphorus
RMSE	Root Mean Square Error

References

1. Li, Y.; Niu, S.L.; Yu, G.R. Aggravated phosphorus limitation on biomass production under increasing nitrogen loading: A meta-analysis. *Glob. Chang. Biol.* **2016**, *22*, 934–943. [CrossRef]
2. Tripler, C.E.; Kaushal, S.S.; Likens, G.E.; Walter, M.T. Patterns in potassium dynamics in forest ecosystems. *Ecol. Lett.* **2006**, *9*, 451–466. [CrossRef]
3. Marklein, A.R.; Houlton, B.Z. Nitrogen inputs accelerate phosphorus cycling rates across a wide variety of terrestrial ecosystems. *New Phytol.* **2012**, *193*, 696–704. [CrossRef]
4. Du, M.Y.; Fan, S.H.; Liu, G.L.; Feng, H.Y.; Guo, B.H.; Tang, X.L. Stoichiometric characteristics of carbon, nitrogen and phosphorus in *Phyllostachys pubescens* forests of China. *Chin. J. Plant Ecol.* **2016**, *40*, 760–774. (In Chinese with English Abstract)
5. Patil, R.H.; Laegdsmand, M.; Olesen, J.E.; Porter, J.R. Effect of soil warming and rainfall patterns on soil N cycling in Northern Europe. *Agric. Ecosyst. Environ.* **2010**, *139*, 195–205. [CrossRef]
6. Lin, J.S.; Shi, X.Z.; Lu, X.X.; Yu, D.S.; Wang, H.J.; Zhao, Y.C.; Sun, W.X. Storage and spatial variation of phosphorus in paddy soils of China. *Pedosphere* **2009**, *19*, 790–798. [CrossRef]
7. Rezaei, S.A.; Gilkes, R.J. The effects of landscape attributes and plant community on soil chemical properties in rangelands. *Geoderma* **2005**, *125*, 167–176. [CrossRef]
8. Rodríguez, A.; Durán, J.; Fernández-Palacios, J.M.; Gallardo, A. Spatial pattern and scale of soil N and P fractions under the influence of a leguminous shrub in a *Pinus canariensis* forest. *Geoderma* **2009**, *151*, 303–310. [CrossRef]
9. Gami, S.K.; Lauren, J.G.; Duxbury, J.M. Influence of soil texture and cultivation on carbon and nitrogen levels in soils of the eastern Indo-Gangetic Plains. *Geoderma* **2009**, *153*, 304–311. [CrossRef]
10. Ross, D.J.; Tate, K.R.; Scott, N.A.; Feltham, C.W. Land-use change: Effects on soil carbon, nitrogen and phosphorus pools and fluxes in three adjacent ecosystems. *Soil Biol. Biochem.* **1999**, *31*, 803–813. [CrossRef]
11. Liu, Z.P.; Shao, M.A.; Wang, Y.Q. Spatial patterns of soil total nitrogen and soil total phosphorus across the entire Loess Plateau region of China. *Geoderma* **2013**, *197–198*, 67–78. [CrossRef]
12. Jennings, E.; Allott, N.; Pierson, D.C.; Schneiderman, E.M.; Lenihan, D.; Samuelsson, P.; Taylor, D. Impact of climate change on phosphorus loading from a grassland catchment: Implication for future management. *Water Res.* **2009**, *43*, 4316–4326. [CrossRef]
13. Yasrebi, J.; Saffari, M.; Fathi, H.; Karimian, N.; Moazallahi, M.; Gazni, R. Evaluation and comparison of ordinary kriging and inverse distance weighting methods for prediction of spatial variability of some soil chemical parameters. *Res. J. Biol. Sci.* **2009**, *4*, 93–102.
14. Wang, Z.; Liu, G.B.; Xu, M.X.; Zhang, J.; Wang, Y.; Tang, L. Temporal and spatial variations in soil organic carbon sequestration following revegetation in the hilly Loess Plateau, China. *Catena* **2012**, *99*, 26–33. [CrossRef]
15. Liu, Z.P.; Shao, M.A.; Wang, Y.Q. Large-scale spatial interpolation of soil pH across the Loess Plateau, China. *Environ. Earth Sci.* **2012**, *69*, 2731–2741. [CrossRef]

16. Kravchenko, A.N. Influence of spatial structure on accuracy of interpolation methods. *Soil Sci. Soc. Am. J.* **2003**, *67*, 1564–1571. [CrossRef]
17. Martín, J.A.R.; Álvaro-Fuentes, J.; Gonzalo, J.; Gil, C.; Ramos-Miras, J.J.; Corbí, J.M.G.; Boluda, R. Assessment of the soil organic carbon stock in Spain. *Geoderma* **2016**, *264*, 117–125. [CrossRef]
18. Zhou, G.M.; Xu, J.M.; Jiang, P.K. Effect of management practices on seasonal dynamics of organic carbon in soils under bamboo plantations. *Pedosphere* **2006**, *16*, 525–531. [CrossRef]
19. Du, M.; Liu, G.; Fan, S.; Feng, H.; Tang, X.; Mao, C. Effects of fertilization on the distribution patterns of biomass and carbon storage in Moso bamboo forest, western Fujian province, China. *Chin. J. Trop. Crops* **2015**, *36*, 1–7. (In Chinese with English Abstract)
20. Liu, J.; Jiang, P.K.; Wang, H.L.; Zhou, G.M.; Wu, J.S.; Yang, F.; Qian, X.B. Seasonal soil CO_2 efflux dynamics after land use change from a natural forest to Moso bamboo plantations in subtropical China. *For. Ecol. Manag.* **2011**, *262*, 1131–1137. [CrossRef]
21. Xu, Q.F.; Jiang, P.K.; Xu, Z.H. Soil microbial functional diversity under intensively managed bamboo plantations in southern China. *J. Soils Sed.* **2008**, *8*, 177–183. [CrossRef]
22. Li, Y.F.; Zhang, J.J.; Chang, S.X.; Jiang, P.K.; Zhou, G.M.; Fu, S.L.; Yan, E.R.; Wu, J.S.; Lin, L. Long-term intensive management effects on soil organic carbon pools and chemical composition in Moso bamboo (*Phyllostachys pubescens*) forests in subtropical China. *For. Ecol. Manag.* **2013**, *303*, 121–130. [CrossRef]
23. Wang, Y.Q.; Zhang, X.C.; Huang, C.Q. Spatial variability of soil total nitrogen and soil total phosphorus under different land uses in a small watershed on the Loess Plateau, China. *Geoderma* **2009**, *150*, 141–149. [CrossRef]
24. Morales, L.A.; Vázquez, E.V.; Paz-Ferreiro, J. Spatial distribution and temporal variability of ammonium-nitrogen, phosphorus, and potassium in a rice field in Corrientes, Argentina. *Sci. World J.* **2014**, *2014*. [CrossRef]
25. Liu, G.L.; Fan, S.H.; Qi, L.H.; Xiao, F.M.; Huang, Y.N. Nutrient distribution and biological cycle characteristics in different types of *Phyllostachys pubescens* forest in Northwest Fujian. *Chin. J. Eco.* **2010**, *29*, 2155–2161. (In Chinese with English Abstract)
26. He, P.; Liu, J.; Yu, K.Y.; Yang, S.P.; Yao, X.; Yu, X.Y.; Deng, Y.B.; Chen, Z.H. Research on spatial heterogeneity of soil organic carbon in the southern bamboo forest. *Chin. J. Soil Sci.* **2016**, *47*, 36–44. (In Chinese with English Abstract)
27. Yong'an Forest Bureau, Forest Management Inventory; Yong'an, Sanming, Fujian, China. Unpublished work, 2014.
28. Tang, X.L.; Fan, S.H.; Qi, L.H.; Liu, G.L.; Guan, F.Y.; Du, M.Y.; Shen, C.X. Effect of different managements on carbon storage and carbon allocation in Moso bamboo forest (*Phyllostachys pubescen*). *Acta Agric. Univ. Jiangxi* **2012**, *34*, 736–742, (In Chinese with English abstract).
29. State Forestry Adiministration. *Forestry Standard of People's Republic of China—Methods of Forest Soil Analysis*; Chinese Standard Press: Beijing, China, 1999.
30. Zhang, L.H.; Xie, Z.K.; Zhao, R.F.; Wang, Y.J. The impact of land use change on soil organic carbon and labile organic carbon stocks in the Longzhong region of Loess Plateau. *J. Arid Land* **2012**, *4*, 241–250. [CrossRef]
31. Xie, Z.B.; Zhu, J.G.; Liu, G.; Cadisch, G.; Hasegawa, T.; Chen, C.M.; Sun, H.F.; Tang, H.Y.; Zeng, Q. Soil organic carbon stocks in China and changes from 1980s to 2000s. *Glob. Chang. Biol.* **2007**, *13*, 1989–2007. [CrossRef]
32. Guan, F.Y.; Tang, X.L.; Fan, S.H.; Zhao, J.C.; Peng, C. Changes in soil carbon and nitrogen stocks followed the conversion from secondary forest to Chinese fir and Moso bamboo plantations. *Catena* **2015**, *133*, 455–460. [CrossRef]
33. R Development Core Team. *R: A Language and Environment for Statistical Computing*; R Foundation for Statistical Computing: Vienna, Austria, 2011. Available online: http://www.R-project.Org/ (accessed on 6 March 2015).
34. Gammadesign. Available online: www.gammadesign.com (accessed on 10 January 2015).
35. Esri. Available online: www.esri.com (accessed on 14 June 2013).
36. Krige, D.G. A statistical approach to some basic mine valuation problems on the witwatersrand. *J. Chem. Metall. Min. Soc. S. Afr.* **1951**, *52*, 119–139.
37. Goovaerts, P. Geostatistics in soil science: State-of-the-art and perspectives. *Geoderma* **1999**, *89*, 1–45. [CrossRef]

38. Schöning, I.; Totsche, K.U.; Kögel-Knabner, I. Small scale spatial variability of organic carbon stocks in litter and solum of a forested Luvisol. *Geoderma* **2006**, *136*, 631–642. [CrossRef]
39. Lv, J.S.; Liu, Y.; Zhang, Z.L.; Dai, J.R.; Dai, B.; Zhu, Y.C. Identifying the origins and spatial distributions of heavy metals in soils of Ju country (Eastern China) using multivariate and geostatistical approach. *J. Soil Sed.* **2015**, *15*, 163–178. [CrossRef]
40. Reza, S.K.; Sarkar, D.; Baruah, U.; Das, T.H. Evaluation and comparison of ordinary kriging and inverse distance weighting methods for prediction of spatial variability of some chemical parameters of Dhalai district, Tripura. *Agropedology* **2010**, *20*, 38–48.
41. Yang, Y.H.; Ma, W.H.; Mohammat, A.; Fang, J.Y. Storage, patterns and controls of soil nitrogen in China. *Pedosphere* **2007**, *17*, 776–785. [CrossRef]
42. Liu, J.; Yang, Q.P.; Yu, D.K.; Song, Q.N.; Zhao, G.D.; Wang, B. Contribution of fine root to soil nutrient heterogeneity at two sides of the bamboo and broadleaved forest interface. *Chin. J. Plant Ecol.* **2013**, *37*, 739–749. (In Chinese with English Abstract) [CrossRef]
43. Wang, T.; Yang, Y.H.; Ma, W.H. Storage, patterns and environmental controls of soil phosphorus in China. *Acta Sci. Nat. Univ. Pekin.* **2008**, *44*, 945–951.
44. Chen, J.; Yang, N. Effects of five plantations on soil properties in subtropical red soil hilly region. *J. Northwest A & F Univ. (Nat. Sci. Ed.)* **2013**, *41*, 167–173,178. (In Chinese with English Abstract)
45. Fang, X.; Xue, Z.J.; Li, B.C.; An, S.S. Soil organic carbon distribution in relation to land use and its storage in a small watershed of the Loess Plateau, China. *Catena* **2012**, *88*, 6–13. [CrossRef]
46. Cambardella, C.A.; Moorman, T.B.; Novak, J.M.; Parkin, T.B.; Karlen, D.L.; Turco, R.F.; Konopka, A.E. Field-scale variability of soil properties in central Iowa soils. *Soil Sci. Soc. Am. J.* **1994**, *58*, 1501–1511. [CrossRef]
47. Fu, W.J.; Jiang, P.K.; Zhao, K.L.; Zhou, G.M.; Li, Y.F.; Wu, J.S.; Du, H.Q. The carbon storage in Moso bamboo plantation and its spatial variation in Anji County of southeastern China. *J. Soils Sed.* **2014**, *14*, 320–329. [CrossRef]
48. Veronesi, F.; Corstanje, R.; Mayr, T. Landscape scale estimation of soil carbon stock using 3D modelling. *Sci. Total Environ.* **2014**, *487*, 578–586. [CrossRef]

9

Nitrogen and Phosphorus Resorption in Relation to Nutrition Limitation along the Chronosequence of Black Locust (*Robinia pseudoacacia* L.) Plantation

Jian Deng [1,2,*], Sha Wang [2], Chengjie Ren [3], Wei Zhang [3], Fazhu Zhao [4], Xianfang Li [1,2], Dan Zhang [2], Xinhui Han [3] and Gaihe Yang [3]

1. Shaanxi Key Laboratory of Chinese Jujube, Yan'an University, Yan'an 716000, China; lixianfang810702@163.com
2. College of Life Science, Yan'an University, Yan'an 716000, China; wwang97625@163.com (S.W.); xtmy1008@126.com (D.Z.)
3. College of Agronomy, Northwest A&F University, Yangling 712100, China; rencj1991@nwsuaf.edu.cn (C.R.); zwgwyd@163.com (W.Z.); hanxinhui@nwsuaf.edu.cn (X.H.); ygh@nwsuaf.edu.cn (G.Y.)
4. College of Urban and Environmental Science, Northwest University, Xi'an 710069, China; zhaofazhu@nwu.edu.cn

* Correspondence: dengjian@yau.edu.cn

Abstract: Plant nitrogen (N) and phosphorus (P) resorption is an important strategy to conserve N and P in the face of nutrient limitation. However, little is known about the variation of N and P resorption efficiency (NRE and PRE) and their correlation with leaves and soil C:N:P stoichiometry in black locust forests (*Robinia pseudoacacia* L.) of different ages. In this study, we measured C, N, and P concentrations in soil, green leaves, and senesced leaves from black locust forests of different ages (i.e, 10-, 20-, 30-, 36-, and 45-year-old), and calculated the NRE, PRE, and C:N:P stoichiometry ratios. The NRE and PRE tended to increase and then decrease with stand age, ranging from 46.8% to 57.4% and from 37.4% to 58.5%, with averages of 52.61 and 51.89, respectively. The PRE:NRE decreased with increased stand ages. The C:P and N:P of soil and green leaves increased with stand ages, indicating the increase of P limitation. In the senesced leaves, C:P and N:P were lower than in green leaves and first increased and then decreased with stand age. The PRE was significantly negatively correlated with the C:P and N:P of soil and green leaves. The NRE was significantly correlated with the C concentration of green leaves, P of the senesced leaves, and C:N. Results suggested that the NRE and PRE responded differently to soil and plant nutrients in black locust forests of different ages. In addition, the black locust plantations would alter the conservation and use strategy of nutrients in the ecosystem through a plant-mediated pathway. Future studies should elucidate the central nutrient utilization strategy of black locust in response to a nutrient-poor environment and determine how it is involved in regulating nutrient resorption.

Keywords: nutrition resorption; ecological stoichiometry; plant-soil feedback; stand age; *Robinia pseudoacacia* L.

---

1. Introduction

Plant nutrient resorption is the process by which a plant transfers nutrients to its growing tissues before the branches and leaves wither, effectively lengthening the retention time of nutrients in plants [1,2]. This process reduces the reliance of plants on soil nutrients, helping to maintain the ecological stoichiometry balance in plants; it also reduces nutrient leaching after litter decomposition, minimizing the ecosystem nutrient loss [2,3]. Nutrient resorption from senescing organs is one of the most important mechanisms for plants to adapt to nutrient limitation and improve nutrient utilization efficiency through conserving and optimizing the use of nutrients. This is also an important strategy to regulate nutrient cycling within plant–soil systems. Nutrient resorption can be defined as nutrient resorption efficiency (NuRE), i.e, the proportion of nutrients withdrawn from green leaves prior to senescence [4]. Therefore, research on the plant NuRE is helpful to reveal the response mechanism of plants to a nutrient-poor environment.

Nitrogen (N) and phosphorus (P), strong limiting factors of natural ecosystems, play an important role in plant growth and metabolism [5,6]. Previous studies have reported that natural forests in low latitudes areas (e.g., tropical forests) are usually limited by N [7] and those in high latitudes are limited by P [6]. In addition, chronosequence studies have shown that young forests are usually vulnerable to N limitation, while ageing forests are usually vulnerable to P limitation, especially in areas with nutrient deficiency or imbalanced N and P input [8,9]. Plant N and P resorption efficiency (NRE and PRE, respectively) attract ample attention because they play an important role in revealing the adaptability of plants to a nutrient limiting environment [10,11]. Vergutz estimated that the mean NRE and PRE for global terrestrial plants were 62.1% and 64.9%, respectively [12]. However, the NRE and PRE varies greatly depending on the species [13] and life-form [14]; NRE and PRE are also easily influenced by environmental factors (e.g., soil nutrient availability, elevation, temperature, precipitation, etc.) [13,15,16] or disturbances [17].

In forest plantations, plant photosynthetic characteristics, soil nutrient supply, and plant nutrition demands usually vary with stand age [18]. These changes may have substantial effects on plant nutrient conservation strategies, thus leading to the change of NuRE with stand age [10,19]. However, previous studies have shown inconsistent trends (e.g., increase [18], decrease [20], and no significant change [21]) of NuRE in response to stand age, which may be related to the tree species and nutrition status [10,21]. The measurement of C:N:P stoichiometry is an effective way to investigate plant and soil nutrient status. For instance, leaf N:P was used in a previous study to indicate the N ($N:P < 14$) or P ($N:P > 16$) limitation of the plants [22]. Therefore, studying NRE and PRE characteristics and their possible links with nutrition stoichiometry, together with forest age changes, may help to reveal the adaptability of forests to varying levels of nutrients [18,19]. This is of great significance to the sustainable management of plantations that are subject to nutrient restriction [10].

Due to its nitrogen fixation, fast growth, and tolerance to drought and barren lands, black locust (*Robinia pseudoacacia* L.) has become one of the most widely introduced tree species for ecological restoration and timber production worldwide [23–25]. The Loess Plateau, China, is a very ecologically fragile region that suffers from severe soil erosion and low availability of nutrients [26]. To restore the region, black locust has been introduced as the dominant plantation species with the aim to aid in soil and water conservation [27,28]. Black locust can significantly improve the soil quality by accumulating soil C and N [25,29]. However, black locust plantations grown in the long-term (for approximately 35 years) suffer from withering branches and a decline in productivity, which may indicate forest degradation [30]. Soil nutrient limitation (usually P limitation in the arid areas) [29,31] and water deficits [32,33] also may become serious with stand development. However, few studies have been conducted on how the resorption of N and P vary with the age of the black locust plantation. Therefore, we studied the adaptability of black locust to the changing nutrient status, especially the variation of NuRE with stand age, which may reveal the possible mechanism of degradation of the black locust plantation.

In this study, we examined the changes of nutrients in soil and leaves (both mature leaves and withered leaves) of black locust plantations covering an age range from 10 to 45 years old in the Loess Plateau of China. The objectives of this study were to quantify the variations of N and P resorption across the black locust plantation chronosequence and to find out whether there are any relationships between NuRE and nutrient stoichiometry indexes in leaves and soil. These data were then used to reveal the responses of plant nutrient use strategy to the altered leaf and soil nutrient status in black locust plantations of different ages. Because nutrient resorption is an important nutrient conservation mechanism and the plant nutrient demand may increase with stand age, we hypothesized that the N and P resorption will: (1) increase across the black locust plantation chronosequence, and (2) show significant negative relationships with leaf and soil N and P contents. The results may provide novel insight into the understanding of the nutrient cycling of black locust forests in arid and semi-arid areas.

2. Materials and Methods

2.1. Research Area

The research was conducted in the Loess Plateau at the Wuliwan catchment of Ansai County, Shaanxi Province, China (36°39′–36°52′ N, 109°20′–109°21′ E). The site has a hilly loess landscape and an elevation of 1010–1400 m above sea level. The study area has a typical semiarid climate with a mean annual rainfall of 505 mm, an annual average temperature of 8.8 °C, and a frost-free period of 159 days. Approximately 60% of the precipitation occurs between July and September, and the precipitation varies greatly in different years. The soil in this region is mainly loessal soil, developed from wind-accumulated loess, which has extremely poor resistance to erosion. The proportion of the differently sized soil particles in the 0–20 cm layer was 63.6% sand, 29.2% silt, and 7.2% clay. Artificial forests have been planted at the study site since the 1970s to mitigate the serious soil erosion on steeply sloped crop land [29]; therefore, the large area of plantations have different stand ages. The dominant plantation species is black locust and *Caragana korshinskii* Kom.

2.2. Experimental Design and Sampling

Based on the afforestation records of the local forestry department, black locust forests of five age classes (i.e., 10, 20, 30, 36, and 45 years old during field sampling in 2016, hereafter referred to as RP10, RP20, RP30, RP36, and RP45, respectively) were selected to represent a chronosequence. All of the sample lands were converted from cropland subjected to similar farming practices, namely, mainly planting maize (*Zea mays* L.), foxtail millet (*Setaria italic* L.), and broomcorn millet (*Panicum miliaceum* L.) with watering mainly through the rainfall and a small amount of manure was the main fertilizer. After trees were planted, the sample lands were fenced to prevent grazing and other human disturbance including thinning, wood usage, or other practices. Detailed information of the sampled lands is shown in Table 1. The climatic and edaphic conditions of the forests of five age classes were similar. Three replicate fields were measured for each black locust stand age. The three replicate fields for each stand age were within 1.5 km of one another to ensure consistent climatic variables and other conditions. In each replicate field, we randomly selected three replicated plots (20 m × 20 m) for sampling. All sites were topographically similar (i.e., slope aspect and slope degree) with similar elevation. Finally, we established 45 sampling plots (5 stand ages × 3 replicate fields × 3 sample plots).

Green and senesced leaves were collected in mid-August and mid-October 2016, respectively, when the green leaves were at peak biomass, and senesced leaves were ready to fall. At least five individual trees were selected in each sample plot. At least 20 branches were chosen around each tree from the lower, middle, and upper canopy. Approximately 50 g leaves were sampled from each individual (picked using a long reach pruner), all the leaves from the same sample plots were mixed (about 250–300 g), and then a quantity sufficient for chemical analysis (about 100 g) was taken out for analysis. The leaves with obvious diseases and/or insect pests were excluded. To avoid nutrient decomposition and leaching, the fully senesced leaves were collected from trees rather than from litter in mid-October, when they were completely yellow, dry, without any signs of deterioration, and would fall from the branch with a gentle touch [10,34]. The sample leaves were transfer to the laboratory in paper bags, then oven-dried at 105 °C for 30 min and maintained at 60 °C until a constant mass was reached. Dried sample leaves were ground and sieved with a 0.1 mm sieve.

Table 1. Field site information for *Robinia pseudoacacia* L. plantations with different stand ages.

Stand Ages	10	20	30	36	45
Location	36°52′00″ N 109°21′38″ E	36°52′08″ N 109°21′06″ E	36°51′36″ N 109°21′06″ E	36°51′37″ N 109°21′05″ E	36°52′16″ N 109°20′54″ E
Elevation (m)	1332.31	1245.2	1283.6	1255	1121.34
Slope (°)	37	30	35	30	28
Aspect (°)	ES38	ES20	EN16	EN15	ES10
Bulk density (g cm^{-3})	1.09	1.29	1.05	1.01	0.98
pH	8.56	8.62	8.58	8.51	8.60
Canopy density (%)	48	55	64	63	60
Stand density (trees hm^{-2})	1540	1535	1350	1305	1240
DBH (cm)	6.33	15.32	17.25	18.27	19.10
Degree of herb cover (%)	46.83	59.13	83.87	76.85	73.40
Number of herb species	13	24	21	22	23

ES and EN means East by South and Ease by North; DBH means diameter at breast height.

The soil samples were collected in mid-August 2016. In each sample plot, we selected five sample points along the diagonal. Because the surface soil (0–10 cm) is the most closely related to plant nutrients [10,29], we collected soil samples at each point from the 0–10 cm layer using a soil auger with a 4 cm diameter. The soil samples from the same plot were adequately mixed and reduced by coning and quartering to achieve appropriate quantities (about 400 g). Then, the soil samples were transported to the laboratory, air-dried at room temperature (25–28 °C), and passed through a 0.25 mm sieve in preparation for soil C, N, and P analysis. To avoid the impact of roots and litter, all samples were collected 80 cm from trees after the litter layer was removed and the plant roots, stones and debris were removed before drying through hand picking and sieving. A vegetation survey was also conducted in each plot in August 2016 using five randomly selected 1 m × 1 m small plots.

2.3. Laboratory Analysis and Determination

The TOC analyzer (Total organic carbon analyzer, Shimadzu Corp., Kyoto, Japan) was used to measure the total C from soil and leaf samples (0.1 mol L^{-1} HCl were used to destroy the carbonates in soil). The total nitrogen (TN) content was measured using the semimicro-Kjeldahl method with a Kjeldahl auto-analyzer (KDY-9830, Beijing, China) after digestion in sulfuric acid hydrogen peroxide (for leaves) or sulfuric acid perchloric acid (for soil). Total phosphorus (TP) content was measured using the molybdenum antimony colorimetric method by using an ultraviolet spectrometer subsystem (UV-6100, Shanghai, China) after digestion following the methods as in TN measurement [35].

2.4. Calculations and Statistical Analysis

NuRE was used to quantify the nutrient resorption, which was calculated as follows [19]:

$$\text{NuRE} = (1 - \frac{W_1}{W_2} \times \text{MLCF}) \times 100\% \quad (1)$$

where NuRE is the nutrient resorption efficiency, and W_1 and W_2 are the nutrient concentrations (TN or TP, g kg^{-1}) in senesced and green leaves, respectively. MLCF is the mass loss correction factor to correct the unbiased resorption value, which is 0.784 for broad-leaved deciduous trees as introduced by Vergutz et al. [12].

The soil and leaf C:N:P stoichiometry ratios were calculated as the mass ratio. There were no missing values in our data. All data were distributed normally after the Kolmogorov–Smirnov test ($p > 0.05$ for each null hypothesis). Then, the Bartlett test was used to investigate the homogeneity of variance, and $p > 0.05$ for each null hypothesis. There were no significant differences for each soil and leaf properties between different replicate fields of the same stand ages, or between different sample plots of the same replicate fields ($p > 0.05$). Therefore, one-way analysis of variance (ANOVA) was used to test the differences of NuRE, nutrients and stoichiometric ratios among the plantations of different ages (the significant level was $p < 0.05$). A least significant difference (LSD) test was conducted for post-hoc multiple comparisons. A Pearson correlation analysis was used to estimate the relationship between NuRE and other leaf or soil nutrition properties. The relationships between the stand age and other properties were estimated by linear regression. All of the statistical analyses were conducted using SAS (SAS Institute Inc., Cary, NC, USA), and the figures were plotted in Origin 2016 (OriginLab Corporation, Northampton, MA, USA).

3. Results

3.1. Leaf and Soil C, N, and P Concentrations

The concentrations of the C and N in green leaves significantly increased from RP10 to RP36; however, the concentrations were decreased from RP36 to RP45 ($p < 0.05$). The concentrations of C and N ranged from 380.81–475.38 g kg^{-1} and 22.38–32.49 g kg^{-1} in all stand ages, respectively (Table 2). The P concentration of green leaves showed no significant variation among the different stand ages, except P concentration in RP45 was significantly lower ($p < 0.05$).

Table 2. Leaf and soil C, N, and P concentrations in different stand aged black locust forests.

Field Site	Soil (g kg^{-1})			Green Leaves (g kg^{-1})			Senesced Leaves (g kg^{-1})		
	C	N	P	C	N	P	C	N	P
RP10	4.65 ± 0.27	0.50 ± 0.03	0.55 ± 0.07	380.81 ± 12.89	22.38 ± 0.94	2.07 ± 0.06	352.61 ± 16.54	15.16 ± 0.46	1.18 ± 0.05
RP20	4.97 ± 0.09	0.61 ± 0.05	0.49 ± 0.04	439.85 ± 13.32	25.05 ± 0.54	2.15 ± 0.04	348.77 ± 12.22	13.62 ± 1.09	1.14 ± 0.02
RP30	8.26 ±0.71	0.87 ± 0.07	0.60 ± 0.04	467.31 ± 8.51	32.02 ± 0.75	2.07 ± 0.05	409.41 ± 12.37	17.90 ± 0.82	1.16 ± 0.06
RP36	8.79 ± 0.35	0.87 ± 0.04	0.53 ± 0.03	475.38 ± 6.11	32.49 ± 0.43	2.17 ± 0.11	425.59 ± 11.48	18.38 ± 0.51	1.25 ± 0.05
RP45	14.41 ± 0.91	1.20 ± 0.13	0.57 ± 0.03	421.53 ± 12.06	28.71 ± 0.80	1.72 ± 0.06	453.97 ± 24.16	19.4 ± 0.52	1.37 ± 0.05

The data are shown as mean ± standard error.

The C and N concentrations of senesced leaves increased from RP20 to RP45 with stand age ($p < 0.05$). The concentrations of C and N ranged from 347.27–453.97 g kg^{-1} and 11.64–19.40 g kg^{-1} in all stand ages (Table 2). The P concentration of senesced leaves showed an increasing trend with stand age, with a range of 1.14–1.37 g kg^{-1}. The C, N, and P concentrations were all higher in green leaves than that in senesced leaves in each stand age. The only exception was the C concentrations in RP45, which was higher in senesced leaves ($p < 0.05$).

The C and N concentrations of soil significantly increased with stand ages ($p < 0.05$), ranging from 4.56–14.41 g kg^{-1} and 0.50–1.20 g kg^{-1}, respectively. The soil P concentrations showed no significant changes with stand ages (Table 2).

3.2. Leaf and Soil C:N:P Stoichiometry Ratios

The C:P and N:P in green leaves increased with stand age ($p < 0.05$), ranging from 184.22–245.60 and 10.83–16.72 with an average of 218.50 and 14.11, respectively (Figure 1). The C:N of green leaves declined as a whole with stand age ($p < 0.05$), ranging from 14.60–17.56 with an average of 15.70. The C:N, C:P, and N:P of senesced leaves all increased and then decreased with stand age ($p < 0.05$), with ranges of 22.62–29.95, 274.35–355.60, and 10.73–15.54 with averages of 23.69, 326.77, and 13.84, respectively (Figure 1). The highest C:N values in senesced leaves were measured in RP20; the highest C:P and N:P in senesced leaves were measured in RP30. Generally, the C:N and C:P of senesced leaves were higher than that of green leaves in all stand ages ($p < 0.05$), and the N:P of senesced leaves were higher from RP10 to RP30 and lower from RP36 to RP45 compared with that of green leaves ($p < 0.05$).

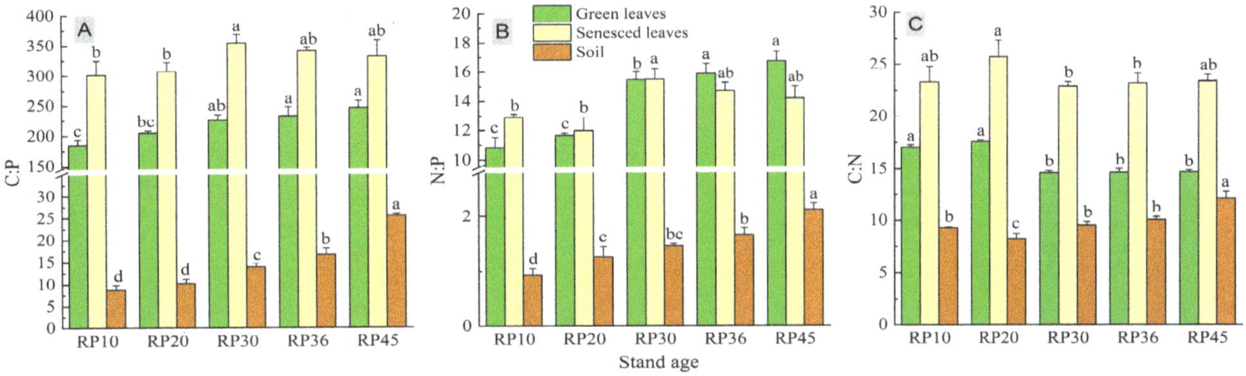

Figure 1. Leaf and soil C:P ratio (**A**), N:P ratio (**B**) and C:N ratio (**C**) in black locust forests with different stand ages. Different lowercase letters represent significant differences among stand ages at $p < 0.05$, and the same letters indicate no significance.

The soil C:P and N:P increased with stand age ($p < 0.05$), ranging from 8.66–25.47 and 0.93–2.11 with averages of 14.98 and 1.48, respectively. The soil C:N decreased by 11.8% from RP10 to RP20, and then increased by 47.8% from RP30 to RP45 ($p < 0.05$), with an average C:N of 9.83 (Figure 1).

3.3. NRE and PRE in Forests of Different Stand Ages and Their Relationship

The NRE and PRE varied among stands of different ages, with the tendency to increase and then decrease with stand age ($p < 0.05$). The NRE and PRE values ranged from 46.8% to 57.4% and from 37.4% to 58.5%, with averages of 52.61 and 51.89, respectively. The highest values for NRE and PRE were obtained in RP20 (Figure 2A). The NRE was higher than the PRE in RP10 and RP20, and lower in other stand ages ($p < 0.05$). A significantly positive correlation was found between NRE and PRE ($R^2 = 0.316$, $p < 0.05$) (Figure 2B). In addition, the PRE:NRE decreased with the age of the stand ($p < 0.05$) (Figure 2C).

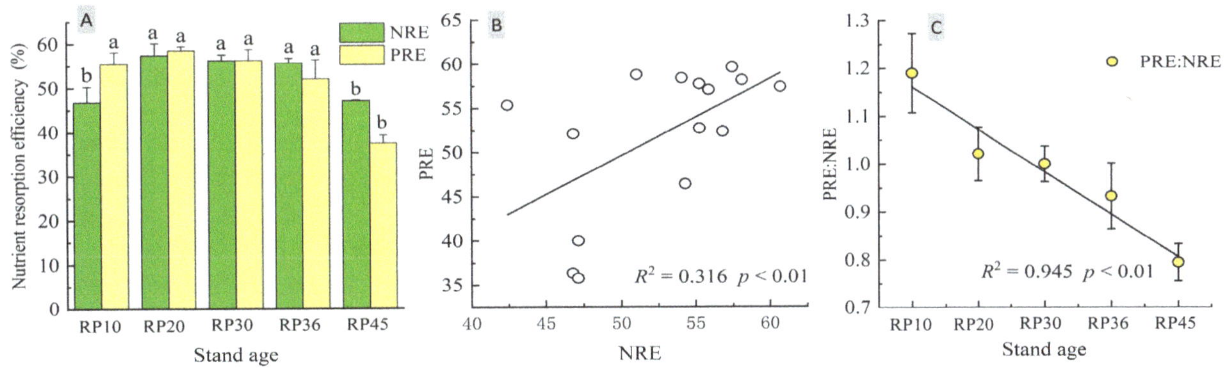

Figure 2. (**A**) Changes of N and P resorption efficiency (NRE and PRE), (**B**) PRE:NRE along the chronosequence of black locust forests, and (**C**) the relationship between NRE and PRE. Data are shown as the mean ± standard deviation. Different lowercase letters represent significant differences among stand ages at $p < 0.05$, and the same letters indication no significant difference.

3.4. Relationships between NuRE, Nutrient Concentrations, and Stoichiometry Ratios in Leaves and Soil

Results from the Pearson correlation analysis showed that the NRE was significantly positively correlated with the C of green leaves and P of senesced leaves ($p < 0.05$); the NRE was significantly negatively correlated with the C:N of senesced leaves among the different stand ages ($p < 0.05$), whereas no significant correlation was found with other indicators (Table 3). The PRE was significantly correlated with most of the nutrient concentration and stoichiometry indicators in leaves and soil ($p < 0.05$). The exception for this was that the soil P; C, N, and C:N of green leaves; and C:N, C:P, and N:P of senesced leaves were not significant. The PRE was positively correlated with the P concentrations of green leaves, whereas it was negatively correlated with other significant related indicators (Table 3).

Table 3. Pearson correlations between nutrient resorption efficiency, nutrient concentrations, and stoichiometry ratios in leaf and soil samples.

		NRE		PRE	
		R	*p*	*R*	*p*
Soil	C	−0.303	0.273	−0.879	<0.001
	N	−0.151	0.592	−0.773	<0.001
	P	−0.116	0.681	-0.258	0.353
	C:N	−0.489	0.064	−0.881	<0.001
	C:P	−0.289	0.296	−0.878	<0.001
	N:P	−0.109	0.700	−0.756	<0.001
Green leaves	C	0.711	<0.001	0.127	0.651
	N	0.432	0.108	−0.188	0.503
	P	0.525	0.044	0.947	<0.001
	C:N	−0.041	0.884	0.489	0.064
	C:P	0.085	0.763	−0.666	<0.001
	N:P	0.058	0.836	−0.645	<0.001
Senesced leaves	C	−0.126	0.656	−0.716	<0.001
	N	−0.292	0.292	−0.675	<0.001
	P	−0.529	0.043	−0.930	<0.001
	C:N	0.504	0.048	0.229	0.413
	C:P	0.345	0.209	−0.045	0.873
	N:P	0.029	0.919	−0.146	0.603

NRE and PRE represent the N and P resorption efficiency, respectively. *R* is the Pearson correlation coefficient.

4. Discussion

4.1. *Leaf and Soil Nutrient Concentration and Stoichiometry of Black Locust Forests of Different Ages*

The nutrient cycle and transformation process of C, N, and P between soil and plants during vegetation restoration is the basis of material circulation in terrestrial ecosystems [36]. Soil provides the necessary nutrients for plant growth, and plants impact the soil nutrients through litter restitution, root growth, and secretions. As a result of plant growth, this correlation would also change with stand ages [37]. Previous studies have shown that the artificial vegetation in the Loess Plateau may face a nutrient imbalance with the increase of stand age, mainly manifesting as P limitation in restored ecosystems [29]. In this study, the soil C and N concentrations increased by 209.9% and 139.2% ($p < 0.05$) from 10-year-old to 45-year-old black locust forests, respectively (Table 1), whereas the soil P concentrations showed no significant variation among the stand ages. The different variation of soil C, N, and P may have occurred because the accumulation of soil C and N are primarily driven by biological factors (e.g., decomposition of plant litter and dead roots), but P transformations in soil are mainly driven by biochemical mineralization (e.g., phosphate decomposition, which takes a long time [29]. Compared to P, the black locust trees can more effectively import N into the soil with symbiosis of N-fixing microbes [38]. Therefore, a significant increase in soil C:P and N:P ratios was found in the older black locust forests ($p < 0.05$) (Figure 1). Our previous study has found a significant decrease of soil-available P and increase of available N along with the increased stand age [39]; therefore, we can speculate that the soil may be relatively deficient in P in older black locust plantations on the Loess Plateau. This provides further support for conclusions from previous studies [29,31,40].

Nutrient concentrations and stoichiometry ratios in leaves change in different growth stages [40,41]. The results showed that the average concentrations of C and N in leaves from black locust forests of different ages were 436.98 g kg^{-1} and 28.13 g kg^{-1}, respectively (Table 2). The C concentration was similar to the average C content of leaves (438 g kg^{-1}) in the Loess Plateau [42], whereas the N content was much higher than the average level in both the Loess Plateau (21.61 g kg^{-1}) [43] and in China (18.60 g kg^{-1}) [44]. The C and N concentration of green leaves increased and then decreased with the increase of stand ages. The concentrations peaked in the RP36 (Table 2). The overall change of C and N in black locust leaves showed the following trend: middle age forest > older forest > young forest, which was consistent with Ma et al.'s results [40]. This may be due to the accumulation of structural substances with more C in the middle- and older-aged plants compared with younger plants. The average P concentrations in green leaves from black locust forests of different ages was 2.01 g kg^{-1} (Table 2), which was close to or slightly less than the global average P content (2.00 g kg^{-1}) [45] and the average of leaves from the Loess Plateau (2.09 g kg^{-1}) [43]. However, P concentration of the green leaves did not show significant differences with the change of stand ages. The N:P ratios of green leaves increased significantly from 10.83 to 16.72 along the chronosequence of black locust forests ($p < 0.05$) (Figure 1). Our results suggested that the black locust plantations would shift from relative N-limitation to relative P-limitation across the chronosequence based on the criteria proposed by Güsewell [22] (i.e., leaves N:P <14, 14–16, and >16 indicating N limitation, N and P co-limitation, and P limitation, respectively). In other words, the young black locust plantations were relatively more N-limited, and the older plantations were relatively more P-limited. Similar findings have also been reported in *Larix olgensis* A. Henry [9], *Larix kaempferi* (Lamb.) Carr. [10], and other plantations [46], which showed that long-term ecosystem development tended to cause a shift from N- to P-limitation. This may be a result of the relatively easy absorption of soil N (with N-fixing microbes in the rhizosphere), but relatively inadequate P absorption by the black locust [38]. However, our research only tested the soil properties in the top layer, thus further study on nutrient effectiveness and root absorption characteristics of deep soil should also be investigated.

Leaf litter is an important way for plants to return nutrients to the soil. Typically, at least some of the nutrients in senesced leaves will be transferred to the branches before falling, resulting in lower levels of nutrients in the senesced leaves compared to the green leaves [12,14,47]. The data obtained in our study also agreed with this rule. The average concentrations of C, N, and P of senesced leaves in stands of different ages were 398.07 g kg^{-1}, 16.89 g kg^{-1}, and 1.22 g kg^{-1} (Table 2), respectively. Compared with the average level of nutrients of senesced leaves in broad-leaved forests in China (the C, N, and P concentrations was 479.9 g kg^{-1}, 13.2 g kg^{-1}, and 1.06 g kg^{-1}, respectively) [48], our results showed relatively high concentrations of N and P, and low concentrations of C. Previous studies have concluded that the abundance of N and P in litter is conducive to microbial (especially bacterial) activity, thereby promoting rapid decomposition of litter, facilitating nutrient release [49,50]. In this study, the C:N and C:P of senesced leaves was 22.62–29.95 and 274.35–355.60 in black locust forests of different ages, respectively (Figure 1). These ratios are significantly lower than the average value of C:N and C:P in senesced leaves in the broad-leaved forests of China (36.36 and 452.73, respectively), demonstrating that the leaf litter in black locust forests are easy to decompose and are beneficial to nutrient release. In addition, our findings showed increasing N and P contents in senesced leaves in older stands, indicating that more N and P would be returned to the soil through litter decomposition.

4.2. N and P Resorption Correlated with Soil and Leaf Nutrients across the Plantation Chronosequence

Plants absorb nutrients from the soil and allocate them to different organs to accumulate or participate in various life activities; the nutrients will be reabsorbed before the leaves fall, thus, prolonging the retention time of nutrients in plants [14]. In this study, the N and P resorption efficiency of black locust forests of different ages were 46.8%–57.4% and 37.4%–58.5%, respectively (Figure 2A); these efficiencies are significantly lower than that of global terrestrial forests (62.1% and 64.9%, respectively) [12], and this may be due to species-dependent differences. The NRE and

PRE were significant correlated with each other, and both of the NRE and PRE showed significantly increasing and then decreasing trends along the chronosequence of black locust forests (Figure 2), which was consistent with previous research in *Pinus massoniana* Lamb and *Metasequoia glyptostroboides* Hu et Cheng plantations of different ages [19,51]. This variation may be because young plants have rapid biomass production, thus need a higher N and P resorption efficiency to supply an appropriate amount of nutrients [19]. Black locust has greater N absorbing efficiency than P absorbing efficiency with symbiosis of N-fixing microbes around roots. Meanwhile, the rapid growth of young trees requires more P for production of genetic material.

These reasons may likely result in a greater PRE than NRE in 10-year-old and 20-year-old forests [40]. A similar trend was also reported in *Eucalyptus urophylla* S.T. Blake × *E. grandis* W. Mill ex Maiden forests [20]. In the older growth stage of black locust forests, the PRE significantly decreased with the increasing stand ages, causing a deficit of P in the soil and trees. This disagrees with our first hypothesis, which may be because the wood biomass became an effective reservoir of P in the later stage of plant growth, thus reducing the P resorption from fallen leaves [52,53].

This indicated that nutrient resorption may not be the main strategy for the preservation of P elements in the older black locust forests. In addition, the NRE decreased with the age of the stand, which is contrary to our first hypothesis. Previous research does not support that tree growth depended more on N resorption to supply its N requirements with the development of the stand [18,37], but it is consistent with the view that decreased NRE was a result of sufficient soil N supply after decomposition [19]. The increased N concentration in soil and senesced leaves with the older stands and relatively low C:N of senesced leaves in our study also supported this view.

The nutrition resorption may be individually or collectively affected by soil nutrient conditions and nutrient contents of green leaves and senesced leaves [12,54]. There is still great uncertainty about the degree of influence and direction of these factors. For example, a study conducted by Aerts reported that there was no significant correlation between NuRE and nutrient concentrations in soil and green leaves, but a negative correlation was found with the nutrient content in senesced leaves [1,55]. However, a study conducted by Vergutz et al. based on data at a global scale concluded that the NuRE depends more on the nutrient state of green leaves than on the senesced leaves [12]. In addition, other studies have also shown that plants subject to N or P limitation should have higher N or P resorption efficiency [56,57]. Killingbeck proposed that nutrient transfer and resorption may be only an inherent feature of species [58], indicating that genetic differences in plants may be the main factor determining nutrient resorption [54,59]. Our results demonstrated that the NRE of black locust forests was significantly correlated with the C concentration in green leaves, P in senesced leaves, and C:N ($p < 0.05$). No significant correlations were found with soil nutrients (Table 3).

This was consistent with a previous study that reported no impact of soil nutrients on NRE [11,55], whereas the PRE was significantly correlated with most of the nutrient content and C:N:P stoichiometry ratios of soil and leaves ($p < 0.05$). Intriguingly, the PRE was negatively correlated with P limitation indicators (i.e., N:P and C:P of soil and leaves) (Figure 3), suggesting that the PRE would be lower with greater P limitation. The decreasing PRE:NRE also indicated relatively lower rates of P resorption with increased plantation age.

This was partly in accordance with our second hypothesis but in contrast with the "relative resorption hypothesis". The "relative resorption hypothesis" revealed that N and P would show an increase in resorption when plants are growing in N- or P-limited conditions [11,60], indicating that the leaves nutrients resorption may be not be the main strategy of black locust to cope the P-limited conditions. Similar results were also observed in *Metasequoia glyptostroboides* and *Larix gmelinii* Rupr. plantations [10,19]. These studies suggested that more efficient nutrient absorption by roots should also be considered.

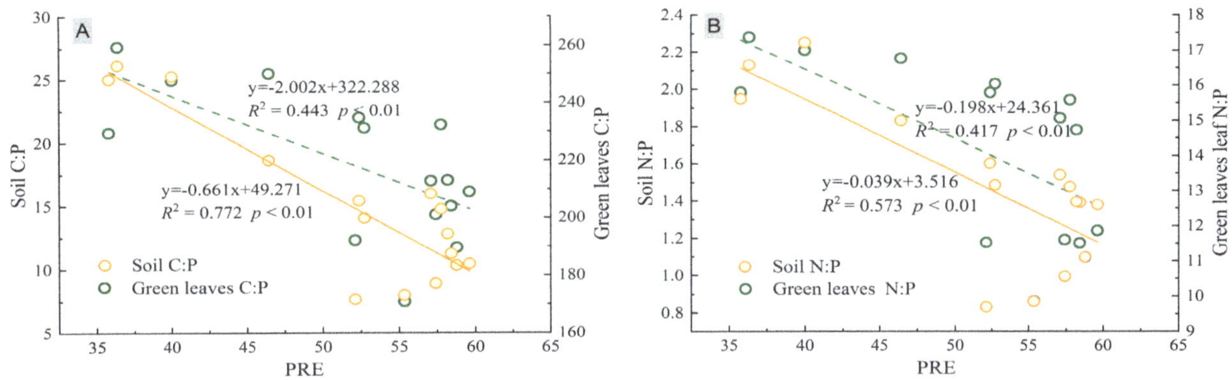

Figure 3. Correlations between PRE and soil and the C:P (**A**) and N:P (**B**) ratios of green leaves in black locust plantations from stands of different ages.

Though the NRE and PRE respond differently to soil and plant nutrients, the results revealed that black locust plantations would alter the conservation and use strategy of nutrients in the ecosystem through a plant-mediated pathway in different stand ages. Based on the present study and previously published studies, we speculate that more nutrients stay in senesced leaves of trees in older forests, possibly prolonging the P cycle and increasing the risk of nutrition loss by leaching in the process of litter decomposition [17,61].

However, litter decomposition may induce increased incorporation of P into the soil microbial biomass, which can promote the recycling speed of nutrients and availability of P in ecosystems [62,63]. Recent studies in P-deficient deciduous forests in Europe have shown that the absorption and return of nutrients also depend on soil P availability [62,64,65].

Our study provided limited information about how the release of litter nutrients and root nutrient absorption affect and react to soil available nutrient change in black locust forests. Since plant nutrient absorption and utilization is a complex process [17,21], future experimental studies considering certain soil nutrient species (e.g., plant available/labile forms, organically-bound forms, etc.), root nutrients absorption, and soil microbes are needed to elucidate the central nutrient utilization strategy of black locust in response to a nutrient-poor environment and how these factors are involved in regulating NuRE in different growth stages.

5. Conclusions

Along with the chronosequence of black locust plantations, the NRE and PRE both increased and then decreased. Imbalanced C, N, and P variation with the increase in stand ages may result in P limitation. The NRE was significantly correlated with the C of green leaves and the P content and C:N of senesced leaves, whereas the PRE was significantly negatively correlated with the C:P and N:P of soil and green leaves.

The PRE:NRE decreased with increased stand ages. In summary, our result showed that the NRE and PRE were both changed as the stand aged. The middle-aged black locust trees (approximately 30 years) were more efficient at resorbing P and N from senesced leaves than younger and older stands, but the NRE and PRE responded differently in soil and plant nutrients. The black locust plantations would alter the conservation and use strategy of nutrients in the ecosystem through a plant-mediated pathway with increasing stand age.

Author Contributions: Conceptualization, J.D., X.L., and X.H.; Formal analysis, S.W. and F.Z.; Investigation, J.D., S.W., C.R., and W.Z.; Writing—original draft, J.D. and D.Z.; Writing—review and editing, C.R., F.Z., X.H., and G.Y.

References

1. Aerts, R. Nutrient resorption from senescing leaves of perennials: Are there general patterns? *J. Ecol.* **1996**, *84*, 597–608. [CrossRef]
2. Deng, M.; Liu, L.; Jiang, L.; Liu, W.; Wang, X.; Li, S.; Yang, S.; Wang, B. Ecosystem scale trade-off in nitrogen acquisition pathways. *Nat. Ecol. Evol.* **2018**, *2*, 1724–1734. [CrossRef] [PubMed]
3. Lü, X.; Reed, S.C.; Yu, Q.; Han, X. Nutrient resorption helps drive intra-specific coupling of foliar nitrogen and phosphorus under nutrient-enriched conditions. *Plant Soil* **2016**, *398*, 111–120. [CrossRef]
4. Wang, L.; Wang, L.; He, W.; An, L.; Xu, S. Nutrient resorption or accumulation of desert plants with contrasting sodium regulation strategies. *Sci. Rep. UK* **2017**, *7*, 17035. [CrossRef]
5. Vitousek, P.M.; Porder, S.; Houlton, B.Z.; Chadwick, O.A. Terrestrial phosphorus limitation: Mechanisms, implications, and nitrogen and phosphorus interactions. *Ecol. Appl.* **2010**, *20*, 5–15. [CrossRef] [PubMed]
6. Elser, J.J.; Bracken, M.E.; Cleland, E.E.; Gruner, D.S.; Harpole, W.S.; Hillebrand, H.; Ngai, J.T.; Seabloom, E.W.; Shurin, J.B.; Smith, J.E. Global analysis of nitrogen and phosphorus limitation of primary producers in freshwater, marine and terrestrial ecosystems. *Ecol. Lett.* **2007**, *10*, 1135–1142. [CrossRef]
7. Campo, J. Shift from ecosystem P to N limitation at precipitation gradient in tropical dry forests at Yucatan, Mexico. *Environ. Res. Lett.* **2016**, *11*, 95006. [CrossRef]
8. Mediavilla, S.; García-Iglesias, J.; González-Zurdo, P.; Escudero, A. Nitrogen resorption efficiency in mature trees and seedlings of four tree species co-occurring in a Mediterranean environment. *Plant Soil* **2014**, *385*, 205–215. [CrossRef]
9. Chen, L.; Zhang, C.; Duan, W. Temporal variations in phosphorus fractions and phosphatase activities in rhizosphere and bulk soil during the development of *Larix olgensis* plantations. *J. Plant Nutr. Soil Sci.* **2015**, *179*, 67–77. [CrossRef]
10. Yan, T.; Lü, X.; Zhu, J.; Yang, K.; Yu, L.; Gao, T. Changes in nitrogen and phosphorus cycling suggest a transition to phosphorus limitation with the stand development of larch plantations. *Plant Soil* **2018**, *422*, 385–396. [CrossRef]
11. Yuan, Z.Y.; Chen, H.Y.H. Changes in nitrogen resorption of trembling aspen (*Populus tremuloides*) with stand development. *Plant Soil* **2010**, *327*, 121–129. [CrossRef]
12. Vergutz, L.; Manzoni, S.; Porporato, A.; Novais, R.F.; Jackson, R.B. Global resorption efficiencies and concentrations of carbon and nutrients in leaves of terrestrial plants. *Ecol. Monogr.* **2012**, *82*, 205–220. [CrossRef]
13. Yan, T.; Zhu, J.; Yang, K. Leaf nitrogen and phosphorus resorption of woody species in response to climatic conditions and soil nutrients: A meta-analysis. *J. For. Res.* **2018**, *29*, 1–9. [CrossRef]
14. Lu, J.; Duan, B.; Yang, M.; Yang, H.; Yang, H. Research progress in nitrogen and phosphorus resorption from senesced leaves and the influence of ontogenetic and environmental factors. *Acta Pratacult. Sin.* **2018**, *27*, 178–188.
15. Du, B.; Ji, H.; Peng, C.; Liu, X.; Liu, C. Altitudinal patterns of leaf stoichiometry and nutrient resorption in *Quercus variabilis* in the Baotianman Mountains, China. *Plant Soil* **2017**, *413*, 193–202. [CrossRef]
16. Zhang, H.; Guo, W.; Yu, M.; Wang, G.G.; Wu, T. Latitudinal patterns of leaf N, P stoichiometry and nutrient resorption of *Metasequoia glyptostroboides* along the eastern coastline of China. *Sci. Total Environ.* **2018**, *618*, 1–6. [CrossRef] [PubMed]
17. Wang, Z.; Lu, J.; Yang, H.; Zhang, X.; Luo, C.; Zhao, Y. Resorption of nitrogen, phosphorus and potassium from leaves of lucerne stands of different ages. *Plant Soil* **2014**, *383*, 301–312. [CrossRef]
18. Sun, Z.; Liu, L.; Peng, S.; Peñuelas, J.; Zeng, H.; Piao, S. Age-Related Modulation of the Nitrogen Resorption Efficiency Response to Growth Requirements and Soil Nitrogen Availability in a Temperate Pine Plantation. *Ecosystems* **2016**, *19*, 698–709. [CrossRef]
19. Zhang, H.; Wang, J.; Wang, J.; Guo, Z.; Wang, G.G.; Zeng, D.; Wu, T. Tree stoichiometry and nutrient resorption along a chronosequence of *Metasequoia glyptostroboides* forests in coastal China. *For. Ecol. Manag.* **2018**, *430*, 445–450. [CrossRef]
20. Qiu, L.; Hu, H.; Lin, Y.; Ge, L.; Wang, K.; He, Z.; Dong, Q. Nutrient resorption effciency and C:N:P stoichiometry of *Eucalyptus urophylla* × *E. grandis* of different ages in a sandy coastal plain area. *Chin. J. Appl. Environ. Biol.* **2017**, *23*, 739–744.

21. Zhou, L.; Addo-Danso, S.D.; Wu, P.; Li, S.; Zou, X.; Zhang, Y.; Ma, X. Leaf resorption efficiency in relation to foliar and soil nutrient concentrations and stoichiometry of *Cunninghamia lanceolata* with stand development in southern China. *J. Soil Sediment* **2016**, *16*, 1–12. [CrossRef]
22. Güsewell, S. N:P ratios in terrestrial plants: Variation and functional significance. *New Phytol.* **2004**, *164*, 243–266. [CrossRef]
23. Meng, F.; Luo, Q.; Wang, Q.; Zhang, X.; Qi, Z.; Xu, F.; Lei, X.; Cao, Y.; Chow, W.S.; Sun, G. Physiological and proteomic responses to salt stress in chloroplasts of diploid and tetraploid black locust (*Robinia pseudoacacia* L.). *Sci. Rep. UK* **2016**, *6*, 23098. [CrossRef]
24. Sitzia, T.; Campagnaro, T.; Dainese, M.; Cierjacks, A. Plant species diversity in alien black locust stands: A paired comparison with native stands across a north-Mediterranean range expansion. *For. Ecol. Manag.* **2012**, *285*, 85–91. [CrossRef]
25. Zhang, G.; Zhang, P.; Cao, Y. Ecosystem carbon and nitrogen storage following farmland afforestation with black locust (*Robinia pseudoacacia*) on the Loess Plateau, China. *J. For. Res.* **2018**, *29*, 761–771. [CrossRef]
26. Hui, S.; Shao, M. Soil and water loss from the Loess Plateau in China. *J. Arid Environ.* **2000**, *45*, 9–20.
27. Ma, C.; Luo, Y.; Shao, M.; Li, X.; Sun, L.; Jia, X. Environmental controls on sap flow in black locust forest in Loess Plateau, China. *Sci. Rep. UK* **2017**, *7*, 13160. [CrossRef]
28. Liu, G.; Shang, G.; Yao, W.; Yang, Q.; Zhao, M.; Dang, X.; Guo, M.; Wang, G.; Wang, B. Ecological Effects of Soil Conservation in Loess Plateau. *Bull. Chin. Acad. Sci.* **2017**, *32*, 11–19.
29. Deng, J.; Sun, P.; Zhao, F.; Han, X.; Yang, G.; Feng, Y.; Ren, G. Soil C, N, P and Its Stratification Ratio Affected by Artificial Vegetation in Subsoil, Loess Plateau China. *PLoS ONE* **2016**, *11*, e151446. [CrossRef] [PubMed]
30. Wei, J.; Li, Z.; Feng, X.; Zhang, Y.; Chen, W.; Wu, X.; Jiao, L.; Wamg, X. Ecological and physiological mechanisms of growth decline of *Robinia pseudoacacia* plantations in the Loess Plateau of China: A review. *Chin. J. Appl. Ecol.* **2018**, *29*, 2433–2444.
31. Cao, Y.; Chen, Y. Coupling of plant and soil C:N:P stoichiometry in black locust (*Robinia pseudoacacia*) plantations on the Loess Plateau, China. *Trees* **2017**, *31*, 1–12. [CrossRef]
32. Jia, X.; Shao, M.; Zhu, Y.; Luo, Y. Soil moisture decline due to afforestation across the Loess Plateau, China. *J. Hydrol.* **2017**, *546*, 113–122. [CrossRef]
33. Shangguan, Z.; Zheng, S. Ecological properties of soil water and effects on forest vegetation in the Loess Plateau. *Int. J. Sustain. Dev. World* **2006**, *13*, 307–314. [CrossRef]
34. Yan, T.; Lü, X.; Yang, K.; Zhu, J. Leaf nutrient dynamics and nutrient resorption: A comparison between larch plantations and adjacent secondary forests in Northeast China. *J. Plant Ecol.* **2015**, *9*, 165–173. [CrossRef]
35. Bao, S. *Analysis of Soil Agrochemical*; Chinese Agricultural Press: Beijing, China, 2000.
36. He, J.; Lu, Y.; Fu, B. *Frontier of Soil Biology*; Science Press: Beijing, China, 2015.
37. Liu, J.; Gu, Z.; Shao, H.; Zhou, F.; Peng, S. N–P stoichiometry in soil and leaves of *Pinus massoniana* forest at different stand ages in the subtropical soil erosion area of China. *Environ. Earth Sci.* **2016**, *75*, 1091. [CrossRef]
38. Rice, S.K.; Westerman, B.; Federici, R. Impacts of the exotic, nitrogen-fixing black locust (*Robinia pseudoacacia*) on nitrogen-cycling in a pine-oak ecosystem. *Plant Ecol.* **2004**, *174*, 97–107. [CrossRef]
39. Zhang, W.; Liu, W.; Xu, M.; Deng, J.; Han, X.; Yang, G.; Feng, Y.; Ren, G. Response of forest growth to C:N:P stoichiometry in plants and soils during *Robinia pseudoacacia* afforestation on the Loess Plateau, China. *Geoderma* **2019**, *337*, 280–289. [CrossRef]
40. Ma, R.; An, S.; Huang, Y. C, N and P stoichiometry characteristics of different-aged *Robinia pseudoacacia* plantations on the Loess Plateau, China. *Chin. J. Appl. Ecol.* **2017**, *28*, 2787–2793.
41. Schreeg, L.A.; Santiago, L.S.; Wright, S.J.; Turner, B.L. Stem, root, and older leaf N:P ratios are more responsive indicators of soil nutrient availability than new foliage. *Ecology* **2016**, *95*, 2062–2068. [CrossRef]
42. Zheng, S.; Shangguan, Z. Spatial patterns of leaf nutrient traits of the plants in the Loess Plateau of China. *Trees* **2007**, *21*, 357–370. [CrossRef]
43. Ma, L.; Chen, Y.; Zhang, X.; Yang, J.; An, S. Characteristics of Leaf Ecological Stoichiometry of *Robinia pseudoacacia* in Loess Plateau. *Res. Soil Water Conserv.* **2014**, *21*, 57–61.
44. Han, W.; Fang, J.; Guo, D.; Zhang, Y. Leaf nitrogen and phosphorus stoichiometry across 753 terrestrial plant species in China. *New Phytol.* **2005**, *168*, 377–385. [CrossRef] [PubMed]
45. Elser, J.J.; Sterner, R.W.; Gorokhova, E.; Fagan, W.F.; Markow, T.A.; Cotner, J.B.; Harrison, J.F.; Hobbie, S.E.; Odell, G.M.; Weider, L.W. Biological stoichiometry from genes to ecosystems. *Ecol. Lett.* **2008**, *3*, 540–550. [CrossRef]

46. Hayes, P.; Turner, B.L.; Lambers, H.; Laliberté, E.; Bellingham, P. Foliar nutrient concentrations and resorption efficiency in plants of contrasting nutrient-acquisition strategies along a 2-million-year dune chronosequence. *J. Ecol.* **2014**, *102*, 396–410. [CrossRef]
47. Chapin, F.S.; Moilanen, L. Nutritional Controls Over Nitrogen and Phosphorus Resorption from Alaskan Birch Leaves. *Ecology* **1991**, *72*, 709–715. [CrossRef]
48. Xie, Y.; Liang, Y.; Xiao, H.; Zhu, R.; Luo, L.; Guo, W.; Cao, Y.; Zhang, Z.; Pan, Y.; Zheng, N.; et al. Pattern and controlling factors of terrestrial leaf litters decomposition in China. *J. East China Univ. Technol. (Nat. Sci.)* **2018**, *41*, 271–276.
49. William, P.; Silver, W.L.; Burke, I.C.; Leo, G.; Harmon, M.E.; Currie, W.S.; King, J.Y.; Adair, E.C.; Brandt, L.A.; Hart, S.C. Global-scale similarities in nitrogen release patterns during long-term decomposition. *Science* **2007**, *315*, 361–364.
50. Gergócs, V.; Hufnagel, L. The effect of microarthropods on litter decomposition depends on litter quality. *Eur. J. Soil Biol.* **2016**, *75*, 24–30. [CrossRef]
51. Li, R.; Wang, S.L.; Wang, Q.K. Nutrient contents and resorption characteristics in needles of different age *Pinus massoniana* (Lamb.) before and after withering. *Chin. J. Appl. Ecol.* **2008**, *19*, 1443.
52. Sardans, J.; Peñuelas, J. Trees increase their P:N ratio with size. *Glob. Ecol. Biogeogr.* **2015**, *24*, 147–156. [CrossRef]
53. Sardans, J.; Peñuelas, J. Tree growth changes with climate and forest type are associated with relative allocation of nutrients, especially phosphorus, to leaves and wood. *Glob. Ecol. Biogeogr.* **2012**, *22*, 494–507. [CrossRef]
54. Luyssaert, S.; Staelens, J.; De Schrijver, A. Does the commonly used estimator of nutrient resorption in tree foliage actually measure what it claims to? *Oecologia* **2005**, *144*, 177–186. [CrossRef]
55. Aerts, R.; Chapin, F.S. The Mineral Nutrition of Wild Plants Revisited: A Re-evaluation of Processes and Patterns. *Adv. Ecol. Res.* **2000**, *30*, 1–67.
56. Güsewell, S. Nutrient Resorption of Wetland Graminoids Is Related to the Type of Nutrient Limitation. *Funct. Ecol.* **2010**, *19*, 344–354. [CrossRef]
57. Agüero, M.L.; Puntieri, J.; Mazzarino, M.J.; Grosfeld, J.; Barroetaveña, C. Seedling response of Nothofagus species to N and P: Linking plant architecture to N/P ratio and resorption proficiency. *Trees* **2014**, *28*, 1185–1195. [CrossRef]
58. Killingbeck, K.T. Nutrients in Senesced Leaves: Keys to the Search for Potential Resorption and Resorption Proficiency. *Ecology* **1996**, *77*, 1716–1727. [CrossRef]
59. De Marco, A.; Arena, C.; Giordano, M.; De Santo, A.V. Impact of the invasive tree black locust on soil properties of Mediterranean stone pine-holm oak forests. *Plant Soil* **2013**, *372*, 473–486. [CrossRef]
60. Han, W.; Tang, L.; Chen, Y.; Fang, J. Relationship between the Relative Limitation and Resorption Efficiency of Nitrogen vs. Phosphorus in Woody Plants. *PLoS ONE* **2013**, *8*, e83366. [CrossRef] [PubMed]
61. Lado-Monserrat, L.; Lidón, A.; Bautista, I. Litterfall, litter decomposition and associated nutrient fluxes in Pinus halepensis: Influence of tree removal intensity in a Mediterranean forest. *Eur. J. For. Res.* **2015**, *134*, 833–844. [CrossRef]
62. Spohn, M.; Zavišić, A.; Nassal, P.; Bergkemper, F.; Schulz, S.; Marhan, S.; Schloter, M.; Kandeler, E.; Polle, A. Temporal variations of phosphorus uptake by soil microbial biomass and young beech trees in two forest soils with contrasting phosphorus stocks. *Soil Biol. Biochem.* **2018**, *117*, 191–202. [CrossRef]
63. Zavišić, A.; Nassal, P.; Yang, N.; Heuck, C.; Spohn, M.; Marhan, S.; Pena, R.; Kandeler, E.; Polle, A. Phosphorus availabilities in beech (*Fagus sylvatica* L.) forests impose habitat filtering on ectomycorrhizal communities and impact tree nutrition. *Soil Biol. Biochem.* **2016**, *98*, 127–137.
64. Netzer, F.; Schmid, C.; Herschbach, C.; Rennenberg, H. Phosphorus-nutrition of European beech (*Fagus sylvatica* L.) during annual growth depends on tree age and P-availability in the soil. *Environ. Exp. Bot.* **2017**, *137*, 194–207. [CrossRef]
65. Pistocchi, C.; Mészáros, É.; Tamburini, F.; Frossard, E.; Bünemann, E.K. Biological processes dominate phosphorus dynamics under low phosphorus availability in organic horizons of temperate forest soils. *Soil Biol. Biochem.* **2018**, *126*, 64–75. [CrossRef]

10

The Effect of Harvest on Forest Soil Carbon

Jason James * and Rob Harrison

School of Environmental and Forest Sciences, University of Washington, Box 352100, Seattle, WA 98195, USA; robh@uw.edu
* Correspondence: jajames@uw.edu

Academic Editors: Scott X. Chang and Xiangyang Sun

Abstract: Forest soils represent a substantial portion of the terrestrial carbon (C) pool, and changes to soil C cycling are globally significant not only for C sequestration but also for sustaining forest productivity and ecosystem services. To quantify the effect of harvesting on soil C, we used meta-analysis to examine a database of 945 responses to harvesting collected from 112 publications from around the world. Harvesting reduced soil C, on average, by 11.2% with 95% CI [14.1%, 8.5%]. There was substantial variation between responses in different soil depths, with greatest losses occurring in the O horizon (−30.2%). Much smaller but still significant losses (−3.3%) occurred in top soil C pools (0–15 cm depth). In very deep soil (60–100+ cm), a significant loss of 17.7% of soil C in was observed after harvest. However, only 21 of the 945 total responses examined this depth, indicating a substantial need for more research in this area. The response of soil C to harvesting varies substantially between soil orders, with greater losses in Spodosol and Ultisol orders and less substantial losses in Alfisols and Andisols. Soil C takes several decades to recover following harvest, with Spodosol and Ultisol C recovering only after at least 75 years. The publications in this analysis were highly skewed toward surface sampling, with a maximum sampling depth of 36 cm, on average. Sampling deep soil represents one of the best opportunities to reduce uncertainty in the understanding of the response of soil C to forest harvest.

Keywords: forest management; harvest; soil carbon; soil order; deep soil; meta-analysis

---

1. Introduction

Forest ecosystems contain 1240 Pg C [1,2], which represents as much as 80% of aboveground terrestrial C and 70% of all soil organic C [3–5]. The relative proportion of forest C found in soils varies among biomes, ranging from roughly 85% of the terrestrial C pool in boreal forests, to 60% in temperate forests, to 50% in tropical rainforests [1,6]. The net balance of soil C in forests relies upon large rates of detrital inputs (61.4 Pg C year^{-1}) and respiratory losses (60 Pg C year^{-1}), which together represent substantial yearly turnover in the soil C pool [7]. By altering the rates of detrital inputs and respiratory outputs in soils, the extent and intensity of forest harvest can have substantial impacts not only on ecosystem function but also on atmospheric chemistry and global climate [6,8,9].

C is one of the principal components of soil organic matter (SOM), a key component of soil that plays an important role in many biological, chemical, and physical properties [10–12]. SOM provides a crucial source of energy and nutrients for soil microbes, buffers soil pH, and helps to stabilize soil structure [12,13]. Along with nitrogen and phosphorus, SOM is considered a critical indicator for soil health and quality.

Thus, soil C is an essential component of forest C accounting, yet many models assume that only surface soil responds to forest management and that soil C returns to equilibrium within 20 years after harvest [14]. Recent national or global assessments of forest C lack any mention of mineral soil

C [15–17], implicitly assuming that soil C remains constant after forest harvest. Furthermore, carbon monitoring programs include soil C inconsistently. For example, the American Carbon Registry [18] and the Verified Carbon Standard [19] do not require or specify protocols for soil C measurements. The Intergovernmental Panel on Climate Change (IPCC) inventory standards [20] assume constant mineral soil C in Tier 1, with an option for inclusion of national soil C inventories only if preferred by a particular agency, and the U.S. Forest Service Inventory and Analysis Program [21] specifically limits soil sampling to 20 cm depth. The inclusion of soil in models of ecosystem C following harvest can have significant effects. For example, in a model of the forest C pool change following intensive bioenergy harvest, Zanchi et al. [22] show that the inclusion of soil increases the C payback period by approximately 25 years when substituting forest bioenergy for coal. Thus, the inclusion or exclusion of soil in ecosystem C models and ecological monitoring programs can have a major impact on forest policy when attempting to mitigate climate change through forest management [14].

Ambiguity about the effect of forest harvesting on soil C has persisted in the literature, likely exacerbated by the inherent spatial and temporal variability in soil measurements that can obscure the results of even the most well-designed studies [23]. By gathering the results from many studies that apply similar treatments, meta-analysis can overcome the high levels of spatial and temporal variability to provide cumulative answers that may not have been evident within individual sites [23,24]. Previous meta-analyses on the effect of harvesting on soil C have found either minimal effects on soil C pools [25] or substantial (30%) loss to O horizon pools with little change to mineral soil C [9]. Variation in soil C response has been shown to significantly differ among soil types and different harvesting strategies [9].

Studies of soil C change due to harvest have historically been strongly biased toward surface sampling [26]. Nave et al. [9] reported a mixed response to harvest in deeper soil (20–100 cm depth), ranging from a slight average decrease (−5%) in studies that reported C pools to a large average increase (+20%) in studies that reported only C concentration. Several recent reviews have highlighted the need for greater sampling of deep soil [26–28], especially as the shifting paradigm of SOM research has come to reject the assumption that deep soil C cannot not change on timescales relevant to anthropogenic C emissions [29–31]. Resolving the response of deep soil horizons to harvesting is important because these horizons occupy a much greater volume than surface O and A horizons. Even small changes in subsurface C can exacerbate or compensate for changes in surface soil C, and neither the magnitude nor direction of subsoil C change is clear from previous research.

The process of meta-analysis is necessarily cumulative, with each iteration updating previous analyses to further constrain the error in effect size estimates and to extend the scope of analysis. Thus, the objective of our meta-analysis is to update and extend the findings of Nave et al. [9] with respect to five major research questions:

(1) What is the overall effect of forest harvesting on soil C pools?
(2) How does the effect of forest harvest on soil C change with soil depth?
(3) To what extent does the effect of harvesting differ among soil orders?
(4) Do site pretreatment strategies or increasing harvesting intensity (i.e., whole tree harvest) moderate or accentuate harvesting impacts on soil C?
(5) How long does soil C take to recover from harvest across different soil types?

2. Materials and Methods

Meta-analysis is a cumulative activity which builds upon previous research and meta-analyses on similar research questions. Our meta-analysis builds upon the work of Nave et al. [9] and Johnson and Curtis [25] by updating their results with studies published between 2008 and 2016. The database published by Nave et al. [9] was independently recreated from each of 75 references. Metadata for each study was verified, and additional metadata such as the sampling depth of each response ratio was gathered. A total of 8 effect sizes differed in our dataset from the Nave et al. [9] database, all of which were either additional data for mineral soils or a split of one effect size into two based upon sampling depth.

To add to this database with studies published between 2008 and 2016, we searched the peer-reviewed literature for relevant studies using the online database Web of Science with combinations of the terms: forest, timber, harvest, logging, soil C, soil organic matter, and management. No climate criteria was used to screen studies. To be included in the meta-analysis, publications had to report both a control as well as harvested treatments. Both pretreatment soil C and unharvested reference stands were considered acceptable controls, and measurements of reference stands were considered the superior control. For forest chronosequence studies, soil C data from the oldest stand was used as the control. A minimum stand age of 30 years was considered acceptable for control stands, although most studies used controls of considerably greater age. Nave et al. [9] found that studies reporting only soil C concentration data yielded different conclusions about the direction of harvest effects than those studies reporting soil C pool data. Consequently, soil C pool data was used in our meta-analysis when both concentration and pool data were available.

We collected potentially useful predictor variables of soil C response from each publication, including soil order, geographic region, and time since harvest (Table 1). Binning of continuous predictor variables (such as precipitation) was carried out in the same intervals as Nave et al. [9] for ease of comparison. Each study was categorized by harvest, residue management, and site preparation strategies. Harvesting technique was categorized as sawlog when only the merchantable bole (stem) was removed from the site or whole tree harvest (WTH) when the tops, limbs, and foliage were removed in addition to the bole. To test the response of soil C at different depths, data from each study was separated into one of five groups: O horizon, top soil (0–15 cm), mid soil (15–30 cm), deep soil (30–60 cm), and very deep soil (60–100+ cm). A sixth group called whole soil was assigned to studies that aggregated mineral soil samples instead of reporting results at separate depths. Several studies aggregated soil data from 0–100 cm, which reduced the number of unique deep and very deep soil observations even though these depths were separately sampled.

Table 1. Factors gathered as potential predictor variables in this meta-analysis.

Factor	**Levels**	
Reporting units	Pool (Mg·ha^{-1}), concentration (% or mg·g^{-1})	
Soil Depth	O horizon	Forest Floor
	Top Soil	0–15 cm
	Mid Soil	15–30 cm
	Deep Soil	30–60 cm
	Very Deep	60–100+ cm
Overstory species	Hardwood, conifer/mixed	
Soil order	Alfisol, Andisol, Entisol, Inceptisol, Mollisol, Spodosol, Oxisol, Ultisol	
Geographic group	NE North America, NW North America, SE North America, SW North America, Europe, Asia, Pacific (Australia, New Zealand)	
Harvest type	Clearcut, thin	
Harvest intensity	Stem only, whole tree	
Residue management	Removed, spread	
Site preparation	Broadcast burn, tillage/scarification	
Soil texture	Fine (mostly silt or clay), coarse (mostly sand), organic	
Time since harvest	Continuous	
Mean Annual Temperature	0–5, 5–7.5, 7.6–10, 10.1–15, 15.1–20, >20 (°C)	
Mean Annual Precipitation	<500, 500–750, 751–1000, 1001–1400, 1401–1800, >1800 (mm)	

Our meta-analysis estimates the magnitude of change in soil C using the ln-transformed response ratio R, which is defined as

$$\ln(R) = \ln\left(\frac{\overline{X_T}}{\overline{X_C}}\right) \quad (1)$$

where $\overline{X_T}$ is the mean soil C value of treatment (harvested) observations, and $\overline{X_C}$ is the mean soil C value of control observations for a given set of experimental conditions at a specific site and depth. Multiple response ratios were recorded for each publication, with the number of response ratios (k) depending upon the number of experimental conditions imposed and the number of samples taken by depth. For example, a publication that reports the results of two thinning treatments and two clear-cut treatments at three depth increments (forest floor, top soil, and mid soil) versus a control would yield 12 response ratios. R is a unit-less measure of effect size, which allows comparison among studies that report data in different units [24]. By back transforming $\ln(R)$, $[(e^{(\ln(R))} - 1) \times 100]$, mean response ratios can be interpreted as the percentage change in soil C relative to the control. Estimates of the standard deviation and sample size for each X^T and X^C were not available in several publications.

Consequently, an accurate estimate of total heterogeneity (Q_T) for the dataset was not possible. Subsequent partitioning of Q_T into within- and among-group heterogeneity (Q_W and Q_A, respectively) for random and mixed effect models (as is customary for meta-analyses) was not possible [24]. Instead, we used nonparametric resampling techniques (bootstrapping) to estimate confidence intervals around mean effect sizes in an unweighted meta-analysis [9].

Adams et al. [32] recommend bootstrapping confidence intervals for ecological meta-analyses, and show that confidence bounds based on this method are more conservative than standard meta-analyses. Bootstrapping was implemented using the bootES package [33] in *R* [34]. For all statistical tests in our analysis, $\alpha = 0.05$.

Although not exhaustive, the database we compiled from the literature search contained 945 soil C response ratios from 112 publications published between 1979 and 2016. Roughly half the dataset was comprised of response ratios analyzed by Nave et al. [9]. The full dataset is available as Supplementary Material, including maximum sampling depth and the number of response ratios from each paper (Appendix A).

3. Results

3.1. Overall Effect and Change with Depth

Across all studies, harvesting led to a significant average decrease in soil C of 11.2% relative to control (Figure 1). Whether the response to harvest was reported as pools or concentrations had a large impact on the estimated effect of harvest on soil C, with mean response for studies reporting C concentration units (%, $mg \cdot g^{-1}$, etc.) 16.2% higher (with a 95% CI [20.9%, 11.8%]) than studies reporting C pool units ($Mg \cdot ha^{-1}$, $tons \cdot ha^{-1}$, etc.). Concentration responses are higher than pool responses at all soil depths, except for very deep and whole mineral soil, which did not have enough concentration response ratios to construct separate confidence intervals (Figure 1). Consequently, all subsequent analyses focused on the subset of data reporting soil C pools.

Several different soil layers show significant losses of C due to harvesting. Overall, O horizons lost 30.2% of their carbon as a result of harvesting. Losses from top soil were much smaller, although the estimated loss when reported in pool units was significant (−3.3%). In mid (15–30 cm) and deep soil (30–60 cm), the average loss of soil C was greater than topsoil, although the smaller number of response ratios for these depths resulted in more poorly constrained estimates. Studies only reporting C concentration observed a 14.5% increase in deep soil (30–60 cm), although the sample size was relatively small. The overall effect in very deep soil (60–100+ cm) was significant, with an average loss of 17.7%. Unfortunately, this region of the profile was not frequently sampled (21 response ratios out of 945 total), and consequently the 95% confidence interval is quite wide.

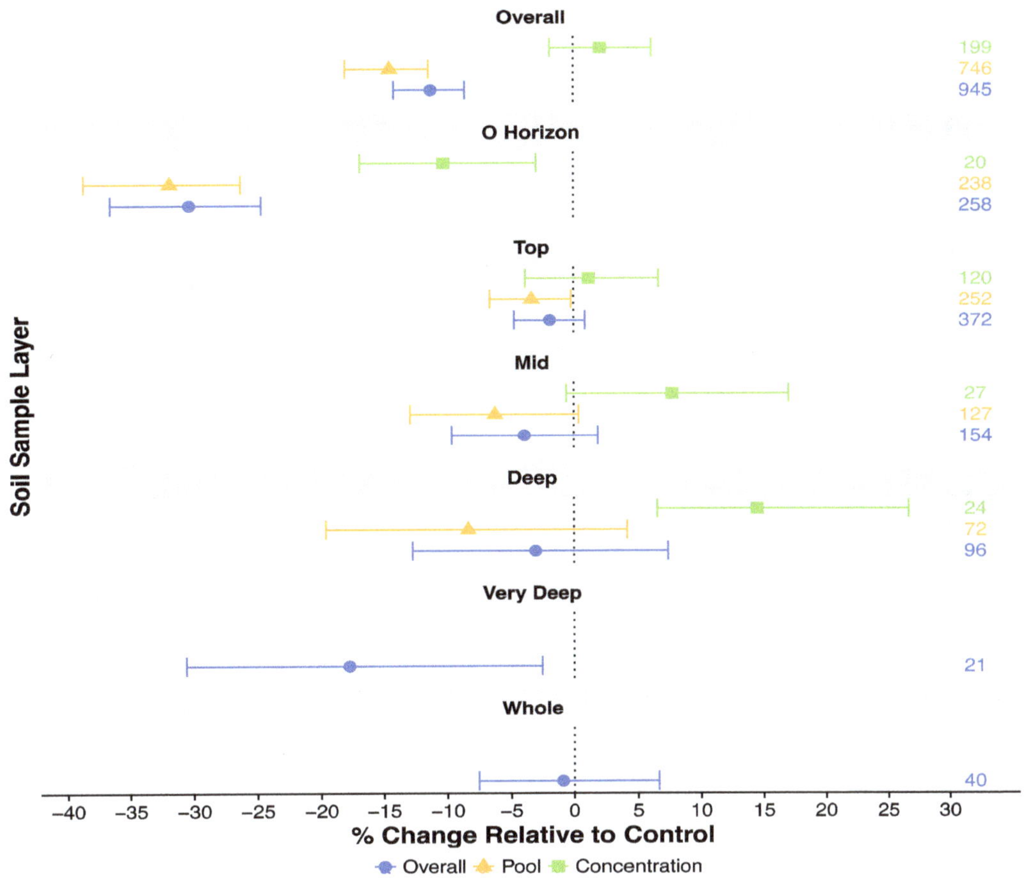

Figure 1. Response of soil C to forest harvesting, overall and faceted by soil depth. All points are back-transformed mean effect sizes ± 95% confidence intervals calculated by nonparametric bootstrap. The number of response ratios (k) that make up each mean effect is listed on the right. Mean effects with confidence intervals overlapping the dashed line (0%) show no significant change in soil C due to harvesting. Within each facet, mean effect sizes are shown for the overall effect as well as separately for studies reporting C pool units or concentration units.

3.2. *Effect of Harvesting across Soil Orders*

The effect of harvesting on soil C differs between soil orders (Figure 2). For the Alfisols and Inceptisols, there are significant losses in O horizon C pools (−12.0% and −45.4%, respectively), but no significant loss in the mineral soil. Mollisols lost an average of 17.7%, although neither O horizon nor mineral soil responses were significantly different from 0. In several cases, small samples sizes made separate testing of organic and mineral soil impossible within a single order (Andisols, Entisols, Oxisols). However, in each of these cases the overall effect was significant. Soil C increased by 24.5% on average in Andisols, but decreased by 18.8% in Entisols and 30.9% in Oxisols. The number of response ratios was more concentrated in the Alfisol, Inceptisol, Spodosol, and Ultisol orders, although a large number of publications did not report information on soil classification. The response to harvesting in Spodosols is substantial (−19.0% overall), with significant losses in both the O horizon (−36.4%) and moderately less in the mineral soil (−9.1%). Likewise, Ultisols lost significant soil C in response to harvesting (−24.7% overall), with the most substantial losses occurring in the O horizon (−66.0%) rather than in the mineral soil (−11.9%).

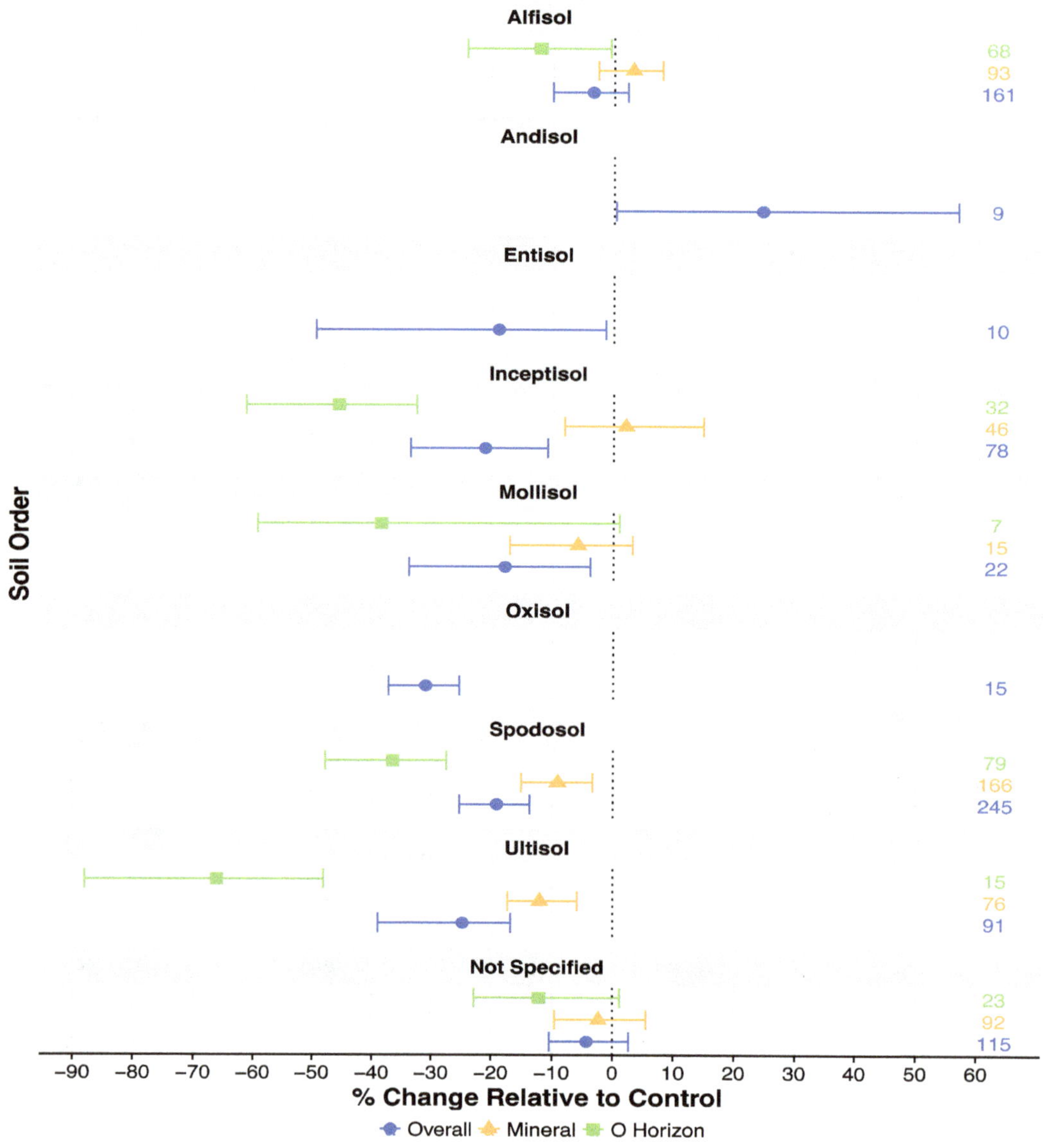

Figure 2. Response of soil C to harvesting in different soil orders. Mean effect sizes ± 95% confidence intervals calculated by nonparametric bootstrap are shown for all response ratios in each soil order (Overall) and broken out into mineral soil or O horizon. The number of response ratios (k) comprising each mean effect are listed on the right. Effect sizes were calculated only on response ratios reported in pool units ($k = 746$).

3.3. Differences in Response to Harvest between Forest Types

The response of soil C to harvest differs between hardwood and coniferous/mixed forest types (Figure 3). The decline in O horizon C pools is significantly greater in conifer/mixed forests (−38.1%) compared to hardwood forests (−25.4%). Differences between forest types were not significant for any mineral soil layer. However, the decline in soil C after harvest was significant for hardwood forests but not conifer/mixed forests in deep soil (30–60 cm) and in studies reporting whole mineral soil C pools. Also in these studies, the difference between hardwood and conifer/mixed forest response to harvest is marginally significant ($p < 0.1$). The number of observations are highly concentrated in O horizon and top soil, consequently limiting the precision of mean effect size estimates in deeper layers. No observations for hardwood forest were made in very deep soil (60–100+ cm).

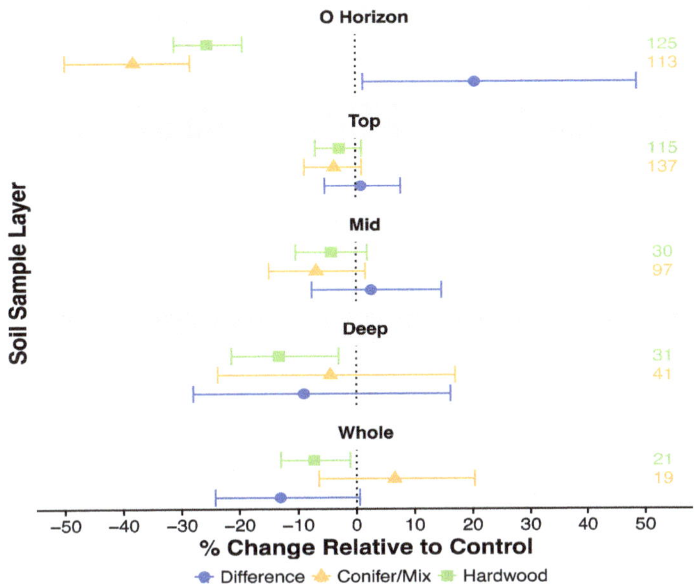

Figure 3. Response of soil C harvesting at different depths in soil, broken down by hardwood or conifer/mixed forest types. Mean effect sizes ± 95% confidence intervals calculated by nonparametric bootstrap are shown for hardwood and conifer/mixed forests. Blue circles show the mean difference between these forest types (Hardwood–Conifer/Mix) ± 95% confidence interval for the difference. Differences are calculated on the logarithmic effect size scale, and then back-transformed to % change, and thus do not necessarily add up on the % change scale. The number of response ratios (k) in each forest type at each depth is listed on the right. Data for very deep soil is not shown because there were no observations for this soil layer in hardwood forests.

3.4. Harvest Intensity, Residue Management and Site Pretreatment

Differences in forest management strategies can significantly impact the response of soil C to harvesting (Figure 4). While there was no significant overall difference observed between thinning and clear-cut harvesting, less C was lost from mineral soils under clear-cut harvesting compared to thinning (+9.3%). Likewise, harvest intensity significantly changed the response of mineral soil C, with soils undergoing whole tree harvesting losing 13.3% less C than bole-only harvesting. Possible mechanisms for these counter-intuitive results are considered in Discussion Section 4.5.

The practice of broadcast burning sites in preparation for planting after a harvest leads to significant additional losses of soil C, with burned soils losing 15.2% more C than soils with no pretreatment. This effect is especially severe in the O horizon (40.9% additional loss than if sites were not burned), and somewhat curtailed in the mineral soil (8.3% additional loss). The wide 95% CI for the estimate of differences in O horizon responses due to burning reflects disparities in burn severity and treatment implementation among different studies.

Spreading of residual materials across harvested sites (by chipping tops and limbs or other methods) resulted in significant additional loss of soil C (−10.9%), with these extra losses occurring mostly in the mineral soil (−17.5%). On the other hand, residue removal resulted in no significant additional losses to soil C.

Tillage is sometimes used to prepare soils for planting after harvest, either to create raised planting beds or to prepare the soil seed bed. This intensive style of site preparation did not result in significant losses in soil C, especially in the mineral soil. However, very large losses were reported in the O horizon (mean effect = −37.1%) with a very wide confidence interval due to a small number of observations. Additional study of the effect of tillage would help to reduce this error.

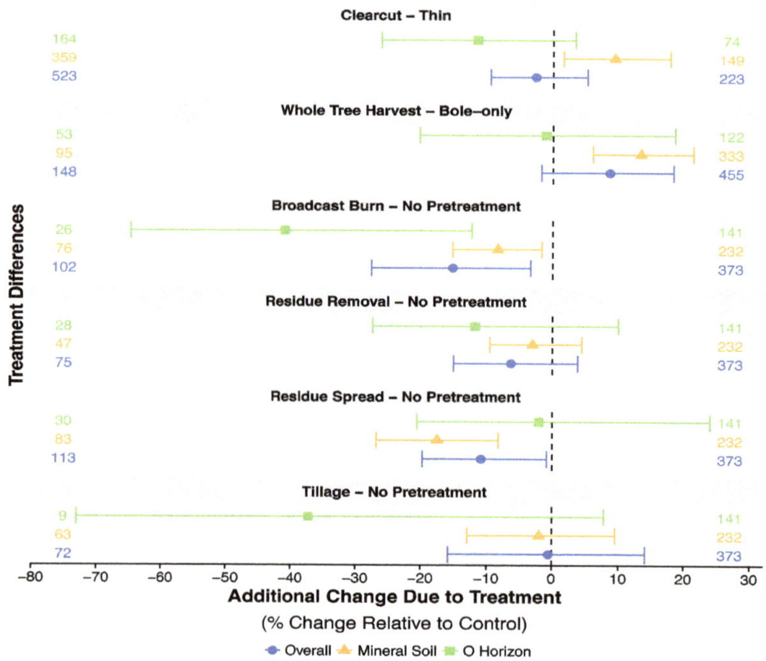

Figure 4. Differences in response of soil C to harvesting between treatment strategies. Differences are calculated by subtracting [more intensive treatment] − [less intensive treatment], such that positive differences represent reduced loss of C due to more intensive treatment, and negative differences represent increased losses of C due to more intensive treatment. Point estimates are back-transformed differences between mean effect sizes ± 95% confidence intervals calculated by nonparametric bootstrap. Mean effect differences with confidence intervals overlapping the dashed line (0%) show no significant difference between the harvesting, residual management, or pretreatment strategies. The number of response ratios (k) for the intensive treatment in each comparison appear on the left and for the less intensive treatment on the right.

3.5. Recovery of Soil C after Harvest

The recovery time for soil C following harvest differs among soil orders (Figure 5). Only 4 soil orders contained enough observations over time to model recovery times: Alfisols, Inceptisols, Spodosols, and Ultisols. We modeled time as a second degree polynomial (Time + Time^2) separately for O horizons and mineral soils for each soil order (Table 2).

Table 2. Linear regression coefficients and significance for second degree polynomial model of response of soil C to harvesting over time.

Coefficient	Estimate	SE	t-Value	p-Value
Intercept (Alfisol, mineral soil)	12.702	3.587	3.541	0.0004
O horizon	−21.475	3.766	−5.703	<0.0001
Inceptisol	−10.876	5.717	−1.902	0.0577
Spodosol	−14.717	4.320	−3.407	0.0007
Ultisol	−24.776	5.391	−4.596	<0.0001
Time	−67.834	41.56	−1.632	0.10325
Time^2	120.412	40.361	2.983	0.0030
Residual SE: 40.24 on 533 *DF*				
F-Statistic: 10.74 on 6 and 533 df, $p < 0.0001$			$R^2 = 0.108$	Adj. $R^2 = 0.098$

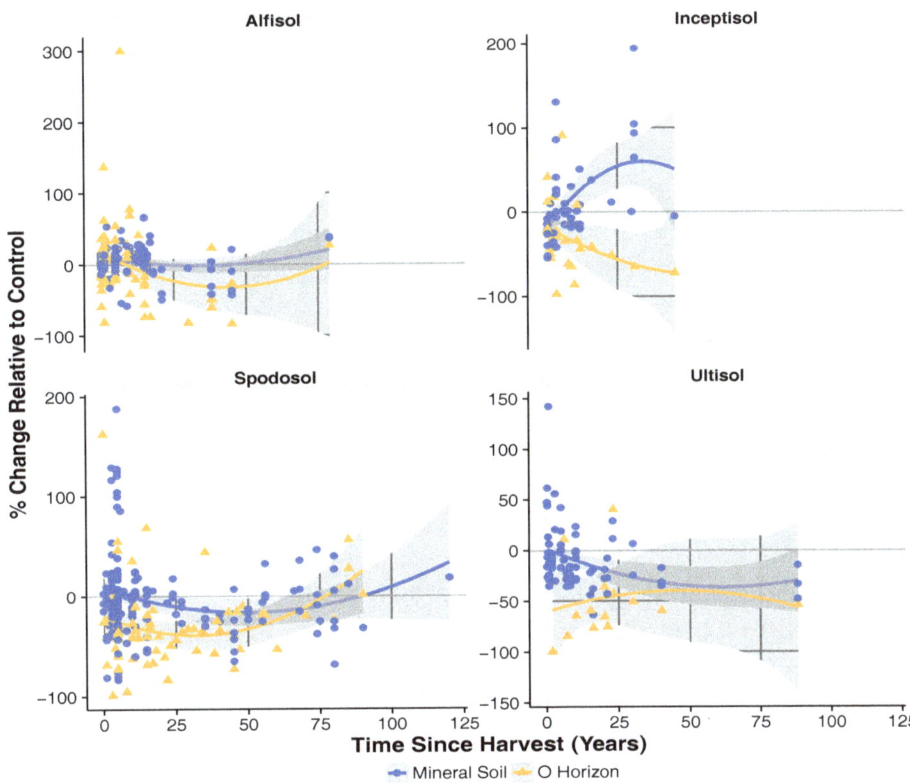

Figure 5. Temporal patterns in both O horizon (yellow triangle) and mineral soil (blue circle) C pools for Alfisol, Inceptisol, Spodosol, and Ultisol orders. Other soil orders are not shown due to an inadequate number of response ratios over time. Regression lines show trends with time using a second order polynomial. For the overall model, $F = 9.205$ on 7 and 532 degrees of freedom, Adj. $R^2 = 0.096$, and $p < 0.0001$ (Table 2).

4. Discussion

4.1. Overall Effect of Harvesting on Soil C

Our results reveal that across many publications in the literature there is a significant loss of soil C in response to harvest (−11.2% overall, −14.4% for studies reporting C pools). This estimate is slightly greater than that found by Nave et al. [9], who reported −8% change relative to control. The difference between these estimates derives from additional losses reported in mineral soil, since the effect of harvesting on O horizon C is identical between this study and Nave et al. [9] (−30%). Indeed, while no significant loss of soil C due to harvesting was reported in previous meta-analyses on the subject [9,25], this analysis reveals significant if small losses in various mineral soil layers. Our meta-analysis has roughly double the number of responses than previous meta-analyses on the subject, and consequently has greater statistical power. In particular, this has allowed us to break down the response of mineral soil C to harvest into more depth increments to better characterize how response is moderated or accentuated by depth.

4.2. Depth Distribution of Soil C Response to Harvest

The response of soil C to harvest differs among depths in the soil. O horizons show the most substantial declines (by percentage), although the O horizon is typically a smaller pool of C than mineral soil horizons. Consequently, smaller absolute declines in O horizon C pools can lead to larger response ratios. Forest type significantly alters the response of O horizons to harvesting, with hardwood forests undergoing less drastic losses than conifer and mixed forests (Figure 3). This result is in contrast with Nave et al. [9], who found that conifer O horizon soil C declines significantly less than hardwood forest floors. Coniferous forest litter is thought to be more chemically recalcitrant to decomposition because of higher C/N and lignin/N, as well as slower N mineralization rates [35,36]. The trend for more soil C loss from coniferous forest floors could be due to differences in the harvesting techniques utilized for each forest type. On the other hand, less change in soil C in coniferous forests in deeper mineral soils could suggest that some of the additional loss in O horizon C pools is the result of translocation of C into mineral soil rather than mineralization to CO_2. Whatever the case, the mechanism for this difference is not clear and warrants additional study.

In mineral soils, the relative response to harvest is typically less than the O horizon, but this small relative loss might correspond to a larger absolute loss of C in the mineral soil in many forests. The major exception to this pattern are Spodosols, which can contain larger proportions of total soil C in deep, acidic O horizons. Declines in top soil C pools were modest (−3.3%) but still significant (Figure 1). Mean effect size estimates become more negative with soil depth, although these estimates are not significant. The overall estimate of change in very deep soil (60–100+ cm) shows substantial and significant loss of C (−17.7%). This estimate, however, only covers a small number of observations (21) from Spodosol, Ultisol, Alfisol, and Inceptisol soil orders and completely excludes hardwood forests. The lack of observations in deeper soil horizons leads to very wide confidence intervals.

On average, the maximum depth of soil sampled by the publications in this meta-analysis was 35.9 cm (Figure 6). The average depth of sampling for each response ratio in the database is even more surface-skewed at 21.3 cm. Many of the observations down to 100 cm in the literature only report treatment differences for the whole mineral soil profile (0–100 cm), which eliminates any possibility of understanding the relative response of different horizons or depths. The scarcity of observations in deep soil is incongruous with the increasing loss of soil C with depth relative to control observed in this analysis. More important than the magnitude or significance of the harvest response in very deep soil is the conclusion that much greater attention should be paid to deep soil C pools in both individual forest manipulation experiments and broad-scale C inventory.

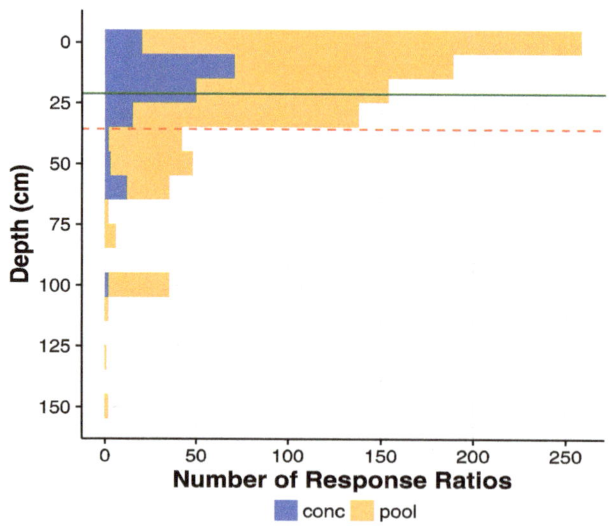

Figure 6. Number of response ratios plotted by the maximum depth of sampling for each observation. Response ratios calculated from concentration are in blue, and from pools in orange. The average depth of all response ratios is denoted by the solid green line (21.3 cm, $n = 945$). The average maximum sampling depth for all 112 publications in the meta-analysis is denoted by the dashed red line (35.9 cm).

While soil C in deep soil is much less concentrated than in O and A horizons, subsurface soil represents a much greater volume of soil than surface soil, especially in older/more well developed soil orders like Alfisols, Ultisols, and Oxisols. Some major regions for forestry contain substantial portions of total soil C in deep horizons. For example, 38% of total soil C was below 50 cm and 24.1% below 1 m in production forest soils in the Pacific Northwest [37]. The imprint of biological activity extends many meters into soil, even into the C horizon [38]. Globally, the average maximum rooting depth for trees is ~7 m [39], far outreaching even the deepest observations in this database. Harvesting disrupts the continued growth and turnover of roots extended deep into soil by mature trees, which in turn disturbs the steady state of C cycling in deep soil by changing environmental conditions (temperature, moisture) as well as the type and rate of C inputs. Furthermore, the flush of nitrate and dissolved organic matter that frequently follows harvest [40,41] could prime the breakdown of older, subsurface C by providing a spike in nutrient availability and labile energy sources [31,42,43]. Alteration of aboveground ecosystems can cause changes in subsurface soils. For example, Mobley et al. [44] observed that, over a period of several decades following afforestation of agricultural land, modest C gains in surface soil were more than offset by large losses in soil C below 30 cm. Neither the response of deep soil C to harvest nor the mechanisms for that change have been sufficiently resolved in the literature, and future work to address these questions are necessary.

4.3. Differences in Soil C Response to Harvest among Soil Orders

Substantial variation in response to harvest was observed among soil orders. Several soil orders had very few response ratios (Andisols, Entisols, Mollisols, and Oxisols), which greatly widens confidence intervals. Nonetheless, significant changes in soil C in response to harvest were observed for all four of these orders. Andisols were the only order to show a significant average increase in soil C in response to harvest. This likely stems from Andisols particular mineralogy, which is often characterized by short-range-order minerals like allophane and imogolite [45–47]. The capacity for these types of minerals to adsorb organic matter makes Andisol soil C especially resistant to loss after harvesting. Alfisols also appear to be resistant to loss of soil C after harvesting, with relatively small loss in O horizons (−12.0%) being the only significant effect. All other soil orders have significant overall losses in soil C, roughly −20% for Entisols, Inceptisols, Mollisols, Spodosols, and Ultisols. The uneven distribution of observations among soil orders (most response ratios in the database are from Alfisols, Inceptisols, Spodosols, and Ultisols) results in substantial differences in the size of confidence intervals among different orders. Unfortunately, many studies did not report soil taxonomic information, and thus 115 response ratios could not be assigned to a soil order. The lack of studies on Andisols is curious given the importance of these soils to forestry in several regions such as the Pacific Northwest, USA and New Zealand. Several studies on Andisol and other under-represented soil orders were excluded from this analysis because of a lack of appropriate controls.

4.4. Recovery of Soil C after Harvest

Recovery of soil C after harvesting can take several decades [9]. O horizon pools decline more severely than mineral soil pools, especially in the first several decades (Figure 5). In Spodosols, O horizons recovered from harvesting after 60–85 years, while mineral soil recovered over a longer period of 75–100+ years. While the response to harvest was less severe in mineral soils, the longer recovery period implies either lagged response time between forest floor and mineral soils or differences in the decay rate constants leading to longer-term changes in mineral soil C compared to the forest floor. In the case of Alfisols and Inceptisols, soil C in mineral pools increased or stayed the same after harvest, while O horizons declined. However, the observations of harvest effects on Alfisols, Inceptisols, and Ultisols were largely confined to within the first 50 years post-harvest. Consequently, an estimate for the recovery period of soil C pools in these soil orders cannot be assessed with much confidence. Continued observation of existing harvesting experiments in other soil orders must be made to better characterize changes in soil C over time. For Andisols, Entisols, Mollisols, and Oxisols, only a few time points have been documented, and much further study will be necessary to understand recovery of soil C after harvest.

The modeled recovery time has a fairly low adjusted R^2 (0.1) and thus a low predictive capacity. Substantial variation in the response to harvest exists within each soil order, reflecting differences in tree species, harvest intensity, and pretreatment strategies, among other factors. Moreover, soil orders are hardly homogeneous, and differences in the response of soil C among lower levels of classification within each order could be as important as order-level differences. Nonetheless, the substantial and significant differences between orders considered in the model suggest that both the resistance of soil C to change and the recovery period of soil C following harvest (resilience) consistently varies among soil types. Compared with 20-year recovery periods assumed by many models [14], our results indicate that soil C recovery takes place over at least triple that time frame for both O horizons and mineral soil in many cases.

While forests >30 years of age were considered acceptable controls for this analysis, the preponderance of data in this meta-analysis show decreases in soil C relative to control at time = 30 years. Consequently, studies that use mature second growth stands barely over this threshold for experimental controls likely underestimate the response of soil C due to harvesting treatments. Depending upon the site conditions and soil order, control stands of at least 50–75+ years since harvest would be recommended, with older stands being more accurate controls.

4.5. The Effect of Harvest Strategies on Soil C

Differences in harvesting and soil pretreatment strategies significantly impact the loss of soil C after harvest. Curiously, despite the greater relative losses of soil C in O horizons, significant differences between harvest intensities and pretreatment strategies were only found in the mineral soil with the exception of broadcast burning (Figure 4). The reduced loss of soil C from mineral soil observed in treatments with greater harvest intensity (+9.3% for clearcut, +13.3% for whole tree harvest) runs counter to the intended effect of these experimental treatments on soil C. One possibility is that increased harvest intensity reduces the quantity of dissolved organic matter and inorganic nutrients leached into the mineral soil, thus reducing the priming [42,48,49] of mineral soil C mineralization through less addition of energy-rich substrates and nutrients. Another possibility is that response of soil C to increased harvest intensity is soil-type specific, and thus an aggregate analysis such as this is subject to bias by unequal sampling of different soil orders. Whatever the case, this dataset cannot identify the specific mechanism(s) driving this difference, and further study is warranted.

Tilling of forest soils prior to planting should intuitively disrupt O horizons to a greater extent than less intensive practices. However, due to the very small number of observations of this practice in the dataset, the large mean treatment effect on soil C was could not be differentiated from 0. By mixing organic material into the surface mineral soil, tilling could increase top soil C in the short term and possibly prime additional breakdown of C over time. In regions where this practice is used, additional research could help to reveal the mechanisms driving change in the soil C of O horizons and mineral following tillage.

Broadcast burning led consistently to additional loss of soil C in both O horizons and mineral soil. The large additional reduction in O horizon C (−40.9%) is expected given that such a treatment is intended to reduce slash on site to facilitate planting. The loss of carbon after harvest extends into deep soil, especially following slash burning (Figure 7). Although there are few observations in very deep soil (60–100+ cm), burning appears to especially exacerbate C losses in this layer. This result is despite the direct effects of fire (such as soil heating and nutrient volatilization) being highly attenuated with depth [50,51]. Levels of mineralized nitrogen (NH_4^+ and NO_3^-) and soluble sugars spike within the first year following fire, leading to increased microbial biomass N and N leaching loss [52]. Thus, the flush of nutrients and organic matter into deeper mineral soil following post-harvest broadcast burning has the potential to impact soil C dynamics throughout the soil profile. The number of observations in deep and very deep layers is small, and consequently additional research is necessary to better differentiate between harvesting and fire effects in deep soil horizons.

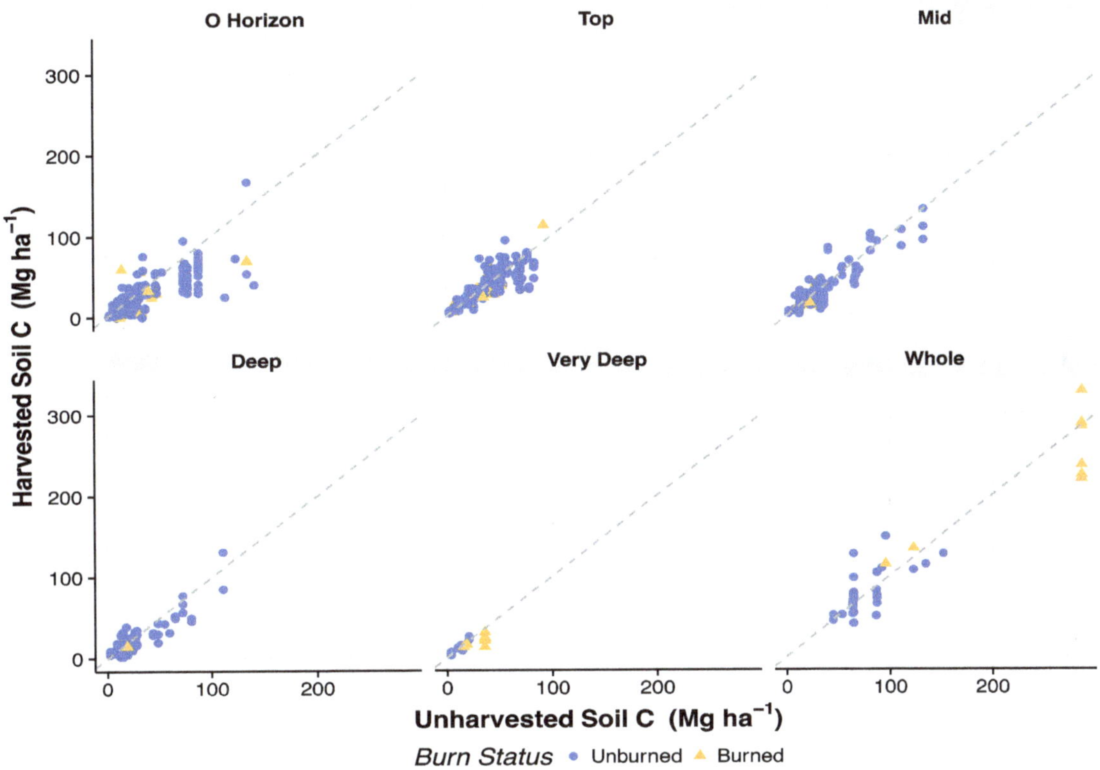

Figure 7. Absolute change in soil C due to harvest for each soil depth in this analysis (O horizon, top, mid, deep, very deep, and whole mineral soil). Different points show burned (yellow triangle) and unburned (blue circle) pretreatment strategies. Dashed 1:1 lines in each facet represent no response due to harvest. The total number of responses shown is $k = 746$.

5. Conclusions

We analyzed 945 studies from 112 publications to examine the effect of harvest on forest soil C around the globe. There is a significant overall reduction in forest soil C following harvest that occurs in both the O horizon and mineral soil. Significant variation in the response to harvesting was observed among different soil depths, among soil orders, between overstory forest types, and between different harvest intensities and pretreatment strategies. Broadcast burning, in particular, appears to exacerbate loss of soil C in both organic and mineral horizons following harvest. The recovery period of soil C following harvest depends upon soil type and takes at least 60 years in many production forests. One of the most important findings of this analysis is a significant loss (−17.7%) of soil C following harvest in very deep soil (60–100+ cm). Deep layers of the soil are greatly under-represented in the literature, and consequently, there is great uncertainty around this estimate. Examination of deep soil horizons in existing manipulative forest studies, in new studies, and in C inventory should be a clear objective for future research.

Acknowledgments: A huge thank you to Bryan James for assistance extracting data from publications and assembling the database. Thanks also to Lucas Nave and an anonymous reviewer for constructive recommendations on the manuscript. This manuscript was supported by funding from the University of Washington Stand Management Cooperative as well as a US Department of Agriculture McIntire-Stennis Grant.

Author Contributions: J.J. and R.H. conceived and designed the research questions for the meta-analysis; J.J. conducted the literature search; J.J. defined the inclusion criteria for the meta-analysis and extracted data from the publications; J.J. analyzed the data; and J.J. wrote the paper in consultation with R.H.

Appendix A. Publications Providing Response Ratios for This Analysis

Reference	Year	k	Max Depth (cm)	Time Since Harvest [a] (years)	Location
Alban and Perala [53]	1992	7	50	35	MN, USA
Bauhus et al. [54]	2004	6	40	9	Germany
Bisbing et al. [55]	2010	6	100	40	MT, USA
Black and Harden [56]	1995	15	20	23	CA, USA
Boerner et al. [57]	2006	4	10	2	SC, USA
Borchers and Perry [58]	1992	4	15	14	OR, USA
Bravo-Oviedo et al. [59]	2015	8	30	15	Spain
Cade-Menun et al. [60]	2000	12	26	5	BC, Canada
Carter et al. [61]	2002	8	15	2	LA, TX, USA
Chatterjee et al. [62]	2009	19	54	21	WY, USA
Chen et al. [63]	2013	24	100	29	China
Chiti et al. [64]	2016	24	100	24	Ghana, Cameroon, Gabon
Christophel et al. [65]	2013	6	30	15	Germany
Christophel et al. [66]	2015	18	30	33	Germany
Cromack et al. [67]	1999	1	100	10	OR, USA
Dai et al. [68]	2001	3	70	14	NH, USA
DeByle et al. [69]	1980	10	5	3	WY, USA
Deluca and Zouhar [52]	2000	6	8	5	MT, USA
Diochon et al. [70]	2009	28	50	35	NS, Canada
Edmonds and McColl [71]	1989	4	20	3	Australia
Edwards and Ross-Todd [72]	1983	6	45	1	TN, USA
Elliott and Knoepp [73]	2005	3	15	3	NC, USA
Ellis et al. [73]	1982	4	10	2	Tasmania
Ellis and Graley [74]	1983	2	10	1	Tasmania
Esquilin et al. [75]	2008	1	10	14	CO, USA
Falsone et al. [76]	2012	3	130	5	Russia
Fraterrigo et al. [77]	2005	1	15	30	NC, USA
Frazer et al. [78]	1990	4	14	12	CA, USA
Gartzia-Bengoetxea et al. [79]	2009	2	5	10	Spain
Gillon et al. [80]	1999	2	0	1	France
Goh and Phillips [81]	1991	4	60	2	New Zealand
Goodale and Aber [82]	2001	2	10	85	NH, USA
Gough et al. [83]	2007	15	80	41	MI, USA
Grady and Hart [84]	2006	2	15	12	AZ, USA
Grand and Lavkulich [85]	2012	6	80		BC, Canada
Gresham [86]	2002	6	30	10	SC, USA
Griffiths and Swanson [87]	2001	3	10	20	OR, USA
Gundale et al. [88]	2005	4	10	3	MT, USA
Gupta and DeLuca [89]	2012	12	50	5	Wales
Hart et al. [90]	2006	2	15	1	AZ, USA
Hendrickson and Chattarpaul [91]	1989	6	20	3	ON, Canada
Herman et al. [92]	2003	2	9	8	CA, USA
Holscher et al. [93]	2001	2	20	22	Germany
Hwang and Son [94]	2006	2	30	2	Korea
Jang and Page-Dumroese [95]	2015	8	30	38	MT, USA
Johnson [96]	1991	3	20	3	NH, USA
Johnson and Todd [97]	1998	6	45	15	TN, USA
Johnson [98]	1995	12		7	NH, USA
Johnson et al. [99]	1997	14	53	6	NH, USA
Johnson et al. [100]	2014	4	60	1	CA, USA
Jones et al. [101]	2011	12	30	15	New Zealand
Kaye and Hart [102]	1998	2	15	1	AZ, USA
Keenan et al. [103]	1994	1	20	4	BC, Canada
Kelliher et al. [104]	2004	4	50	22	OR, USA

Reference	Year	k	Max Depth (cm)	Time Since Harvest [a] (years)	Location
Kishchuk et al. [105]	2014	4	7	6	AB, Canada
Klockow et al. [106]	2013	9	20	1	MN, USA
Klopatek [107]	2002	6	20	30	WA, USA
Knoepp and Swank [108]	1997	4	30	33	NC, USA
Korb et al. [109]	2004	1	10	1	AZ, USA
Kraemer and Hermann [110]	1979	2	10	26	WA, USA
Kurth et al. [111]	2014	72	30	8	MI, MN, USA
Laiho et al. [112]	2003	5	22	5	NC, LA, USA
Latty et al. [113]	2004	2	15	90	NY, USA
Law et al. [114]	2001	3	100	21	OR, USA
Law et al. [115]	2003	9	100	62	OR, USA
Leduc and Rothstein [116]	2007	1	10	5	MI, USA
Maassen and Wirth [117]	2004	2	5		Germany
Mattson and Smith [118]	1993	30	10	11	WV, USA
Mattson and Swank [119]	1989	8	60	5	NC, USA
May and Attiwill [120]	2003	2	10	5	Australia
McLaughlin and Phillips [121]	2006	2	50	17	ME, USA
McKee et al. [122]	2013	8	60	24	AL, USA
McLaughlin [123]	1996	10	50	5	MI, USA
Merino and Edeso [124]	1999	6	15	1	Spain
Moreno-Fernandez et al. [125]	2015	54	50	60	Spain
Mu et al. [126]	2013	18	50	5	China
Murphy et al. [127]	2006	20	60	1	CA, USA
Neher et al. [128]	2003	3	20	2	NC, USA
Norris et al. [129]	2009	15	100	16	SK, Canada
O'Brien et al. [130]	2003	6	50	18	Australia
Powers et al. [131]	2011	20	30	13	MN, WI, USA
Prest et al. [132]	2014	5	50	35	NS, Canada
Prietzel et al. [133]	2004	4	0	1	WA, USA
Puhlick et al. [134]	2016	10	100		ME, USA
Rab [135]	1996	8	10	1	Australia
Riley and Jones [136]	2003	3	10	1	SC, USA
Roaldson et al. [137]	2014	16	20	5	CA, USA
Rothstein and Spaulding [138]	2010	6	30		MI, USA
Sanchez et al. [139]	2007	6	105	2	SC, USA
Sanscrainte et al. [140]	2003	4	70		WA, USA
Saynes et al. [141]	2012	8	5	11	Mexico
Selig et al. [142]	2008	3	30	14	VA, USA
Shelburne et al. [143]	2004	4	10	1	SC, USA
Sheng et al. [144]	2015	5	100	8	China
Skovsgaard et al. [145]	2006	12	30	0	Denmark
Slesak et al. [146]	2012	12	60	5	OR, WA, USA
Small and McCarthy [147]	2005	3	10	7	OH, USA
Stone et al. [148]	1999	1	15	1	AZ, USA
Stone and Elioff [149]	1998	4	30	5	MN, USA
Strong [150]	1997	8	40	18	MN, USA
Strukelj et al. [151]	2015	12	10	5	QC, Canada
Tang et al. [152]	2009	12	60	29	MI, WI, USA
Trettin et al. [153]	2011	6	150	11	MI, USA
Ussiri and Johnson [154]	2007	15	60	8	NH, USA
Vario et al. [155]	2014	6	60	49	NH, USA
Vesterdal et al. [156]	1995	9	0		Denmark
Waldrop et al. [157]	2003	3	0	1	CA, USA
Wu et al. [158]	2010	1	20	10	China
Xiang et al. [159]	2009	8	30	0	China
Yanai et al. [160]	2000	35	0	29	NH, USA
Zabowski et al. [161]	2008	2	20	25	OR, WA, USA
Zhong and Makeshin [162]	2003	2	10	16	Germany
Zummo and Friedland [163]	2011	15	60	3	NH, USA

[a] For chronosequence studies, time since harvest in this table is averaged across all response ratios for that study.

References

1. Dixon, R.K.; Brown, S.; Houghton, R.A.; Solomon, A.M.; Trexler, M.C.; Wisniewski, J. Carbon Pools and Flux of Global Forest Ecosystems. *Science* **1994**, *263*, 185–190. [CrossRef] [PubMed]
2. Prentice, I.C.; Farquhar, G.; Fasham, M.J.R.; Goulden, M.L.; Heimann, M.; Jarmillo, V.J.; Kheshgi, H.S.; Le Quere, C.; Scholes, R.J.; Wallace, D.W.R. *The Carbon Cycle and Atmospheric Carbon Dioxide*; Cambridge University Press: Cambridge, UK, 2001; pp. 183–237.
3. Batjes, N.H. Total carbon and nitrogen in the soils of the world. *Eur. J. Soil Sci.* **1996**, *47*, 151–163. [CrossRef]
4. Jobbagy, E.G.; Jackson, R.B. The vertical distribution of soil organic carbon and its relation to climate and vegetation. *Ecol. Appl.* **2000**, *10*, 423–436. [CrossRef]
5. Six, J.; Conant, R.T.; Paul, E.A.; Paustian, K. Stabilization mechanisms of soil organic matter: Implications for C-saturation of soils. *Plant Soil* **2002**, *241*, 155–176. [CrossRef]
6. Lal, R. Forest soils and carbon sequestration. *For. Ecol. Manag.* **2005**, *220*, 242–258. [CrossRef]
7. Schimel, D.S. Terrestrial ecosystems and the carbon cycle. *Glob. Chang. Biol.* **1995**, *1*, 77–91. [CrossRef]
8. Jandl, R.; Lindner, M.; Vesterdal, L.; Bauwens, B.; Baritz, R.; Hagedorn, F.; Johnson, D.W.; Minkkinen, K.; Byrne, K.A. How strongly can forest management influence soil carbon sequestration? *Geoderma* **2007**, *137*, 253–268. [CrossRef]
9. Nave, L.E.; Vance, E.D.; Swanston, C.W.; Curtis, P.S. Harvest impacts on soil carbon storage in temperate forests. *For. Ecol. Manag.* **2010**, *259*, 857–866. [CrossRef]
10. Schlesinger, W.H.; Bernhardt, E.S. *Biogeochemistry: An Analysis of Global Change*; Academic Press: Oxford, UK, 2013.
11. Brady, N.C.; Weil, R.R. *The Nature and Properties of Soils*; Macmillan Publ. Co.: New York, NY, USA, 2002; Volume 13, p. 960.
12. Krull, E.S.; Skjemstad, J.O.; Baldock, J.A. *Functions of Soil Organic Matter and the Effect on Soil Properties*; Final Rep. CSO00029; Grains Research and Development Corporation: Glen Osmond, South Australia, 2009; p. 128.
13. Six, J.; Paustian, K. Aggregate-associated soil organic matter as an ecosystem property and a measurement tool. *Soil Biol. Biochem.* **2014**, *68*, A4–A9. [CrossRef]
14. Buchholz, T.; Friedland, A.J.; Hornig, C.E.; Keeton, W.S.; Zanchi, G.; Nunery, J. Mineral soil carbon fluxes in forests and implications for carbon balance assessments. *Glob. Chang. Biol. Bioenergy* **2014**, *6*, 305–311. [CrossRef]
15. Ryan, M.G.; Harmon, M.E.; Birdsey, R.A.; Giardina, C.P.; Heath, L.S.; Houghton, R.A.; Jackson, R.B.; McKinley, D.C.; Morrison, J.F.; Murray, B.C.; et al. A synthesis of the science on forests and carbon for U.S. Forests. *Issues Ecol.* **2010**, *13*, 1–16.
16. McKinley, D.C.; Ryan, M.G.; Birdsey, R.A.; Giardina, C.P.; Harmon, M.E.; Heath, L.S.; Houghton, R.A.; Jackson, R.B.; Morrison, J.F.; Murray, B.C.; et al. A synthesis of current knowledge on forests and carbon storage in the United States. *Ecol. Appl.* **2011**, *21*, 1902–1924. [CrossRef] [PubMed]
17. Fahey, T.J.; Woodbury, P.B.; Battles, J.J.; Goodale, C.L.; Hamburg, S.P.; Ollinger, S.V.; Woodall, C.W. Forest carbon storage: Ecology, management, and policy. *Front. Ecol. Environ.* **2010**, *8*, 245–252. [CrossRef]
18. American Carbon Registry. *The ACR Forest Carbon Project Standard v 2.1*; Winrock International: Arlington, VA, USA, 2011.
19. Verified Carbon Standard. *Agriculture, Forestry and other Land Use (AFOLU) Requirements*; VCS Version 3 Requirements Doc., v3.4 2013; Verified Carbon Standard: Washington, DC, USA, 2013.
20. IPCC. 2006 IPCC Guidelines for national greenhouse gas inventories. In *Agriculture, Forestry and Other Land Use*; Institute for Global Environmental Strategies: Hayama, Japan, 2006; Volume 4, pp. 1–29.
21. Bechtold, W.A.; Patterson, P.L. The Enhanced Forest Inventory and Analysis Program—National Sampling Design and Estimation Procedures. In *USDA General Technical Report SRS-80*; U.S. Department of Agriculture, Forest Service, Southern Research Station: Asheville, NC, USA, 2005; p. 85.
22. Zanchi, G.; Pena, N.; Bird, N. Is woody bioenergy carbon neutral? A comparative assessment of emissions from consumption of woody bioenergy and fossil fuel. *GCB Bioenergy* **2012**, *4*, 761–772. [CrossRef]
23. Homann, P.S.; Bormann, B.T.; Boyle, J.R. Detecting treatment differences in soil carbon and nitrogen resulting from forest manipulations. *Soil Sci. Soc. Am. J.* **2001**, *65*, 463–469. [CrossRef]
24. Hedges, L.V.; Gurevitch, J.; Curtis, P.S. The meta-analysis of response ratios in experimental ecology. *Ecology* **1999**, *80*, 1150–1156. [CrossRef]
25. Johnson, D.W.; Curtis, P.S. Effects of forest management on soil C and N storage: Meta analysis. *For. Ecol. Manag.* **2001**, *140*, 227–238. [CrossRef]

26. Harrison, R.B.; Footen, P.W.; Strahm, B.D. Deep soil horizons: Contribution and importance to soil carbon pools and in assessing whole-ecosystem response to management and global change. *For. Sci.* **2011**, *57*, 67–76.
27. Schmidt, M.W.I.; Torn, M.S.; Abiven, S.; Dittmar, T.; Guggenberger, G.; Janssens, I.A.; Kleber, M.; Koegel-Knabner, I.; Lehmann, J.; Manning, D.A.C.; et al. Persistence of soil organic matter as an ecosystem property. *Nature* **2011**, *478*, 49–56. [CrossRef] [PubMed]
28. Deb Richter, D.; Yaalon, D.H. "The Changing Model of Soil" Revisited. *Soil Sci. Soc. Am. J.* **2012**, *76*, 76–778. [CrossRef]
29. Marin-Spiotta, E.; Gruley, K.E.; Crawford, J.; Atkinson, E.E.; Miesel, J.R.; Greene, S.; Cardona-Correa, C.; Spencer, R.G.M. Paradigm shifts in soil organic matter research affect interpretations of aquatic carbon cycling: Transcending disciplinary and ecosystem boundaries. *Biogeochemistry* **2014**, *117*, 279–297. [CrossRef]
30. Lehmann, J.; Kleber, M. The contentious nature of soil organic matter. *Nature* **2015**, *528*, 60–68. [CrossRef] [PubMed]
31. Fontaine, S.; Barot, S.; Barre, P.; Bdioui, N.; Mary, B.; Rumpel, C. Stability of organic carbon in deep soil layers controlled by fresh carbon supply. *Nature* **2007**, *450*, 277–280. [CrossRef] [PubMed]
32. Adams, D.C.; Gurevitch, J.; Rosenberg, M.S. Resampling tests for meta-analysis of ecological data. *Ecology* **1997**, *78*, 1277–1283. [CrossRef]
33. Gerlanc, D.; Kirby, K. *bootES: Bootstrap Effect Sizes*; R Package, Version 1.2; R Foundation for Statistical Computing: Vienna, Austria, 2015.
34. R Development Core Team. *A Language and Environment for Statistical Computing*; R Foundation for Statistical Computing: Vienna, Austria, 2016.
35. Currie, W.S. Responsive C and N biogeochemistryof the temperate forest floor. *Trends Ecol. Evol.* **1999**, *14*, 316–320. [CrossRef]
36. Silver, W.; Miya, R. Global patterns in root decomposition: Comparisons of climate and litter quality effects. *Oecologia* **2001**, *129*, 407–419. [CrossRef]
37. James, J.; Devine, W.; Harrison, R.; Terry, T. Deep soil carbon: quantification and modeling in subsurface layers. *Soil Sci. Soc. Am. J.* **2014**, *78*, S1–S9. [CrossRef]
38. Richter, D.D.; Markewitz, D. How Deep Is Soil? *Bioscience* **1995**, *45*, 600–609. [CrossRef]
39. Canadell, J.; Jackson, R.B.; Ehleringer, J.R.; Mooney, H.A.; Sala, O.E.; Schulze, E.D. Maximum rooting depth of vegetation types at the global scale. *Oecologia* **1996**, *108*, 583–595. [CrossRef]
40. Strahm, B.D.; Harrison, R.B.; Terry, T.A.; Flaming, B.L.; Licata, C.W.; Petersen, K.S. Soil solution nitrogen concentrations and leaching rates as influenced by organic matter retention on a highly productive Douglas-fir site. *For. Ecol. Manag.* **2005**, *218*, 74–88. [CrossRef]
41. Likens, G.E.; Bormann, F.H.; Johnson, N.M. Nitrification—Importance to nutrient losses from a cutover forested ecosystem. *Science* **1969**, *163*, 1205–1206. [CrossRef] [PubMed]
42. Kuzyakov, Y. Priming effects: Interactions between living and dead organic matter. *Soil Biol. Biochem.* **2010**, *42*, 1363–1371. [CrossRef]
43. Blagodatskaya, E.; Kuzyakov, Y. Mechanisms of real and apparent priming effects and their dependence on soil microbial biomass and community structure: Critical review. *Biol. Fertil. Soils* **2008**, *45*, 115–131. [CrossRef]
44. Mobley, M.L.; Lajtha, K.; Kramer, M.G.; Bacon, A.R.; Heine, P.R.; Richter, D.D. Surficial gains and subsoil losses of soil carbon and nitrogen during secondary forest development. *Glob. Chang. Biol.* **2015**, *21*, 986–996. [CrossRef] [PubMed]
45. Soil Survey Staff. *Keys to Soil Taxonomy*; Department of Agriculture, Natural Resources Conservation Service: Washington, DC, USA, 2014.
46. Wada, K. Minerals formed and mineral formation from volcanic ash by weathering. *Chem. Geol.* **1987**, *60*, 17–28. [CrossRef]
47. Nanzyo, M.; Shoji, S.; Dahlgren, R. *Volcanic Ash Soils—Genesis, Properties and Utilization*; Elsevier: Amsterdam, The Netherlands, 1993; Volume 21.
48. Blagodatskaya, E.V.; Blagodatsky, S.A.; Anderson, T.H.; Kuzyakov, Y. Priming effects in Chernozem induced by glucose and N in relation to microbial growth strategies. *Appl. Soil Ecol.* **2007**, *37*, 95–105. [CrossRef]
49. Kuzyakov, Y.; Friedel, J.K.; Stahr, K. Review of mechanisms and quantification of priming effects. *Soil Biol. Biochem.* **2000**, *32*, 1485–1498. [CrossRef]
50. Valette, J.; Gomendy, V.; Marechal, J.; Houssard, C.; Gillon, D. Heat Transfer in the Soil During Very Low-intensity Experimental Fires: The Role of Duff and Soil Moisture Content. *Int. J. Wildland Fire* **1994**, *4*, 225–237. [CrossRef]

51. Certini, G. Effects of fire on properties of forest soils: A review. *Oecologia* **2005**, *143*, 1–10. [CrossRef] [PubMed]
52. DeLuca, T.H.; Zouhar, K.L. Effects of selection harvest and prescribed fire on the soil nitrogen status of ponderosa pine forests. *For. Ecol. Manag.* **2000**, *138*, 263–271. [CrossRef]
53. Alban, D.H.; Perala, D.A. Carbon storage in lake states aspen ecosystems. *Can. J. For. Res.* **1992**, *22*, 1107–1110. [CrossRef]
54. Bauhus, J.; Vor, T.; Bartsch, N.; Cowling, A. The effects of gaps and liming on forest floor decomposition and soil C and N dynamics in a Fagus sylvatica forest. *Can. J. For. Res.* **2004**, *34*, 509–518. [CrossRef]
55. Bisbing, S.M.; Alaback, P.B.; DeLuca, T.H. Carbon storage in old-growth and second growth fire-dependent western larch (*Larix occidentalis* Nutt.) forests of the Inland Northwest, USA. *For. Ecol. Manag.* **2010**, *259*, 1041–1049. [CrossRef]
56. Black, T.A.; Harden, J.W. Effect of timber harvest on soil carbon storage at Blodgett Experimental Forest, California. *Can. J. For. Res.* **1995**, *25*, 1385–1396. [CrossRef]
57. Boerner, R.E.J.; Waldrop, T.A.; Shelburne, V.B. Wildfire mitigation strategies affect soil enzyme activity and soil organic carbon in loblolly pine (*Pinus taeda*) forests. *Can. J. For. Res.* **2006**, *36*, 3148–3154. [CrossRef]
58. Borchers, J.G.; Perry, D.A. The influence of soil texture and aggregation on carbon and nitrogen dynamics in southwest Oregon forest and clearcuts. *Can. J. For. Res.* **1992**, *22*, 298–305. [CrossRef]
59. Bravo-Oviedo, A.; Ruiz-Peinado, R.; Modrego, P.; Alonso, R.; Montero, G. Forest thinning impact on carbon stock and soil condition in Southern European populations of *P. sylvestris* L. *For. Ecol. Manag.* **2015**, *357*, 259–267. [CrossRef]
60. Cade-Menun, B.J.; Berch, S.M.; Preston, C.M.; Lavkulich, L.M. Phosphorus forms and related soil chemistry of Podzolic soils on northern Vancouver Island. II. The effects of clear-cutting and burning. *Can. J. For. Res.* **2000**, *30*, 1726–1741. [CrossRef]
61. Carter, M.C.; Dean, T.J.; Zhou, M.; Messina, M.G.; Wang, Z. Short-term changes in soil C, N, and biota following harvesting and regeneration of loblolly pine (*Pinus taeda* L.). *For. Ecol. Manag.* **2002**, *164*, 67–88. [CrossRef]
62. Chatterjee, A.; Vance, G.F.; Tinker, D.B. Carbon pools of managed and unmanaged stands of ponderosa and lodgepole pine forests in Wyoming. *Can. J. For. Res.* **2009**, *39*, 1893–1900. [CrossRef]
63. Chen, L.-C.; Liang, M.-J.; Wang, S.-L. Carbon stock density in planted versus natural *Pinus massoniana* forests in sub-tropical China. *Ann. For. Sci.* **2016**, *73*, 461–472. [CrossRef]
64. Chiti, T.; Perugini, L.; Vespertino, D.; Valentini, R. Effect of selective logging on soil organic carbon dynamics in tropical forests in central and western Africa. *Plant Soil* **2015**, *399*, 283–294. [CrossRef]
65. Christophel, D.; Spengler, S.; Schmidt, B.; Ewald, J.; Prietzel, J. Customary selective harvesting has considerably decreased organic carbon and nitrogen stocks in forest soils of the Bavarian Limestone Alps. *For. Ecol. Manag.* **2013**, *305*, 167–176. [CrossRef]
66. Christophel, D.; Höllerl, S.; Prietzel, J.; Steffens, M. Long-term development of soil organic carbon and nitrogen stocks after shelterwood- and clear-cutting in a mountain forest in the Bavarian Limestone Alps. *Eur. J. For. Res.* **2015**, *134*, 623–640. [CrossRef]
67. Cromack, K.; Miller, R.E.; Anderson, H.W.; Helgerson, O.T.; Smith, R.B. Soil Carbon and Nutrients in a Coastal Oregon Douglas-Fir Plantation with Red Alder. *Soil Sci. Soc. Am. J.* **1999**, *63*, 232–239. [CrossRef]
68. Dai, K.H.; Johnson, C.E.; Driscoll, C.T. Organic matter chemistry and dynamics in clear-cut and unmanaged hardwood forest ecosystems. *Biogeochemistry* **2001**, *54*, 51–83. [CrossRef]
69. DeByle, N.V. Harvesting and Site Treatment Influences on the Nutrient Status of Lodgepole Pine Forests in Western Wyoming. In *General Technical Report INT-90*; USDA Forest Service Intermountain Forest And Range Experimental Station: Ogden, UT, USA, 1979; pp. 137–156.
70. Diochon, A.; Kellman, L.; Beltrami, H. Looking deeper: An investigation of soil carbon losses following harvesting from a managed northeastern red spruce (*Picea rubens* Sarg.) forest chronosequence. *For. Ecol. Manag.* **2009**, *257*, 413–420. [CrossRef]
71. Edmonds, R.L.; McColl, G.J. Effects of forest management on soil nitrogen in Pinus radiata stands in the Australian Capital Territory. *For. Ecol. Manag.* **1989**, *29*, 199–212. [CrossRef]
72. Edwards, N.T.; Ross-Todd, B.M. Soil carbon dynamics in a mixed deciduous forest following clear-cutting with and without residue removal. *Soil Sci. Soc. Am. J.* **1983**, *47*, 1014–1021. [CrossRef]
73. Elliott, K.J.; Knoepp, J.D. The effects of three regeneration harvest methods on plant diversity and soil characteristics in the southern Appalachians. *For. Ecol. Manag.* **2005**, *211*, 296–317. [CrossRef]

74. Ellis, R.; Graley, A.M. Gains and losses in soil nutrients associated with harvesting and burning eucalypt rainforest. *Plant Soil* **1983**, *74*, 437–450. [CrossRef]
75. Esquilín, J.; Aida, E.; Stromberger, M.E.; Shepperd, W.D. Soil scarification and wildfire interactions and effects on microbial communities and carbon. *Soil Sci. Soc. Am. J.* **2008**, *72*, 111–118. [CrossRef]
76. Falsone, G.; Celi, L.; Caimi, A.; Simonov, G.; Bonifacio, E. The effect of clear cutting on podzolisation and soil carbon dynamics in boreal forests (Middle Taiga zone, Russia). *Geoderma* **2012**, *177–178*, 27–38. [CrossRef]
77. Fraterrigo, J.; Turner, M.; Pearson, S.M.; Dixon, P. Effects of past land use on spatial heterogeneity of soil nutrients in southern Appalachian forests. *Ecol. Monogr.* **2005**, *75*, 215–230. [CrossRef]
78. Frazer, D.W.; McColl, J.G.; Powers, R.F. Soil Nitrogen Mineralization in a Clearcutting Chronosequence in a Northern California Conifer Forest. *Soil Sci. Soc. Am. J.* **1990**, *54*, 1145. [CrossRef]
79. Gartzia-Bengoetxea, N.; González-Arias, A.; Merino, A.; De Arano, I.M. Soil organic matter in soil physical fractions in adjacent semi-natural and cultivated stands in temperate Atlantic forests. *Soil Biol. Biochem.* **2009**, *41*, 1674–1683. [CrossRef]
80. Gillon, D.; Houssard, C.; Valette, J.C.; Rigolot, E. Nitrogen and phosphorus cycling following prescribed burning in natural and managed Aleppo pine forests. *Can. J. For. Res.* **1999**, *29*, 1237–1247. [CrossRef]
81. Goh, K.M.; Phillips, M.J. Effects of clearfell logging and clearfell logging and burning of a Nothofagus forest on soil nutrient dynamics in South Island, New Zealand—Changes in forest floor organic matter and nutrient status. *N. Z. J. Bot.* **1991**, *29*, 367–384. [CrossRef]
82. Goodale, C.L.; Aber, J.D. The Long-term effects of land-use history on nitrogen cycling in northern hardwood forests. *Ecol. Appl.* **2001**, *11*, 253–267. [CrossRef]
83. Gough, C.M.; Vogel, C.S.; Harrold, K.H.; George, K.; Curtis, P.S. The legacy of harvest and fire on ecosystem carbon storage in a north temperate forest. *Glob. Chang. Biol.* **2007**, *13*, 1935–1949. [CrossRef]
84. Grady, K.C.; Hart, S.C. Influences of thinning, prescribed burning, and wildfire on soil processes and properties in southwestern ponderosa pine forests: A retrospective study. *For. Ecol. Manag.* **2006**, *234*, 123–135. [CrossRef]
85. Grand, S.; Lavkulich, L.M. Effects of forest harvest on soil carbon and related variables in Canadian spodosols. *Soil Sci. Soc. Am. J.* **2012**, *76*, 1816–1827. [CrossRef]
86. Gresham, C.A. Sustainability of intensive loblolly pine plantation management in the South Carolina Coastal Plain, USA. *For. Ecol. Manag.* **2002**, *155*, 69–80. [CrossRef]
87. Griffiths, R.P.; Swanson, A.K. Forest soil characteristics in a chronosequence of harvested Douglas-fir forests. *Can. J. For. Res.* **2001**, *31*, 1871–1879. [CrossRef]
88. Gundale, M.J.; DeLuca, T.H.; Fiedler, C.E.; Ramsey, P.W.; Harrington, M.G.; Gannon, J.E. Restoration treatments in a *Montana ponderosa* pine forest: Effects on soil physical, chemical and biological properties. *For. Ecol. Manag.* **2005**, *213*, 25–38. [CrossRef]
89. Das Gupta, S.; DeLuca, T.H. Short-term changes in belowground C, N stocks in recently clear felled Sitka spruce plantations on podzolic soils of North Wales. *For. Ecol. Manag.* **2012**, *281*, 48–58. [CrossRef]
90. Hart, S.C.; Selmants, P.C.; Boyle, S.I.; Overby, S.T. Carbon and nitrogen cycling in southwestern ponderosa pine forests. *For. Sci.* **2006**, *52*, 683–693.
91. Hendrickson, O.; Chatarpaul, L. Nutrient cycling following whole-tree and conventional harvest in a northern mixed forest. *Can. J. For. Res.* **2016**, *19*, 725–735. [CrossRef]
92. Herman, D.J.; Halverson, L.J.; Firestone, M.K. Nitrogen dynamics in an annual grassland: Oak canopy, climate, and microbial population effects. *Ecol. Appl.* **2003**, *13*, 593–604. [CrossRef]
93. Hölscher, D.; Schade, E.; Leuschner, C. Effects of coppicing in temperate deciduous forests on ecosystem nutrient pools and soil fertility. *Basic Appl. Ecol.* **2001**, *2*, 155–164. [CrossRef]
94. Hwang, J.; Son, Y. Short-term effects of thinning and liming on forest soils of pitch pine and Japanese larch plantations in central Korea. *Ecol. Res.* **2006**, *21*, 671–680. [CrossRef]
95. Jang, W.; Page-Dumroese, D.S.; Keyes, C.R. Long-term soil changes from forest harvesting and residue management in the Northern Rocky Mountains. *Soil Sci. Soc. Am. J.* **2016**. [CrossRef]
96. Johnson, C.E. Whole-tree clear-cutting effects on soil horizons and organic matter pools. *Soil Sci. Soc. Am. J.* **1991**, *55*, 497–502. [CrossRef]
97. Johnson, D.; Todd, D.E. Harvesting effects on long-term changes in nutrient pools of mixed oak forest. *Soil Sci. Soc. Am. J.* **1998**, *62*, 1725–1735. [CrossRef]
98. Johnson, C.E. Soil nitrogen status 8 years after whole-tree clear-cutting. *Can. J. For. Res.* **1995**, 1346–1355. [CrossRef]

99. Johnson, C.E.; Romanowicz, R.B.; Siccama, T.G. Conservation of exchangeable cations after clear-cutting of a northern hardwood forest. *Can. J. For. Res.* **1997**, *27*, 859–868. [CrossRef]
100. Johnson, D.W.; Walker, R.F.; Glass, D.W.; Stein, C.M.; Murphy, J.B.; Blank, R.R.; Miller, W.W. Effects of thinning, residue mastication, and prescribed fire on soil and nutrient budgets in a Sierra Nevada mixed-conifer forest. *For. Sci.* **2014**, *60*, 170–179. [CrossRef]
101. Jones, H.S.; Beets, P.N.; Kimberley, M.O.; Garrett, L.G. Harvest residue management and fertilisation effects on soil carbon and nitrogen in a 15-year-old Pinus radiata plantation forest. *For. Ecol. Manag.* **2011**, *262*, 339–347. [CrossRef]
102. Kaye, J.P.; Hart, S.C. Restoration and canopy-type effects on soil respiration in a ponderosa pine-bunchgrass ecosystem. *Soil Sci. Soc. Am. J.* **1998**, *62*, 1062–1072. [CrossRef]
103. Keenan, R.J.; Messier, C.; Kimmins, J.P. (Hamish) Effects of clearcutting and soil mixing on soil properties and understorey biomass in western red cedar and western hemlock forests on northern Vancouver Island, Canada. *For. Ecol. Manag.* **1994**, *68*, 251–261. [CrossRef]
104. Kelliher, F.M.; Ross, D.J.; Law, B.E.; Baldocchi, D.D.; Rodda, N.J. Limitations to carbon mineralization in litter and mineral soil of young and old ponderosa pine forests. *For. Ecol. Manag.* **2004**, *191*, 201–213. [CrossRef]
105. Kishchuk, B.E.; Thiffault, E.; Lorente, M.; Quideau, S.; Keddy, T.; Sidders, D. Decadal soil and stand response to fire, harvest, and salvage-logging disturbances in the western boreal mixedwood forest of Alberta, Canada. *Can. J. For. Res.* **2015**, *45*, 141–152. [CrossRef]
106. Klockow, P.A.; D'Amato, A.W.; Bradford, J.B. Impacts of post-harvest slash and live-tree retention on biomass and nutrient stocks in *Populus tremuloides* Michx.-dominated forests, northern Minnesota, USA. *For. Ecol. Manag.* **2013**, *291*, 278–288. [CrossRef]
107. Klopatek, J.M. Belowground carbon pools and processes in different age stands of Douglas-fir. *Tree Physiol.* **2002**, *22*, 197–204. [CrossRef] [PubMed]
108. Knoepp, J.D.; Swank, W.T. Forest management effects on surface soil carbon and nitrogen. *Soil Sci. Soc. Am. J.* **1997**, *61*, 928–935. [CrossRef]
109. Korb, J.E.; Johnson, N.C.; Covington, W.W. Slash pile burning effects on soil biotic and chemical properties and plant establishment: Recommendations for amelioration. *Restor. Ecol.* **2004**, *12*, 52–62. [CrossRef]
110. Kraemer, J.F.; Hermann, R.K. Broadcast burning: 25-year effects on forest soils in the western flanks of the Cascade Mountains. *For. Sci.* **1979**, *25*, 427–439.
111. Kurth, V.J.; D'Amato, A.W.; Palik, B.J.; Bradford, J.B. Fifteen-year patterns of soil carbon and nitrogen following biomass harvesting. *Soil Sci. Soc. Am. J.* **2014**, *78*, 624–633. [CrossRef]
112. Laiho, R.; Sanchez, F.; Tiarks, A.; Dougherty, P.M.; Trettin, C.C. Impacts of intensive forestry on early rotation trends in site carbon pools in the southeastern US. *For. Ecol. Manag.* **2003**, *174*, 177–189. [CrossRef]
113. Latty, E.F.; Canham, C.D.; Marks, P.L. The effects of land-use history on soil properties and nutrient dynamics in northern hardwood forests of the Adirondack Mountains. *Ecosystems* **2004**, *7*, 193–207. [CrossRef]
114. Law, B.E.; Thornton, P.E.; Irvine, J.; Anthoni, P.M.; Van Tuyl, S. Carbon storage and fluxes in *Ponderosa pine* forests at different developmental stages. *Glob. Chang. Biol.* **2001**, *7*, 755–777. [CrossRef]
115. Law, B.E.; Sun, O.J.; Campbell, J.; Van Tuyl, S.; Thornton, P.E. Changes in carbon storage and fluxes in a chronoseuence of *Ponderosa pine*. *Glob. Chang. Biol.* **2003**, *4*, 510–524. [CrossRef]
116. LeDuc, S.D.; Rothstein, D.E. Initial recovery of soil carbon and nitrogen pools and dynamics following disturbance in jack pine forests: A comparison of wildfire and clearcut harvesting. *Soil Biol. Biochem.* **2007**, *39*, 2865–2876. [CrossRef]
117. Maassen, S.; Wirth, S. Soil microbiological monitoring of a pine forest after partial thinning for stand regeneration with beech seedlings. *Soil Sci. Plant Nutr.* **2004**, *50*, 815–819. [CrossRef]
118. Mattson, K.G.; Smith, H.C. Detrital organic matter and soil CO_2 efflux in forests regenerating from cutting in West Virginia. *Soil Biol. Biochem.* **1993**, *25*, 1241–1248. [CrossRef]
119. Mattson, K.G.; Swank, W.T. Soil and detrital carbon dynamics following forest cutting in the Southern Appalachians. *Biol. Fertil. Soils* **1989**, *7*, 247–253. [CrossRef]
120. May, B.M.; Attiwill, P.M. Nitrogen-fixation by Acacia dealbata and changes in soil properties 5 years after mechanical disturbance or slash-burning following timber harvest. *For. Ecol. Manag.* **2003**, *181*, 339–355. [CrossRef]

121. McLaughlin, J.W.; Phillips, S.A. Soil carbon, nitrogen, and base cation cycling 17 years after whole-tree harvesting in a low-elevation red spruce (*Picea rubens*)-balsam fir (*Abies balsamea*) forested watershed in central Maine, USA. *For. Ecol. Manag.* **2006**, *222*, 234–253. [CrossRef]
122. McKee, S.E.; Seiler, J.R.; Aust, W.M.; Strahm, B.D.; Schilling, E.B.; Brooks, S. Carbon pools and fluxes in a tupelo (*Nyssa aquatica*)-baldcypress (*Taxodium distichum*) swamp24-years after harvest disturbances. *Biomass Bioenergy* **2013**, *55*, 130–140. [CrossRef]
123. McLaughlin, J.W.; Liu, G.; Jurgensen, M.F.; Gale, M.R. Organic carbon characteristics in a spruce swamp five years after harvesting. *Soil Sci. Soc. Am. J.* **1996**, *60*, 1228–1236. [CrossRef]
124. Merino, A.; Edeso, J.M. Soil fertility rehabilitation in young *Pinus radiata* D. Don. plantations from northern Spain after intensive site preparation. *For. Ecol. Manag.* **1999**, *116*, 83–91. [CrossRef]
125. Moreno-Fernandez, D.; Diaz-Pines, E.; Barbeito, I.; Sanchez-Gonzalez, M.; Montes, F.; Rubio, A.; Canellas, I. Temporal carbon dynamics over the rotation period of two alternative management systems in Mediterranean mountain Scots pine forests. *For. Ecol. Manag.* **2015**, *348*, 186–195. [CrossRef]
126. Mu, C.C.; Lu, H.C.; Bao, X.; Wang, B.; Cui, W. Effects of selective cutting on vegetation carbon storage of boreal Larix gmelinii-Carex schmidtii forested wetlands in Daxing'anling, China. *Shengtai Xuebao/Acta Ecol. Sin.* **2013**, *33*, 5286–5298.
127. Murphy, J.D.; Johnson, D.W.; Miller, W.W.; Walker, R.F.; Blank, R.R. Prescribed fire effects on forest floor and soil nutrients in a Sierra Nevada Forest. *Soil Sci.* **2006**, *171*, 181–199. [CrossRef]
128. Neher, D.A.; Barbercheck, M.E.; El-Allaf, S.M.; Anas, O. Effects of disturbance and ecosystem on decomposition. *Appl. Soil Ecol.* **2003**, *23*, 165–179. [CrossRef]
129. Norris, C.E.; Quideau, S.A.; Bhatti, J.S.; Wasylishen, R.E.; MacKenzie, M.D. Influence of fire and harvest on soil organic carbon in jack pine sites. *Can. J. For. Res.* **2009**, *39*, 642–654. [CrossRef]
130. O'Brien, N.D.; Attiwill, P.M.; Weston, C.J. Stability of soil organic matter in *Eucalyptus regnans* forests and *Pinus radiata* plantations in south eastern Australia. *For. Ecol. Manag.* **2003**, *185*, 249–261. [CrossRef]
131. Powers, M.; Kolka, R.; Palik, B.; McDonald, R.; Jurgensen, M. Long-term management impacts on carbon storage in Lake States forests. *For. Ecol. Manag.* **2011**, *262*, 424–431. [CrossRef]
132. Prest, D.; Kellman, L.; Lavigne, M.B. Mineral soil carbon and nitrogen still low three decades following clearcut harvesting in a typical Acadian Forest stand. *Geoderma* **2014**, *214–215*, 62–69. [CrossRef]
133. Prietzel, J.; Wagoner, G.L.; Harrison, R.B. Long-term effects of repeated urea fertilization in Douglas-fir stands on forest floor nitrogen pools and nitrogen mineralization. *For. Ecol. Manag.* **2004**, *193*, 413–426. [CrossRef]
134. Puhlick, J.J.; Fernandez, I.J.; Weiskittel, A.R. Evaluation of forest management effects on the mineral soil carbon pool of a lowland, mixed-species forest in Maine, USA. *Can. J. Soil Sci.* **2016**, *96*, 207–218. [CrossRef]
135. Rab, M.A. Soil physical and hydrological properties following logging and slash burning in the *Eucalyptus regnans* forest of southeastern Australia. *For. Ecol. Manag.* **1996**, *84*, 159–176. [CrossRef]
136. Riley, J.M.; Jones, R.H. Factors limiting regeneration of *Quercus alba* and *Cornus florida* in formerly cultivated coastal plain sites, South Carolina. *For. Ecol. Manag.* **2003**, *177*, 571–586. [CrossRef]
137. Roaldson, L.M.; Johnson, D.W.; Miller, W.W.; Murphy, J.D.; Walker, R.F.; Stein, C.M.; Glass, D.W.; Blank, R.R. Prescribed Fire and Timber Harvesting Effects on Soil Carbon and Nitrogen in a Pine Forest. *Soil Sci. Soc. Am. J.* **2014**, *78*, S48. [CrossRef]
138. Rothstein, D.E.; Spaulding, S.E. Replacement of wildfire by whole-tree harvesting in jack pine forests: Effects on soil fertility and tree nutrition. *For. Ecol. Manag.* **2010**, *260*, 1164–1174. [CrossRef]
139. Sanchez, F.G.; Scott, D.A.; Ludovici, K.H. Negligible effects of severe organic matter removal and soil compaction on loblolly pine growth over 10 years. *For. Ecol. Manag.* **2006**, *227*, 145–154. [CrossRef]
140. Sanscrainte, C.L.; Peterson, D.L.; McKay, S. Carbon storage and soil properties in late-successional and second-growth subalpine forests in the North Cascade Range, Washington. *Northwest Sci.* **2003**, *77*, 297–307.
141. Saynes, V.; Etchevers, J.D.; Galicia, L.; Hidalgo, C.; Campo, J. Soil carbon dynamics in high-elevation temperate forests of Oaxaca (Mexico): Thinning and rainfall effects. *Bosque (Valdivia)* **2012**, *33*, 3–11. [CrossRef]
142. Selig, M.F.; Seiler, J.R.; Tyree, M.C. Soil carbon and CO_2 efflux as influenced by the thinning of loblolly pine (*Pinus taeda* L.) plantations on the Piedmont of Virginia. *For. Sci.* **2008**, *54*, 58–66.
143. Shelburne, V.B.; Boyle, M.F.; Lione, D.J.; Waldrop, T.A. Preliminary effects of prescribed burning and thinning as fuel reduction treatments on the piedmont soils of the clemson experimental forest. In *General Technical Report SRS-71*; U.S. Department of Agriculture, Forest Service, Southern Research Station: Asheville, NC, USA, 2004; pp. 35–38.

144. Sheng, H.; Zhou, P.; Zhang, Y.; Kuzyakov, Y.; Zhou, Q.; Ge, T.; Wang, C. Loss of labile organic carbon from subsoil due to land-use changes insubtropical China. *Soil Biol. Biochem.* **2015**, *88*, 148–157. [CrossRef]
145. Skovsgaard, J.P.; Stupak, I.; Vesterdal, L. Distribution of biomass and carbon in even-aged stands of Norway spruce (*Picea abies* (L.) Karst.): A case study on spacing and thinning effects in northern Denmark. *Scand. J. For. Res.* **2006**, *21*, 470–488. [CrossRef]
146. Slesak, R.A.; Schoenholtz, S.H.; Harrington, T.B. Soil carbon and nutrient pools in Douglas-fir plantations 5 years after manipulating biomass and competing vegetation in the Pacific Northwest. *For. Ecol. Manag.* **2011**, *262*, 1722–1728. [CrossRef]
147. Small, C.J.; McCarthy, B.C. Relationship of understory diversity to soil nitrogen, topographic variation, and stand age in an eastern oak forest, USA. *For. Ecol. Manag.* **2005**, *217*, 229–243. [CrossRef]
148. Stone, J.E.; Kolb, T.E.; Covington, W.W. Effects of restoration thinning on presettlement *Pinus ponderosa* in northern Arizona. *Restor. Ecol.* **1999**, *7*, 172–182. [CrossRef]
149. Stone, D.; Elioff, J. Soil properties and aspen development five years after compaction and forest floor removal. *Can. J. Soil Sci.* **1998**, *78*, 51–58. [CrossRef]
150. Strong, T.F. Harvesting intensity influences the carbon distribution in a northern hardwood ecosystem. In *Research Paper NC-329*; Department of Agriculture, Forest Service, North Central Research Station: St. Paul, MN, USA, 1997.
151. Strukelj, M.; Brais, S.; Paré, D. Nine-year changes in carbon dynamics following different intensities of harvesting in boreal aspen stands. *Eur. J. For. Res.* **2015**, *134*, 737–754. [CrossRef]
152. Tang, J.; Bolstad, P.V.; Martin, J.G. Soil carbon fluxes and stocks in a Great Lakes forest chronosequence. *Glob. Chang. Biol.* **2009**, *15*, 145–155. [CrossRef]
153. Trettin, C.C.; Jurgensen, M.F.; Gale, M.R.; McLaughlin, J.W. Recovery of carbon and nutrient pools in a northern forested wetland 11 years after harvesting and site preparation. *For. Ecol. Manag.* **2011**, *262*, 1826–1833. [CrossRef]
154. Ussiri, D.A.N.; Johnson, C.E. Organic matter composition and dynamics in a northern hardwood forest ecosystem 15 years after clear-cutting. *For. Ecol. Manag.* **2007**, *240*, 131–142. [CrossRef]
155. Vario, C.L.; Neurath, R.A.; Friedland, A.J. Response of mineral soil carbon to clear-cutting in a northern hardwood forest. *Soil Sci. Soc. Am. J.* **2014**, *78*, 309–318. [CrossRef]
156. Vesterdal, L.; Dalsgaard, M.; Felby, C.; Raulund-Rasmussen, K.; Jørgensen, B.B. Effects of thinning and soil properties on accumulation of carbon, nitrogen and phosphorus in the forest floor of Norway spruce stands. *For. Ecol. Manag.* **1995**, *77*, 1–10. [CrossRef]
157. Waldrop, M.P.; McColl, J.G.; Powers, R.F. Effects of forest postharvest management practices on enzyme activities in decomposing litter. *Soil Sci. Soc. Am. J.* **2003**, *67*, 1250–1256. [CrossRef]
158. Wu, J.-S.; Jiang, P.-K.; Chang, S.X.; Xu, Q.-F.; Lin, Y. Dissolved soil organic carbon and nitrogen were affected by conversion of native forests to plantations in subtropical China. *Can. J. Soil Sci.* **2010**, *90*, 27–36. [CrossRef]
159. Xiang, W.; Chai, H.; Tian, D.; Peng, C. Marginal effects of silvicultural treatments on soil nutrients following harvest in a Chinese fir plantation. *Soil Sci. Plant Nutr.* **2009**, *55*, 523–531. [CrossRef]
160. Yanai, R.D.; Arthur, M.A.; Siccama, T.G.; Federer, C.A. Challenges of measuring forest floor organic matter dynamics: Repeated measures from a chronosequence. *For. Ecol. Manag.* **2000**, *138*, 273–283. [CrossRef]
161. Zabowski, D.; Chambreau, D.; Rotramel, N.; Thies, W.G. Long-term effects of stump removal to control root rot on forest soil bulk density, soil carbon and nitrogen content. *For. Ecol. Manag.* **2008**, *255*, 720–727. [CrossRef]
162. Zhong, Z.; Makeschin, F. Soil biochemical and chemical changes in relation to mature spruce (*Picea abies*) forest conversion and regeneration. *J. Plant Nutr. Soil Sci.* **2003**, *166*, 291–299. [CrossRef]
163. Zummo, L.M.; Friedland, A.J. Soil carbon release along a gradient of physical disturbance in a harvested northern hardwood forest. *For. Ecol. Manag.* **2011**, *261*, 1016–1026. [CrossRef]

11

Divergent Responses of Foliar N:P Stoichiometry During Different Seasons to Nitrogen Deposition in an Old-Growth Temperate Forest, Northeast China

Dongxing Yang, Hongrui Mao and Guangze Jin *

Center for Ecological Research, Northeast Forestry University, Harbin 150040, China; ydx19930719@sina.com (D.Y.); damao731@126.com (H.M.)

* Correspondence: kwpine@nefu.edu.cn

Abstract: Atmospheric nitrogen (N) deposition has rapidly increased during the last few decades; however, the seasonal responses of leaf N:P stoichiometry to N deposition remain unclear. In 2008, a simulated N deposition experiment (0, 30, 60, and 120 $kg \cdot N \cdot ha^{-1} \cdot yr^{-1}$) was conducted in an old-growth temperate forest in Northeast China. In 2014, the leaves of 17 woody species and soil were sampled in spring, summer, and autumn in each treatment, and N:P stoichiometry was assessed. Community N and P in summer were significantly lower than that in spring and autumn. Unlike broadleaved species, conifers showed no significant variation among the three seasons. N addition significantly enhanced community N and soil available P but decreased soil total P in summer and autumn, and decreased community P, as well as the P concentration of three life forms (conifer, tree, and shrub), in autumn. Our results emphasize the importance of multiple sampling across seasons in temperate forests. Arguing against the traditional consensus, the productivity of the old-growth temperate forests is limited by both N and P.

Keywords: nitrogen deposition; N and P colimitation; leaf N:P stoichiometry; soil N:P stoichiometry; seasonal variations

1. Introduction

Increased atmospheric nitrogen (N) deposition due to the combustion of fossil fuels and the increase in agricultural production during the last few decades, has captured the attention of many ecologists [1–3]. N deposition also has rapidly increased in China since the 1980s because of the country's rapid industrial development [4,5]. In addition, N and P are commonly considered the two most limiting elements of net primary productivity (NPP) in terrestrial ecosystems [6,7], increased N deposition may result in an imbalance in the input of N vs. P in ecosystems [8,9] and further alter foliar N:P, the community nutrient limitation, and species composition [10–12].

Although positive effects of increased foliar N by N deposition are often reported [8,13], neutral and even inhibitory effects have also been observed [14,15]. Studies of foliar P under N deposition are relatively less common than those of foliar N [16], and foliar N:P ratio variations under N deposition remain uncertain. In fact, the determination of a limiting status is important for predicting how a change in foliar N:P stoichiometry under N deposition occurs. The effects of N deposition on community production and on N:P stoichiometry occur in three stages [16–18]. During the first stage, plant growth under N limitation will show a rapid increase [19]. During the second stage, the response of plant growth will decrease compared to that during the first stage as the status of N saturation is approached [17]. When N limitation changes to P limitation during the third stage, increased N input may result in the loss of soil available nutrients, and plant growth will be inhibited in response to N deposition [14]. In addition, P is a major limiting nutrient in tropical regions with old-aged soils, whereas temperate regions in the Northern Hemisphere with younger soils are often limited by N [7]. Arguing against the traditional consensus, Elser et al. [6] suggested that P limitation has an equivalent status to that of N limitation across terrestrial systems. Through a fertilization experiment in 13 boreal forests, Goswami et al. [20] found evidence that most middle-aged and mature forests exhibit P limitation rather than N limitation. Other studies have described a shift from N limitation to P limitation with succession, as N-cycling properties recover and the dominance of a conservative P cycle re-emerges, in mature and old-growth forests [21,22]. Overall, as the critical forest type of the northern region of China, further research on the mixed broadleaved-Korean pine (*Pinus koraiensis*) forest, an old-growth temperate forest, regarding its nutrient limitation status is needed.

Initially, fertilization experiments are widely employed to assess community nutrient status and N or P limitation [23–26]. Based on the strong relationship between N and P, Koerselman and Meuleman [27] also proposed a simple method that utilizes the foliar N:P ratio to evaluate the limiting status. However, a one-time sampling may not fully reflect the stoichiometric characteristics of a plant community [28]. Because of the change in foliar development stages and the existence of a retranslocation mechanism in plant leaves, foliar N and P may undergo marked variations through two seasons in just a few months, even in mature green leaves [19,29,30]. Thus, considering the effect of seasonal change is important when studying patterns of change in foliar N:P stoichiometry and the response of foliar N:P stoichiometry to N deposition. To our knowledge, there is no research focusing on the effects of N deposition on foliar N:P stoichiometry in different seasons.

Because plant traits affect internal and external nutrient cycling, different responses to N deposition and seasonal change may appear among species and plant functional groups [7,31]. Additionally, mycorrhizae, as a type of key classification for plants, are also attracting the attention of many biologists and ecologists [20,32,33]. Indeed, mycorrhizal symbiosis has a strong effect on plant growth and plays a key role in nutrient cycling [34,35], and various mycorrhizal types possess different mechanisms and capacities for influencing the process in which plants acquire nutrients from the soil. For example, the hyphae of mycorrhizal fungi can search for limited nutrients, including nitrates and phosphates, to support plant growth, but only ectomycorrhizal (EM) and ericoid mycorrhizal fungi can obtain organically bound nutrients [36]. As plants of old-growth temperate forests are associated with both arbuscular mycorrhizal (AM) and ectomycorrhizal (EM) species, the plants associated with differential mycorrhizal types may have diverse responses to altered surroundings.

This study investigated the effects of seasonal changes on the response pattern of leaf N:P stoichiometry to N deposition through a simulated N addition experiment. We sought to determine (1) how leaf stoichiometric characteristics change with seasonal change, (2) whether the responses of the leaf N:P stoichiometry to N deposition show differences in three seasons, and (3) the nature of the nutrient limitation status of an old-growth temperate forest in Northeast China.

2. Materials and Methods

2.1. Site Description

The study site is in the Heilongjiang Liangshui National Nature Reserve (47°10′50″ N, 128°53′20″ E) in Dailing, Heilongjiang Province, Northeast China. The annual mean temperature is −0.3 °C, and the annual mean maximum temperature is 7.5 °C. The annual mean minimum temperature is −6.6 °C. The ≥ 0 °C accumulated temperature is 2200 °C–2600 °C. The annual mean precipitation is 676 mm, and >90% of the precipitation falls during the growing season from May to October. There are 100–120 frost-free days and 130–150 snow-cover days. The average elevation is 400 m, and the difference in the relative elevation is 100–200 m. The zonal soil is a dark brown forest soil. The mixed broadleaved-Korean pine (*Pinus koraiensis*) forest is the zonal vegetation of the reserve. The stand age is between 200 and 300 years, with some individual ages older than 400 years.

2.2. Experimental Design and Measurements

The N addition experiment began during May 2008 in the mixed broadleaved-Korean pine (*Pinus koraiensis*) forest, an old-growth temperate forest. Twelve 20-m × 20-m sequential plots were established in a random manner. To prevent mutual interference, 10-m barriers between adjacent plots were retained. In every plot, species with a diameter at breast height (DBH, 1.3 m) >2 cm were surveyed; species, DBH, and coordinates were recorded and labeled in May 2008 and surveyed again in July 2014. The background conditions of each plot surveyed in 2008 are shown in Table 1. Using urea ($CO(NH_2)_2$) as the N source, four treatment groups were established, with three replicates—control (N0, 0 kg·ha^{-1}·yr^{-1}), low N (N1, 30 kg·ha^{-1}·yr^{-1}), medium N (N2, 60 kg·ha^{-1}·yr^{-1}), and high N (N3, 120 kg·ha^{-1}·yr^{-1}). The background N deposition (inorganic and organic N) was 12.9 kg·N·ha^{-1}·yr^{-1}, as recorded by a dry–wet deposition collector (New Star Environmental LLC, Roswell, GA, USA). N additions were applied at the beginning of June, July, and August every year and began in June 2008. The $CO(NH_2)_2$ for each treatment was dissolved in 20 L of water, and the solutions were sprayed onto the soil surface using a backpack sprayer. The control group was sprayed with the same amount of water.

Table 1. Natural conditions in each plot before the N addition experiment in 2008.

Plots	Treatments	Average N Addition (kg·ha^{-1}·yr^{-1})	Repeat	Density (trees·ha^{-1})	Mean Breast Diameter (cm)	Slope Aspect	Slope Degree
1	N0	0	N0-1	1325	10.70	W	<6°
2	N0	0	N0-2	1625	8.60	W	<6°
3	N1	30	N1-1	1575	8.46	W	<6°
4	N3	120	N3-1	2925	7.68	W	<6°
5	N3	120	N3-2	1825	9.48	W	<6°
6	N3	120	N3-3	1100	14.02	W	<6°
7	N0	0	N0-3	1650	9.43	W	<6°
8	N1	30	N1-2	1550	11.47	W	<6°
9	N1	30	N1-3	1575	10.41	W	<6°
10	N2	60	N2-1	2300	7.62	W	<6°
11	N2	60	N2-2	1800	11.44	W	<6°
12	N2	60	N2-3	1525	11.45	W	<6°

When collecting samples from tall trees, a branch with many leaves and sufficient sunlight was cut by a person specialized in climbing trees. The healthy and green leaves were selected and placed into a plastic vacuum packaging bag. Seventeen species were collected randomly in 3 plots for each treatment in the middle of May, July, and September during 2014. However, insufficient material was collected for a proportion of the species because of natural and anthropogenic reasons; for example, most of the *Fraxinus mandshurica* individuals had not renewed their foliage in May, and their leaves had already fallen by September, or a few trees were so high that samples could not be obtained. In this study, May, July, and September represented spring, summer, and autumn, respectively. The 17 species were of three types— deciduous broad-leaved trees, deciduous broad-leaved shrubs and evergreen conifers (hereafter trees, shrubs and conifers, respectively). Conifers were assigned their own group because they have unique traits compared to those of other trees [7]. The mycorrhizal types of species in this study were according to Shi et al. [37], and species belonging to the same family were classified into the same mycorrhizal type. The detailed sampling information is listed in Table 2.

Soil samples were also collected at the same time as leaf samples. In each plot, soils were randomly sampled by taking three 5-cm-diameter soil cores from depths of 0–10, 10–20, 20–30, and 30–40 cm. The litter on the soil surface was removed before sampling. The samples were placed in plastic bags and transported to the lab after being air-dried.

Table 2. Sampling condition of each species in 2014.

Scientific Name of Species	Life Form	Ectomycorrhizal (EM) or Arbuscular Mycorrhizal (AM)	*The important value* (%) under Control Group in 2014	Uncollected Samples in Different Treatments		
				Spring	Summer	Autumn
Pinus koraiensis	Conifer	EM	27.67	N3		
Acer mono	Tree	EM	11.19		N1	
Corylus mandshurica	Shrub	EM	9.60			
Tilia amurensis	Tree	EM	8.04	N1		N1
Betula costata	Tree	EM	7.05	N1 N2 N3		N0 N1 N2
Ulmus laciniata	Tree	EM	5.66			
Ulmus japonica	Tree	EM	4.13	N2		
Acer tegmentosum	Tree	AM	3.75		N1	N0 N1 N2 N3
Fraxinus mandshurica	Tree	AM	3.03	N1 N2 N3		N1 N2
Prunus padus	Tree	AM	1.98			
Syringa reticulate var. *mandshurica*	Tree	AM	1.54			
Acer ukurunduense	Tree	EM	1.19	N1	N1	
Euonymus pauciflorus	Shrub	AM	1.13			
Picea koraiensis	Conifer	EM	0	N2		
Deutzia gladata	Shrub	AM	0			
Lonicera chrysantha	Shrub	AM	0			
Ribes mandschuricum	Shrub	AM	0			

Each leaf sample was ground to a fine powder, passed through a 0.149-mm mesh, and enveloped. The sample was oven-dried at 65 °C to a constant weight for further analysis. The soil sample was passed through a 2-mm mesh; a portion was conserved in an envelope, and the remaining material was conserved after passing through a 0.149-mm mesh. In October 2014, leaf N concentration was analyzed using a Kjeltec KTM 2300 analyzer (Foss Teactor AB, Hoganas, Sweden) after H_2SO_4 + K_2SO_4 + $CuSO_4$ digestion. Soil total N (TN) in every layer was analyzed using the same method but with H_2SO_4 + H_2O_2 digestion. Leaf and soil total P (TP) were measured using the molybdenum antimony colorimetric spectrophotometer method after H_2SO_4 + H_2O_2 digestion. These samples were oven-dried again at 65 °C for 2 h before digestion. For soil available P (AP), 2-mm of soil from every layer was pooled and fully mixed, and the air-dried soil was extracted using 0.05 mol·L^{-1} HCl + 0.025 mol·L^{-1} H_2SO_4. The P concentration in the extract was measured using the molybdenum antimony colorimetric spectrophotometer method. N and P concentrations are expressed as mg·g^{-1}, and a mass basis was used to calculate stoichiometric ratios.

2.3. Statistical Analyses

The important value (*IV*) of *i*-th species in each treatment group in 2008 and 2014 were calculated using the following equation:

$$IV_{(i)} = (\text{relative density}_{(i)} + \text{relative frequency}_{(i)} + \text{relative prominence}_{(i)}) \ / \ 3$$

The change in *IVs* for each species was directly calculated using the following equation:

$$\text{The changed value}_{(i)} = IV_{(i)(2014)} - IV_{(i)(2008)}$$

Two-way analysis of variance (ANOVA) with season and treatment as fixed factors was applied to examine the effects on leaf and soil N, P and the N:P ratio based on averages. Community N:P stoichiometry (N:P stoichiometry means N, P, and the N:P ratio, community N:P stoichiometry means N:P stoichiometry of the plant community) weighted by $IV_{(i)\,(2014)}$ was also examined in the same manner (hereafter, community N:P stoichiometry is the leaf N:P stoichiometry weighted by the $IV_{(i)\,(2014)}$). Meanwhile, the marginal means of the main factors were calculated and recorded. Three-way ANOVA with season, treatment, and different groups (broadleaves vs. conifers, trees vs. shrubs, ectomycorrhizal species (EM) vs. arbuscular mycorrhizal species (AM), broadleaves including broadleaved trees and shrubs; the data for conifers were not included in tree and EM groups) as fixed factors were used to analyze effects on leaf N, P, and the N:P ratio. Nonsignificant interactions were removed from the analysis. Leaf P, the leaf N:P ratio and soil N:P stoichiometry in our data were apparently of right-skewed distribution, and they were naturally logarithmically transformed before performing ANOVA; untransformed data are shown in the figures.

The data were separated according to (1) different seasons and (2) different seasons and different classified groups. Least squares discrimination (LSD) one-way ANOVA was applied to assess differences between treatment groups, and $IV_{(i)\,(2014)}$ was calculated as weightiness. Differences between N:P stoichiometry averages in our study and previous studies were evaluated using the one-sample *t*-test. The Kolmogorov–Smirnov and Levene's tests were employed to assess the normality of the residuals and the homogeneity of variance, respectively. All figures were created using Origin 9.0, and all data were analyzed using SPSS 20 (SPSS, Inc., Chicago, IL, USA). Significance was determined at the 0.05 level.

3. Results

3.1. N, P, and the N:P Ratio in the Three Seasons Under Ambient (without N Addition) Conditions

Community N, average N, and soil N showed a significant increase in September after an initial decrease from spring to summer (Figure 1a,d and Figure 2a). A significant increase from spring to autumn was found for soil N:P, which exhibited the same trend as soil N and average N:P (Figures 1f and 2a,d). Community N:P in autumn was significantly lower than that in summer but was significantly higher than that in spring. Community P and average P in spring were significantly higher than those in the other two seasons (Figure 1b,c,e,f). Although soil TP showed no significant variation during the three seasons, soil AP had a similar trend to that of community P, tending to increase again in autumn after a significant decrease in summer (Figures 1b and 2b,c). The leaf N:P stoichiometry of conifers was significantly lower than that of trees and shrubs ($p < 0.05$), and the conifer P and N:P ratios displayed no significant seasonal variation, different from the patterns of the trees and shrubs. Moreover, a nearly identical seasonal pattern between the trees and shrub N:P stoichiometry was observed (Figure 3a–i).

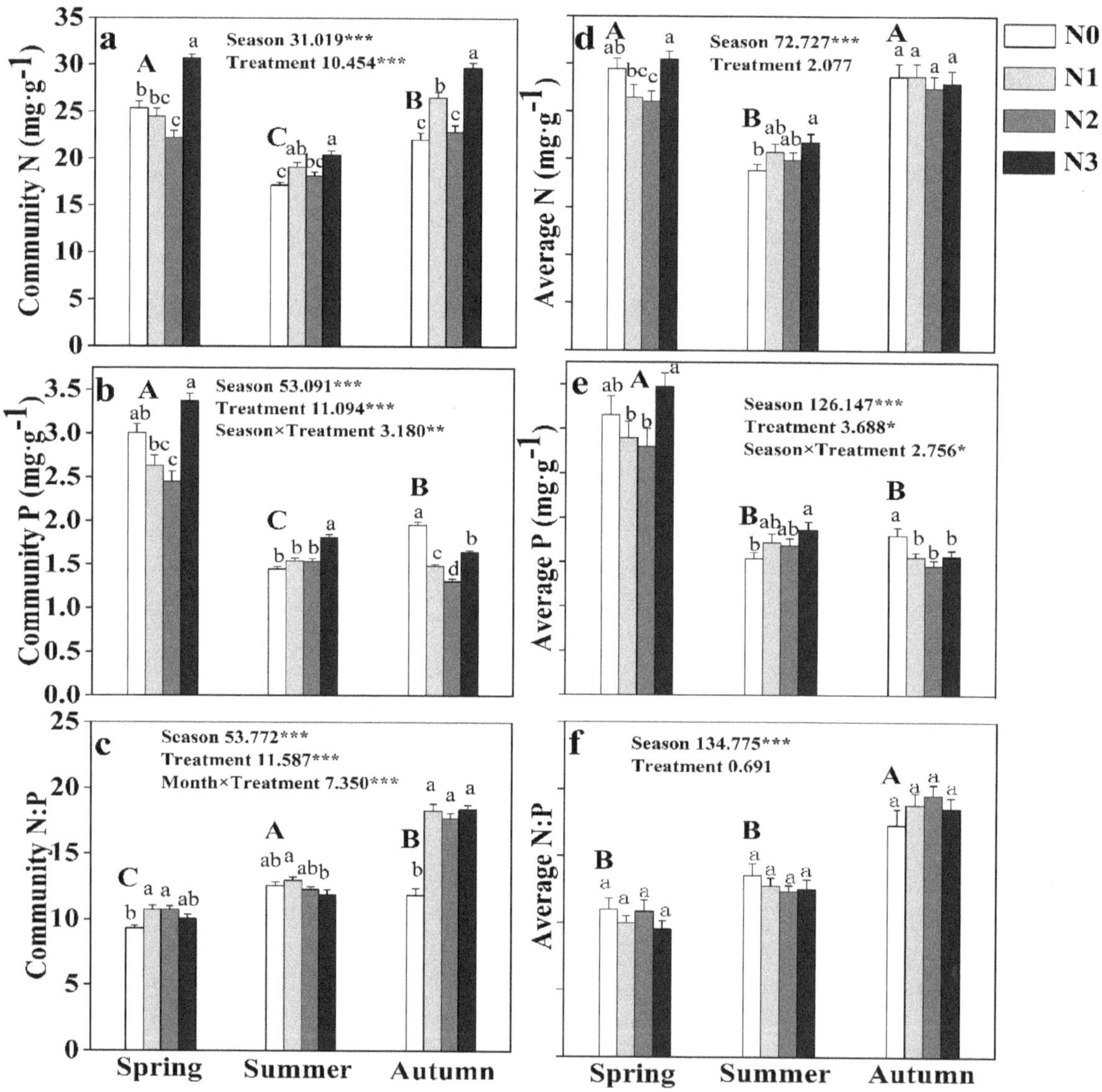

Figure 1. Effects of N addition on community and average N (**a**,**d**), P (**b**,**e**) and the N:P ratio (**c**,**f**) in three seasons. The results of two-way ANOVAs are shown in each panel and interactions without significance were removed from the analysis. The results of least squares discrimination (LSD) one-way ANOVA were shown above the bars. Different uppercase letters indicate a significant difference among the three seasons under ambient conditions. Different lowercase letters indicate a significant difference among treatments in each month. Data are shown as the mean + *SE*. * $p < 0.05$, ** $p < 0.01$, *** $p < 0.001$.

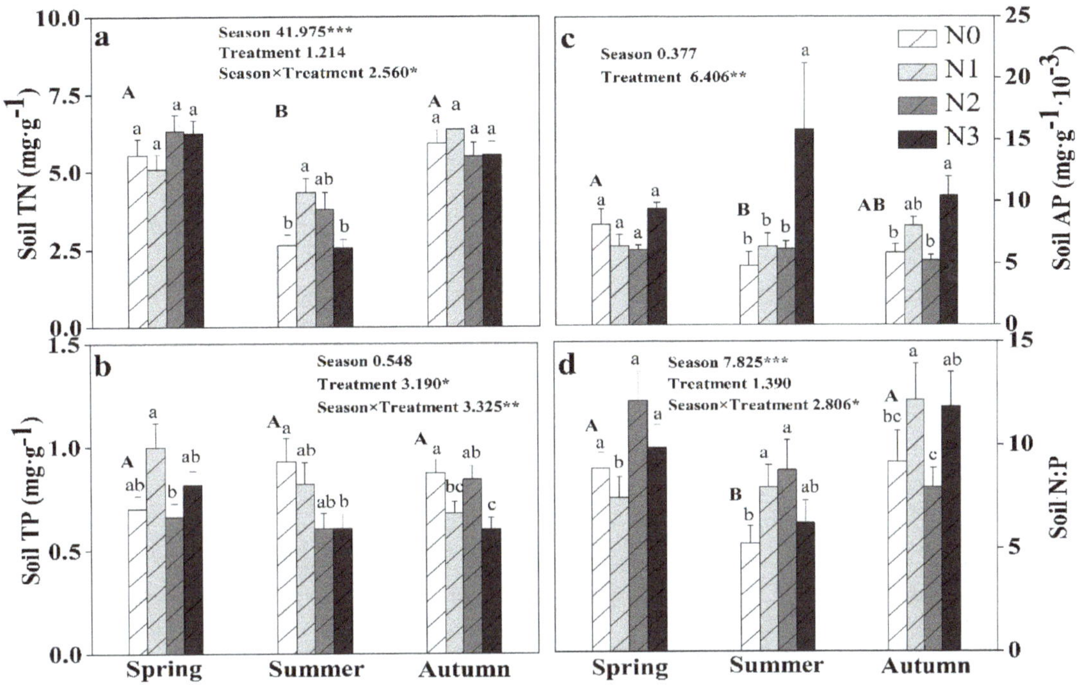

Figure 2. Effects of N addition on soil total N (TN) (**a**), soil total P (TP) (**b**), soil available P (AP) (**c**) and soil total N:P ratio (**d**) in the three seasons. The results of two-way ANOVAs are shown in each panel and interactions without significance were removed from the analysis. The results of least squares discrimination (LSD) one-way ANOVA were shown above the bars. Different uppercase letters indicate a significant difference among the three seasons under ambient conditions. Different lowercase letters indicate a significant difference among the treatments in each month. Data are shown as the mean + *SE*. * $p < 0.05$, ** $p < 0.01$, *** $p < 0.001$.

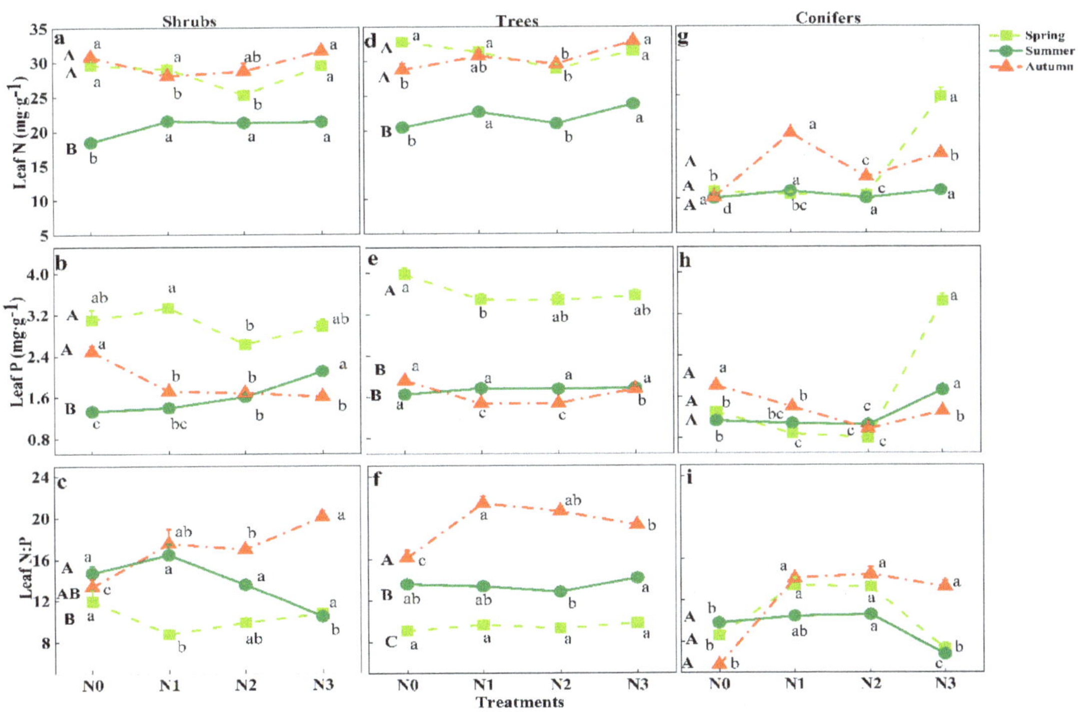

Figure 3. Effects of N addition on leaf N (**a**,**d**,**g**), leaf P (**b**,**e**,**h**) and the leaf N:P ratio (**c**,**f**,**i**) of three life forms in three seasons. Different uppercase letters indicate a significant difference among the three seasons under ambient. Different lowercase letters indicate a significant difference among the treatments in each month. Data are shown as the mean + *SE*.

Summer was the peak season in the survey region. During this season, leaf average N and P in our study was similar to the average for plants in China, though N:P was significantly lower. Leaf P was significantly lower than reported in two global studies, and leaf N was significantly lower than that in the study of Elser et al. [38]. We also found some inconsistent results for leaf N:P stoichiometry in spring and autumn compared to that in summer (Table 3).

Table 3. Comparison of the average leaf N, P, and N:P in our study with China and the world.

Stoichiometry and Seasons	N			P			N:P		
	Spring	Summer	Autumn	Spring	Summer	Autumn	Spring	Summer	Autumn
Han et al. [39]		20.2			1.46			16.3	
(China)	+ *	−	+ *	+ *	+	+ *	− *	− *	+
Reich and Oleksyn [7]		20.1			1.77			13.8	
(World)	+ *	−	+ *	+ *	− *	+	− *	−	+ *
Elser et al. [38]		20.6			1.99			12.7	
(World)	+ *	− *	+ *	+ *	− *	−	−	+	+ *

"+" means the average in this study is higher than average of China and the world. "−" means the value in this study is lower than the tested value. * $p < 0.05$.

3.2. Effect of N Addition on Community and Soil N:P Stoichiometry during Different Seasons

Community N and P showed a similar variation to that of average N and P under N addition in spring and summer, with a similar change pattern for N to that of P (Figure 1a,b,d,e). During spring, community N and average N were significantly increased under N3 after a decreasing trend (Figure 1a,d). Significant increases in community N:P were found for N1 and N2 treatments, but there was no significant variation between the different treatment groups with regard to average N:P (Figure 1c,f). In summer, an upward trend in community and average N and P was observed (Figure 1a,b,d,e); community N:P under N3 was significantly lower than that under N1, and there was no significant effect of N deposition on average N:P (Figure 1c,f). In autumn, N addition significantly increased community N, P, and N:P and significantly decreased average P; however, there was no significant variation in average N and N:P (Figure 1a–f). A two-way ANOVA showed that seasonal change had a significant impact on the community and average N:P stoichiometry (Figure 1a–f). Treatment also had a significant impact on community N:P stoichiometry but no significant impact on average N and N:P (Figure 1c,d,f). In addition, their interaction had a significant influence on community P, N:P and average P (Figure 1b,c,e).

During summer, soil N and N:P showed an increasing trend, with a decrease under N3. In addition, the N3 treatment significantly decreased soil P and increased soil AP. The patterns of variation in soil P and soil AP in autumn were similar to those in summer; soil N:P increased under N1 and N3, and soil N showed no significant change (Figure 2a–d). A two-way ANOVA showed that seasonal change had a significant impact on soil N and soil N:P, and treatment had a significant impact on soil P and soil AP; however, their interaction had no significant impact on soil AP (Figure 2a–d). In addition, the marginal means of main factors (season and treatment) are listed in the Table S1.

3.3. Response of Leaf N:P Stoichiometry in Different Groups to Seasonal Change and N Addition

A three-way ANOVA showed that although only a small variation in the response of N:P stoichiometry between shrubs and trees to N addition was found in the three seasons, the response was significantly different than that of conifers (Table 4; Figure 3a–i). In spring, the leaf N of shrubs, trees and conifers all exhibited a significant decrease under N2 that was increased under

N3 (Figure 3a,d,g). A similar decrease under N2 in leaf P was found (Figure 3b,e,h). In summer, N addition significantly increased N in trees and shrubs, with no remarkable effect on conifer N (Figure 3a,d,g). In autumn, increased leaf N was found in the leaves of the trees and conifers, but their leaf P showed a trend opposite to leaf N under N addition (Figure 3d,e,g,h). N addition significantly increased the N:P of shrub, trees, and conifers in autumn (Figure 3c,f,i).

A three-way ANOVA showed that different mycorrhizal types displayed notable differences in leaf N:P stoichiometry, and there were significant interactions between mycorrhizal type and other factors (Table 4; Figure 4a–f). In particular, EM species in autumn showed a nearly opposite trend to that of AM species in terms of N:P stoichiometry (Figure 4a–f). From 2008 to 2014, the changed *IV* of AM species showed an increasing trend with higher available N. In contrast, the changed *IV* of EM showed a decreased trend under increasing N addition (Figure 5).

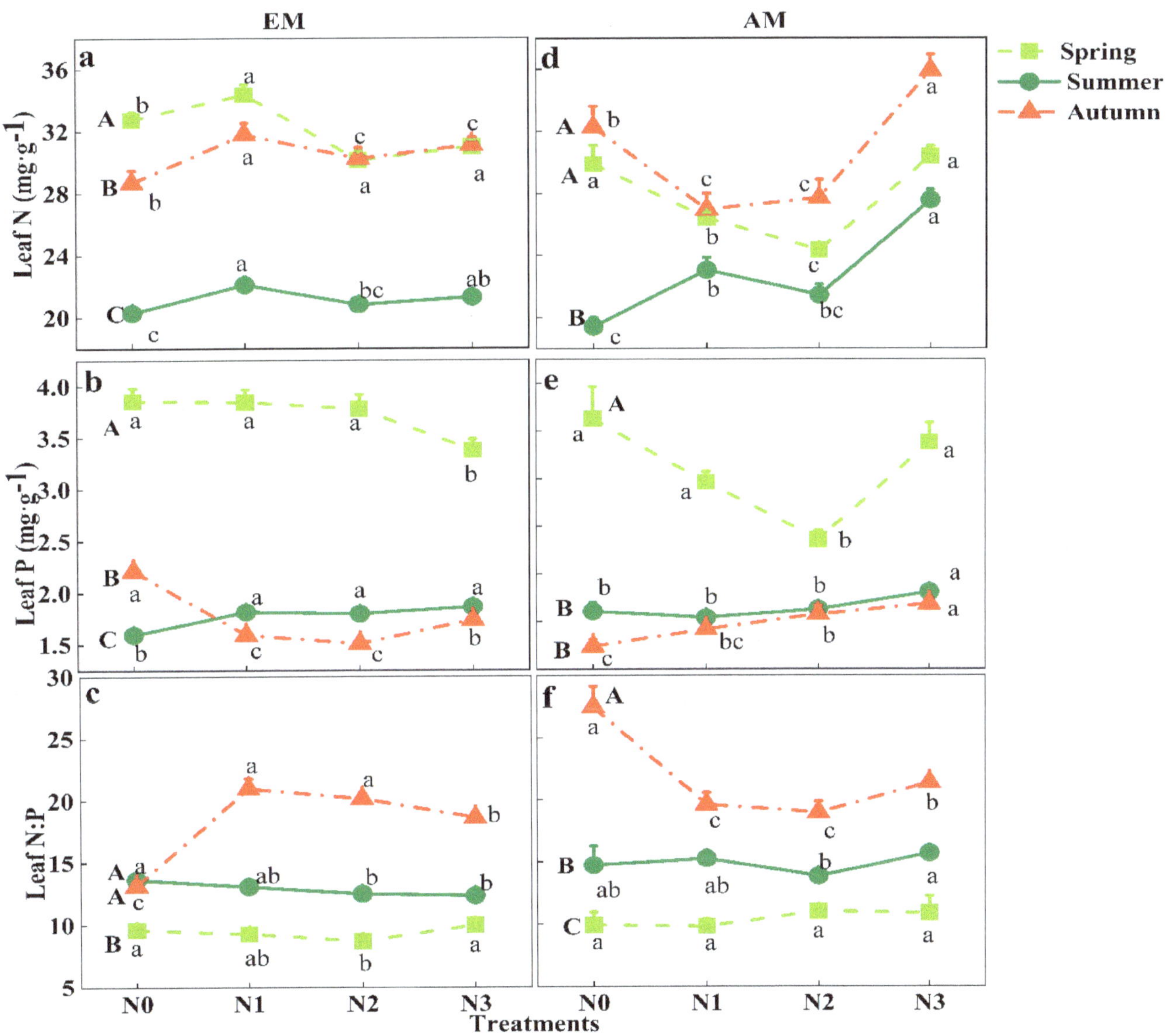

Figure 4. Effect of N addition on leaf N (**a**,**d**), leaf P (**b**,**e**) and the leaf N:P ratio (**c**,**f**) of ectomycorrhizal species (EM) vs. arbuscular mycorrhizal species (AM) in the three seasons. Different uppercase letters indicate a significant difference among the three seasons under ambient conditions. Different lowercase letters indicate a significant difference among the treatments in each month. Data are shown as the mean + *SE*.

Table 4. Results (*F* values) of three-way ANOVA on the effects of leaf N:P stoichiometry in different groups.

Different Groups	Factors	Season	Treatment	Group	Season × Treatment	Season × Group	Treatment × Group	Season × Treatment × Group
Broadleaves vs. conifers	N	53.480 ***	6.269 ***	408.073 ***	2.857 *	8.434 ***		2.220 *
	P	25.034 ***	20.112 ***	134.599 ***	6.762 ***	18.939 ***	9.209 ***	3.286 **
	N:P	33.11 ***	21.895 **	39.339 ***	11.868 **	14.017 **	9.716 **	3.817 **
Trees vs. Shrubs	N	106.743 ***	4.307 **	5.477 *				
	P	104.666 ***	2.415	0.906	2.458 *	3.691 *		
	N:P	131.018 ***	3.085 *	0.005	2.581 *			
EM vs. AM	N	80.135 ***	7.201 ***	1.092		8.889 ***	7.089 ***	
	P	142.893 ***	1.639	17.898 ***				2.250 **
	N:P	127.291 ***	0.433	14.060 ***				2.798 ***

Interactions without significance were removed from the analysis, and the date are not shown in the table. * $p < 0.05$, ** $p < 0.01$, *** $p < 0.001$.

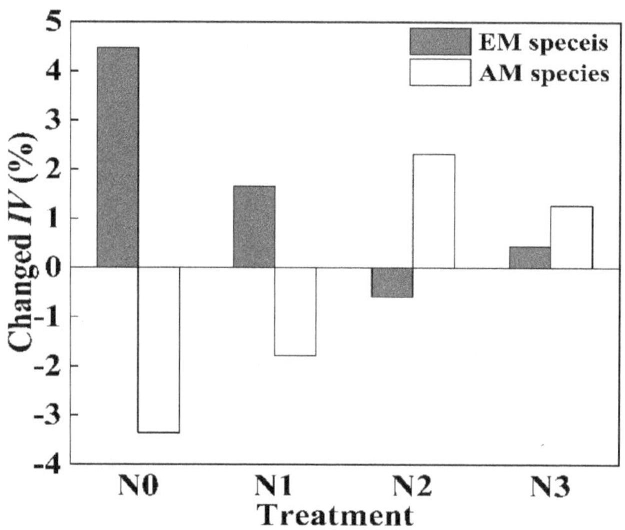

Figure 5. Changed *IV* (%) of AM species and EM species from 2008 to 2014 under four treatments.

4. Discussion

4.1. Seasonal Change in Leaf N:P Stoichiometry

Consistent with previous study [19], among the three seasons, community N and P were highest in spring, which is due to the rapid growth rate of leaves [28,29]. N is the main element of proteins, and P is the main element of P-rich ribosomal RNA (rRNA). In spring, leaves are in the early stage of their growth period, the foliar size is relatively small, and foliar cells show a high degree of splitting and selection to absorb considerable N and P to support their rapid growth. Therefore, foliar nutrient concentrations are relatively high [18,40]. Summer is the peak season: The rate of foliar nutrient absorption is lower than that of growth, leaves are larger than they are in spring, and most of the broad leaves have fully expanded. Thus, the concentrations of nutrients are diluted, which results in rapidly decreasing nutrient concentrations in summer [41]. However, when leaves are no longer growing and are fully expanded, the dilution process stops, and foliar nutrient concentrations increase again because of continuous nutrient absorption; accordingly, community N and P in autumn are higher than in summer.

Because of the rapid decrease in foliar P compared to foliar N in summer, community N:P is of a significantly higher level than that in spring. The sharp decrease in foliar P from the early growing season to the late growing season has also been found in previous studies, but without explanation [19,42]. A possible reason is the decreased demand for P. Because plants are older in summer than in spring, active meristems (e.g., inflorescences, shoot tips, young leaves) will restrict plant growth; in contrast, mature leaves no longer grow but are still photosynthetically active, which results in a greater decrease in the P requirement for rRNA synthesis compared to the decrease in N for protein synthesis [43]. Similarly, soil AP significantly decreases, by 41.2%, from spring to summer, and the percentage of soil AP of soil TP during Spring, Summer, and Autumn is 1.17%, 0.52%, and 0.67%, respectively. Wang and Moore [30] also found a strong increase in available N and K but no change in soil AP in summer and autumn compared to available N, P, and K in spring.

When we divided the foliar data into shrubs, trees, and conifers, remarkably different seasonal patterns and nutrient concentrations between coniferous and broadleaved species were found. Indeed, leaves with high nutrient concentrations (both N and P) tend to be shorter lived and have a high photosynthetic capacity, whereas leaves with low nutrient concentrations tend to have longer lifespans and lower metabolic capacities [44,45] (Evergreen-Deciduous hypothesis). Conifers (mainly consisting of *Pinus koraiensis*, the *IV* of which under ambient conditions is 27.67%) maintain a low leaf nutrient status that may constitute a strategy related to canopy dominance and space occupancy [46]. In addition, unlike broadleaved species that translocate N and P to leaves during the early growing season, conifers gradually translocate these nutrients to new leaves throughout the growing season to maintain a relatively stable nutrient concentration and stoichiometry, thereby decreasing the effect of the external environment [29]. This finding confirms our results that conifers exhibit less seasonal variation than the other two plant groups examined in this study. It was also noted that the weighted N:P stoichiometry of shrubs and trees demonstrated a pattern similar to that of the average N:P stoichiometry, although this was not the case for conifers, which explains the reduced variation in community N:P stoichiometry from summer to autumn compared to the variation in the average N:P stoichiometry. These findings suggest that compared to average values, weighted values may better reflect the N:P stoichiometry at a community level.

Similar seasonal pattern of foliar nutrient concentrations has also been exhibited in the previous study [30]: foliar N concentration of deciduous species showed an increased trend from July to September after a sharp decrease from May to June, though the increase was not statistically significant and there was a sharp decrease again in October due to nutrient resorption. As mentioned earlier, the seasonal pattern of N concentration exhibits the strong effects of foliar development stage on nutrient concentrations of leaves, and exposes the limitation of calculating foliar nutrients based on the mass ratio. In fact, it is a well-known phenomenon in agriculture that nutrient concentrations decrease with ontogenetic development of individual plants, and this decrease is mainly because of the imbalanced rate between increased nutrients and accumulated dry matter [19]. However, the application of mass ratio is still widely accepted in the field of plant stoichiometry [16,19,43], because we are accustomed to thinking about organism body mass or estimating total biomass of species or components in ecosystems [40]. Nonetheless, our study suggests that compared to the foliar concentration ($mg \cdot g^{-1}$), the calculation based on the foliar mass ($mass \cdot leaf^{-1}$) may be a better expression when considered the seasonal effects, because it can directly express the nutrient contents in leaves and ignore the effects of foliar development stage [29].

4.2. Seasonal Variation in the Leaf N:P Stoichiometry Response to N Addition

Seasonal change had a highly significant effect on community N:P stoichiometry. Furthermore, leaf nutrients responded differently to N addition in the three seasons. In fact, although N addition increases the availability of N and promotes its absorption, it can at the same time stimulate the growth of plants, which is accompanied by dilution of nutrient concentrations [19,28]. Thus, the differences among treatments may reflect the relative magnitude of the foliar growth rate and the nutrient absorption rate. During the early growth stage, N addition increases the growth rate of leaves more rapidly than it increased N absorption rate. Accumulated N is diluted by rapidly growing leaves, which results in a decreasing trend under low and medium N addition. When the increase in absorption rate is faster than the increase in growth rate, the N concentration shows an increasing trend, for example: the community N under high N addition during spring. Uniformly, the foliar growth rate gradually slowed or even stopped, but the effect of N addition on the nutrient absorption rate was still effective, which resulted in the increasing trend under N addition during summer.

Studies have suggested that imbalanced N-P loading affects P cycling and may enhance P limitation [12,45,46]. This effect is exhibited in the decreased soil TP in summer and autumn and increased community N:P in autumn under N addition. In the short term, N-induced P limitation may be alleviated by enhanced P absorption due to enhanced activity of soil and root phosphatases or by increased P conservation due to increased P resorption [47–51]. In this study, a similar change pattern between community P and community N during spring and summer was found, and soil AP rose significantly as a result of high N addition, confirming the alleviation of N-induced P limitation. Mitigation of N-induced P limitation under N deposition has also been reported in temperate plantations by Deng et al. [16] and in a semiarid grassland by Long et al. [52], reflecting the capacity of plants to regulate P acquired to maintain homeostasis (*H*) [43,53].

Nonetheless, community P as well as the P concentration of the three life forms (shrub, tree, and conifer) all showed a decreasing trend under N addition in autumn. There are two possible reasons for this variation. One, is that these regulatory mechanisms may not be able to quantitatively supply sufficient P to balance the increase in N in plants under continuous N input [26]; however, the limited variations across seasons in soil TP, increased community P under ambient conditions and decreased P demand of leaves during this stage suggest that this explanation may not be viable. Another, more plausible reason, is internal nutrient retranslocation, i.e., the process by which plants transform nutrients in leaves into other tissues [29,49]. The recycling mechanism occurs not only in senescing leaves but also during the entire life cycle of leaves [54] as a response to N-induced P limitation. Plants thus transfer more P from mature green leaves, which have a relatively low P demand in autumn, to other tissues to support their growth. Unfortunately, we did not collect more data for nutrients in other tissues to support this explanation, and more research needs to be completed in the future. In the long term, the mitigating effect of N addition on P limitation may not be persistent, and the degree of P limitation will still be enhanced by N addition because of the reduced P concentration in leaves, which will slow the P recycling rate and further decrease the amount of available P [12,47,55].

Mycorrhizal associations play a critical role in regulating the acquisition of limited nutrients [34,35]. Mycorrhizal associations exist in nearly all ecosystems, with mycorrhizal plants dominating in many habitats [56,57]. AM-associated plants are often dominant in low-latitude areas, whereas EM-associated plants are often dominant in high-latitude areas. Overall, compared to EM, AM associations result in more effective P acquisition [31,58]. Consistent with Deng et al. [16], who reported an increased AM:EM ratio in a young stand under N addition, we also found that the changed *IV* of AM species increased but that of EM species decreased with increasing N addition, which suggests once again an enhanced P limitation under N addition as well as the regulation of community N-induced P limitation. Furthermore, we did not observe significant differences in responses to N addition between shrubs and trees, but AM species and EM species did display such differences. Studies have shown that EM fungi tend to specialize in N acquisition, whereas AM species might experience secondary N limitation under the enhanced P limitation induced by N addition [20,59]. This may explain the diverse variation in N concentration under N addition between the two mycorrhizal types. Similar to the variation in community P, the P concentration among EM species also decreased under N addition. Zhang et al. [33] found a significantly greater P resorption rate for EM-associated trees than for AM trees in a boreal forest, which may from another perspective bolster our explanation for the decrease in P during autumn. Moreover, the increased P level in AM species demonstrates a stronger capacity to acquire P. In addition, other properties are also important in driving the growth of a plant community, for example: The evolution of soil organic to humic matter, which may interact in the response of foliar N:P stoichiometry to N deposition [60,61].

4.3. Status of Nutrient Limitation

How to assess nutrient limitation in primary productivity is always a concern of ecologists, and a fertilization experiment is a frequently used method that defines nutrient limitation as a requirement for additional nutrients [26,47,62]. Accordingly, the improved foliar N level via N addition across the three seasons showed that N may be a constraint in this forest. Nevertheless, we cannot conclude that N is the sole limiting nutrient. In contrast, we infer that P is also a critical limiting nutrient for this forest because of the initiation of some adaptive strategies that aim to alleviate P limitation under N addition.

In addition to fertilization experiments, community foliar nutrient concentrations and ratios based on averages are widely applied for assessing nutrient limitation. Koerselman and Meuleman [27] found that a foliar N:P ratio <14 indicated that the plant was limited by N alone, that N:P >16 indicated limitation solely by P, and that 14 < N:P < 16 indicated colimitation. Later, Güsewell et al. [63] proposed a more conservative threshold value; i.e., N:P < 10 and N:P > 20 solely represented N limitation and P limitation, respectively. Recently, community N:P weighted by the *IV* or biomass of species is also being applied in some research [13,22]. In this study, the values of N:P across three seasons were all between 10 and 20, regardless of whether they were calculated using the weighted or average value. In addition, through a comparison with previous research during the peak season, our results show that the foliar P in the study region is lower than the worldwide average and that the foliar N:P is lower than the average for China. These results support the opinion of Han et al. [39], who found that foliar P across most areas in China was lower than the global average and that northern regions in China were more limited by N. More evidence regarding the existence of P limitation is reflected in the low-supplied soil AP. Through investigation of pools and distributions of soil P in China, Zhang et al. [64] found that dark brown forest soil, which is the zonal soil in this region, has a relatively high soil TP but a quite low soil AP compared to other soil types. The highest soil AP:TP ratio in this study was approximately 0.012 under ambient conditions. These results further prove our inference that N and P limitation coexist in this forest.

Studies have shown that the relative status of N vs. P limitation changes with succession [21,22]. Inputs of N and P have different sources; P is derived from rock weathering, whereas N can accumulate from the atmosphere through biological N fixation and atmospheric N deposition [65]. Therefore, young ecosystems lacking the input of atmospheric elements are usually limited by the supply of N, and with the development of succession, P will decrease to a limiting level because of the accumulation of soil organic carbon and soil TN [26]. The old-growth temperate forest in the present study has a long history of succession, which may explain the reason for the observed N and P colimitation. Wang and Moore [30] also recently reported colimitation of N and P in an ombrotrophic peatland in eastern Ontario, Canada, and Goswami et al. [20] found evidence of P limitation in most middle-aged and mature stands in central New Hampshire, USA. These two regions are all at latitudes similar to that of the region in this study.

5. Conclusions

This is the first study that explores the divergent responses of foliar N:P stoichiometry during different seasons to N deposition. In this study, plant leaf stoichiometric characteristics showed different variations with seasonal changes, and responses to N deposition in three seasons also differed. Thus, significant variations in leaf stoichiometric characteristics across these three seasons over just four months may be worth considering. Our research emphasizes the importance of multiple sampling across seasons in a temperate forest. Broadleaved species vs. conifers and AM species vs. EM species showed significant differences in seasonal variation and in response to N addition. Arguing against the traditional consensus, the net primary productivity of this mixed broadleaved-Korean pine (*Pinus koraiensis*) forest, an old-growth temperate forest and a critical forest type in Northeast China, is under N and P colimitation. The level of P limitation is enhanced by N deposition, and this N-induced

P limitation may be alleviated by the adjustment of strategies for plant adaption such as nutrient retranslocation and mycorrhizal association.

Author Contributions: G.J. conceived and designed the experiments; D.Y. and H.M. performed the experiments; D.Y. analyzed the data; D.Y. contributed reagents/materials/analysis tools; D.Y. and G.J. wrote the paper.

Acknowledgments: We thank three anonymous reviewers and editor for constructive suggestions to improve the quality of the manuscript.

References

1. Galloway, J.N.; Dentener, F.J.; Capone, D.G.; Boyer, E.W.; Howarth, R.W.; Seitzinger, S.P.; Asner, G.P.; Cleveland, C.C.; Green, P.A.; Holland, E.A.; et al. Nitrogen cycles: Past, present, and future. *Biogeochemistry* **2004**, *70*, 153–226. [CrossRef]
2. Neff, J.C.; Townsend, A.R.; Gleixner, G.; Lehman, S.J.; Turnbull, J.; Bowman, W.D. Variable effects of nitrogen additions on the stability and turnover of soil carbon. *Nature* **2002**, *419*, 915–917. [CrossRef] [PubMed]
3. Phoenix, G.K.; Emmett, B.A.; Britton, A.J.; Caporn, S.J.M.; Dise, N.B.; Helliwell, R.; Jones, L.; Leake, J.R.; Leith, I.D.; Sheppard, L.J.; et al. Impacts of atmospheric nitrogen deposition: Responses of multiple plant and soil parameters across contrasting ecosystems in long-term field experiments. *Glob. Chang. Biol.* **2012**, *18*, 1197–1215. [CrossRef]
4. Liu, X.; Duan, L.; Mo, J.; Du, E.; Shen, J.; Lu, X.; Zhang, Y.; Zhou, X.; He, C.; Zhang, F. Nitrogen deposition and its ecological impact in China: An overview. *Environ. Pollut.* **2011**, *159*, 2251–2264. [CrossRef] [PubMed]
5. Lu, X.; Mao, Q.; Gilliam, F.S.; Luo, Y.; Mo, J. Nitrogen deposition contributes to soil acidification in tropical ecosystems. *Glob. Chang. Biol.* **2014**, *20*, 3790–3801. [CrossRef] [PubMed]
6. Elser, J.J.; Bracken, M.E.; Cleland, E.E.; Gruner, D.S.; Harpole, W.S.; Hillebrand, H.; Ngai, J.T.; Seabloom, E.W.; Shurin, J.B.; Smith, J.E. Global analysis of nitrogen and phosphorus limitation of primary producers in freshwater, marine and terrestrial ecosystems. *Ecol. Lett.* **2007**, *10*, 1135–1142. [CrossRef] [PubMed]
7. Reich, P.B.; Oleksyn, J. Global patterns of plant leaf N and P in relation to temperature and latitude. *Proc. Natl. Acad. Sci. USA* **2004**, *101*, 11001–11006. [CrossRef] [PubMed]
8. Crowley, K.F.; McNeil, B.E.; Lovett, G.M.; Canham, C.D.; Driscoll, C.T.; Rustad, L.E.; Denny, E.; Hallett, R.A.; Arthur, M.A.; Boggs, J.L.; et al. Do nutrient limitation patterns shift from nitrogen toward phosphorus with increasing nitrogen deposition across the northeastern United States? *Ecosystems* **2012**, *15*, 940–957. [CrossRef]
9. Parfitt, R.L.; Ross, D.J.; Coomes, D.A.; Richardson, S.J.; Smale, M.C.; Dahlgren, R.A. N and P in new zealand soil chronosequences and relationships with foliar N and P. *Biogeochemistry* **2005**, *75*, 305–328. [CrossRef]
10. Clark, C.M.; Tilman, D. Loss of plant species after chronic low-level nitrogen deposition to prairie grasslands. *Nature* **2008**, *451*, 712–715. [CrossRef] [PubMed]
11. Peñuelas, J.; Sardans, J.; Rivas-ubach, A.; Janssens, I.A. The human-induced imbalance between C, N and P in earth's life system. *Glob. Chang. Biol.* **2012**, *18*, 3–6. [CrossRef]
12. Penuelas, J.; Poulter, B.; Sardans, J.; Ciais, P.; van der Velde, M.; Bopp, L.; Boucher, O.; Godderis, Y.; Hinsinger, P.; Llusia, J.; et al. Human-induced nitrogen-phosphorus imbalances alter natural and managed ecosystems across the globe. *Nat. Commun.* **2013**, *4*, 2934. [CrossRef] [PubMed]
13. Han, X.; Sistla, S.A.; Zhang, Y.H.; Lü, X.T.; Han, X.G. Hierarchical responses of plant stoichiometry to nitrogen deposition and mowing in a temperate steppe. *Plant Soil* **2014**, *382*, 175–187. [CrossRef]
14. Magill, A.H.; Aber, J.D.; Currie, W.S.; Nadelhoffer, K.J.; Martin, M.E.; McDowell, W.H.; Melillo, J.M.; Steudler, P. Ecosystem response to 15 years of chronic nitrogen additions at the Harvard forest LTER, Massachusetts, USA. *For. Ecol. Manag.* **2004**, *196*, 7–28. [CrossRef]
15. Mayor, J.R.; Wright, S.J.; Turner, B.L.; Austin, A. Species-specific responses of foliar nutrients to long-term nitrogen and phosphorus additions in a lowland tropical forest. *J. Ecol.* **2014**, *102*, 36–44. [CrossRef]
16. Deng, M.; Liu, L.; Sun, Z.; Piao, S.; Ma, Y.; Chen, Y.; Wang, J.; Qiao, C.; Wang, X.; Li, P. Increased phosphate uptake but not resorption alleviates phosphorus deficiency induced by nitrogen deposition in temperate *Larix principis-rupprechtii* plantations. *New Phytol.* **2016**, *212*, 1019–1029. [CrossRef] [PubMed]

17. Aber, J.D.; Nadelhoffer, K.J.; Steudler, P.; Melillo, J.M. Nitrogen saturation in northern forest ecosystem. *Bioscience* **1989**, *39*, 378–386. [CrossRef]
18. Elser, J.J.; Fagan, W.F.; Kerkhoff, A.J.; Swenson, N.G.; Enquist, B.J. Biological stoichiometry of plant production: Metabolism, scaling and ecological response to global change. *New Phytol.* **2010**, *186*, 593–608. [CrossRef] [PubMed]
19. Ågren, G.I. Stoichiometry and nutrition of plant growth in natural communities. *Annu. Rev. Ecol. Evol. Syst.* **2008**, *39*, 153–170. [CrossRef]
20. Goswami, S.; Fisk, M.C.; Vadeboncoeur, M.A.; Garrison-Johnston, M.; Yanai, R.D.; Fahey, T.J. Phosphorus limitation of aboveground production in northern hardwood forests. *Ecology* **2018**, *99*, 438–449. [CrossRef]
21. Davidson, E.A.; de Carvalho, C.J.; Figueira, A.M.; Ishida, F.Y.; Ometto, J.P.; Nardoto, G.B.; Saba, R.T.; Hayashi, S.N.; Leal, E.C.; Vieira, I.C.; et al. Recuperation of nitrogen cycling in amazonian forests following agricultural abandonment. *Nature* **2007**, *447*, 995–998. [CrossRef]
22. Zhang, W.; Zhao, J.; Pan, F.; Li, D.; Chen, H.; Wang, K. Changes in nitrogen and phosphorus limitation during secondary succession in a karst region in southwest China. *Plant Soil* **2015**, *391*, 77–91. [CrossRef]
23. Chapin, F.S. The mineral nutrition of wild plants. *Ann. Rev. Ecol. Syst.* **1980**, *11*, 233–260. [CrossRef]
24. Chapin, F.S.; Vitousek, P.M.; Van Cleve, K. The nature of nutrient limitation in plant communities. *Am. Nat.* **1986**, *127*, 48–58. [CrossRef]
25. Vitousek, P.M.; Howarth, R.W. Nitrogen limitation on land and in the sea: How can it occur? *Biogeochemistry* **1991**, *13*, 87–115. [CrossRef]
26. Vitousek, P.M.; Porder, S.; Houlton, B.Z.; Chadwick, O.A. Terrestrial phosphorus limitation: Mechanisms, implications, and nitrogen–phosphorus interactions. *Ecol. Appl.* **2010**, *20*, 5–15. [CrossRef]
27. Koerselman, W.; Meuleman, A.F.M. The vegetation N:P ratio: A new tool to detect the nature of nutrient limitation. *J. Appl. Ecol.* **1996**, *33*, 1441–1450. [CrossRef]
28. Townsend, A.R.; Cleveland, C.C.; Asner, G.P.; Bustamamte, M.M.C. Controls over foliar N:P ratios in tropical forests. *Ecology* **2007**, *88*, 107–118. [CrossRef]
29. Fife, D.N.; Nambiar, E.K.S.; Saur, E. Retranslocation of foliar nutrients in evergreen tree species planted in a mediterranean environment. *Tree Physiol.* **2008**, *28*, 187–196. [CrossRef]
30. Wang, M.; Moore, T.R. Carbon, nitrogen, phosphorus, and potassium stoichiometry in an ombrotrophic peatland reflects plant functional type. *Ecosystems* **2014**, *17*, 673–684. [CrossRef]
31. Lambers, H.; Raven, J.A.; Shaver, G.R.; Smith, S.E. Plant nutrient-acquisition strategies change with soil age. *Trends Ecol. Evol.* **2008**, *23*, 95–103. [CrossRef]
32. Jeffries, P.; Gianinazzi, S.; Perotto, S.; Turnau, K.; Barea, J.M. The contribution of arbuscular mycorrhizal fungi in sustainable maintenance of plant health and soil fertility. *Biol. Fertil. Soils* **2003**, *37*, 1–16.
33. Zhang, H.Y.; Lu, X.T.; Hartmann, H.; Keller, A.; Han, X.G.; Trumbore, S.; Phillips, R.P. Foliar nutrient resorption differs between arbuscular mycorrhizal and ectomycorrhizal trees at local and global scales. *Glob. Ecol. Biogeogr.* **2018**, *27*, 875–885. [CrossRef]
34. Smith, S.E.; Smith, F.A. Roles of arbuscular mycorrhizas in plant nutrition and growth: New paradigms from cellular to ecosystem scales. *Annu. Rev. Plant Biol.* **2011**, *62*, 227–250. [CrossRef]
35. Van der Heijden, M.G.; Martin, F.M.; Selosse, M.A.; Sanders, I.R. Mycorrhizal ecology and evolution: The past, the present, and the future. *New Phytol.* **2015**, *205*, 1406–1423. [CrossRef]
36. Read, D.J.; Perez-Moreno, J. Mycorrhizas and nutrient cycling in ecosystems—A journey towards relevance? *New Phytol.* **2003**, *157*, 475–492. [CrossRef]
37. Shi, W.; Wang, Z.Q.; Liu, J.L.; Gu, J.C.; Guo, D.L. Fine root morphology of twenty hardwood species in maoershan natural secondary forest northeastern China. *J. Plant Ecol.* **2008**, *32*, 1217–1226.
38. Elser, J.J.; Sterner, R.W.; Gorokhova, E.; Fagan, W.F.; Markow, T.A.; Cotner, J.B.; Harrison, J.F.; Hobbie, S.E.; Odell, G.M.; Weider, L.J. Biological stoichiometry from genes to ecosystems. *Ecol. Lett.* **2000**, *3*, 540–550. [CrossRef]
39. Han, W.; Fang, J.; Guo, D.; Zhang, Y. Leaf nitrogen and phosphorus stoichiometry across 753 terrestrial plant species in China. *New Phytol.* **2005**, *168*, 377–385. [CrossRef]
40. Sterner, R.W.; Elser, J.J. *Ecological Stoichiometry: The Biology of Elements from Molecules to the Biosphere*; Princeton University Press: Princeton, NJ, USA, 2002.
41. Aronsson, A.; Elowson, S.; Persson, T. Effects of irrigation and fertilization on mineral nutrients in scots pine needles. *Ecol. Bull.* **1980**, *32*, 219–228.

42. Son, Y.; Lee, I.K.; Ryu, S.R. Nitrogen and phosphorus dynamics in foliage and twig of pitch pine and Japanese larch plantations in relation to fertilization. *J. Plant Nutr.* **2000**, *23*, 697–710. [CrossRef]
43. Güsewell, S. N:P ratios in terrestrial plants: Variation and functional significance. *New Phytol.* **2004**, *164*, 243–266. [CrossRef]
44. Reich, P.B.; Uhl, C.; Walters, M.B.; Ellsworth, D.S. Leaf lifespan as a determinant of leaf structure and function among 23 amazonian tree species. *Oecologia* **1991**, *86*, 16–24. [CrossRef]
45. Reich, P.B.; Walters, M.B.; Ellsworth, D.S. From tropics to tundra: Global convergence in plant functioning. *Proc. Natl. Acad. Sci. USA* **1997**, *94*, 13730–13734. [CrossRef]
46. Westoby, M.; Falster, D.S.; Moles, A.T.; Vesk, P.A.; Wright, I.J. Plant ecological strategies: Some leading dimensions of variation between species. *Annu. Rev. Ecol. Syst.* **2002**, *33*, 125–159. [CrossRef]
47. Li, Y.; Niu, S.; Yu, G. Aggravated phosphorus limitation on biomass production under increasing nitrogen loading: A meta-analysis. *Glob. Chang. Biol.* **2016**, *22*, 934–943. [CrossRef]
48. Perring, M.P.; Hedin, L.O.; Levin, S.A.; McGroddy, M.; de Mazancourt, C. Increased plant growth from nitrogen addition should conserve phosphorus in terrestrial ecosystems. *Proc. Natl. Acad. Sci. USA* **2008**, *105*, 1971–1976. [CrossRef]
49. Cleveland, C.C.; Houlton, B.Z.; Smith, W.K.; Marklein, A.R.; Reed, S.C.; Parton, W.; Del Grosso, S.J.; Running, S.W. Patterns of new versus recycled primary production in the terrestrial biosphere. *Proc. Natl. Acad. Sci. USA* **2013**, *110*, 12733–12737. [CrossRef]
50. Phoenix, G.K.; Booth, R.E.; Leake, J.R.; Read, D.J.; Grime, J.P.; Lee, J.A. Simulated pollutant nitrogen deposition increases P demand and enhances root-surface phosphatase activities of three plant functional types in a calcareous grassland. *New Phytol.* **2003**, *161*, 279–290. [CrossRef]
51. Treseder, K.K. A meta-analysis of mycorrhizal responses to nitrogen, phosphorus, and atmospheric CO_2 in field studies. *New Phytol.* **2004**, *164*, 347–355. [CrossRef]
52. Long, M.; Wu, H.H.; Smith, M.D.; La Pierre, K.J.; Lü, X.T.; Zhang, H.Y.; Han, X.G.; Yu, Q. Nitrogen deposition promotes phosphorus uptake of plants in a semi-arid temperate grassland. *Plant Soil* **2016**, *408*, 475–484. [CrossRef]
53. Yu, Q.; Chen, Q.; Elser, J.J.; He, N.; Wu, H.; Zhang, G.; Wu, J.; Bai, Y.; Han, X. Linking stoichiometric homoeostasis with ecosystem structure, functioning and stability. *Ecol. Lett.* **2010**, *13*, 1390–1399. [CrossRef] [PubMed]
54. Nambiar, E.K.S.; Fife, D.N. Nutrient retranslocation in temperate conifers. *Tree Physiol.* **1991**, *9*, 185–207. [CrossRef]
55. Knorr, M.; Frey, S.D.; Curtis, P.S. Nitrogen additions and litter decomposition: A meta-analysis. *Ecology* **2005**, *86*, 3252–3257. [CrossRef]
56. Brundrett, M.C. Mycorrhizal associations and other means of nutrition of vascular plants: Understanding the global diversity of host plants by resolving conflicting information and developing reliable means of diagnosis. *Plant Soil* **2009**, *320*, 37–77. [CrossRef]
57. Read, D.J.; Lewis, D.H.; Fitter, A.H.; Alexander, I.J.; Battley, E.H. Mycorrhizas in ecosystems. *Experientia* **1991**, *47*, 376–391. [CrossRef]
58. Averill, C.; Turner, B.L.; Finzi, A.C. Mycorrhiza-mediated competition between plants and decomposers drives soil carbon storage. *Nature* **2014**, *505*, 543–545. [CrossRef]
59. Smith, S.; Read, D. Mycorrhizal symbiosis. *Q. Rev. Biol.* **2008**, *3*, 273–281.
60. Pizzeghello, D.; Nicolini, G.; Nardi, S. Hormone-like activity of humic substances in *Fagus sylvaticae* forests. *New Phytol.* **2001**, *151*, 647–657. [CrossRef]
61. Pizzeghello, D.; Nicolini, G.; Nardi, S. Hormone-like activities of humic substances in different forest ecosystems. *New Phytol.* **2002**, *155*, 393–402. [CrossRef]
62. Gilbert, J.B.; Lawesj, H. Agricultural, botanical, and chemical results of experiments on the mixed herbage of permanent meadow, conducted for more than twenty years in succession on the same land. Part I. *Philos. Trans. R. Soc. Lond.* **1880**, *171*, 289–416.
63. Güsewell, S.; Koerselman, W.; Verhoeven, J.T.A. Biomass N:P ratios as indicators of nutrient limitation for plant populations in wetland. *Ecol. Appl.* **2003**, *13*, 372–384. [CrossRef]
64. Zhang, C.; Tian, H.Q.; Liu, J.Y.; Wang, S.Q.; Liu, M.L.; Pan, S.F.; Shi, X.Z. Pools and distributions of soil phosphorus in China. *Glob. Biogeochem. Cycles* **2005**, *19*. [CrossRef]
65. Walker, T.W.; Syers, J.K. The fate of phosphorus during pedogenesis. *Geoderma* **1976**, *15*, 1–19. [CrossRef]

12

Effects of Hurricane-Felled Tree Trunks on Soil Carbon, Nitrogen, Microbial Biomass and Root Length in a Wet Tropical Forest

D. Jean Lodge [1,*], Dirk Winter [2], Grizelle González [3] and Naomi Clum [4]

1 United States Department of Agriculture, Forest Service, Northern Research Station, Luquillo 00773-1377, Puerto Rico
2 Department of Psychiatry, Columbia University, 180 Ft. Washington Ave., New York, NY 10032, USA; dcw4@caa.columbia.edu
3 United States Department of Agriculture, Forest Service, International Institute of Tropical Forestry, Jardín Botánico Sur, 1201 Ceiba St.-Río Piedras 00926, Puerto Rico; ggonzalez@fs.fed.us
4 Carmel Valley Middle School, 3800 Mykonos Lane, San Diego, CA 92007, USA; naomi.clum@sdhusd.net
* Correspondence: dlodge@fs.fed.us or dlodgester@gmail.com or djlodge@caribe.net

Academic Editors: Scott X. Chang and Xiangyang Sun

Abstract: Decaying coarse woody debris can affect the underlying soil either by augmenting nutrients that can be exploited by tree roots, or by diminishing nutrient availability through stimulation of microbial nutrient immobilization. We analyzed C, N, microbial biomass C and root length in closely paired soil samples taken under versus 20–50 cm away from large trunks of two species felled by Hugo (1989) and Georges (1998) three times during wet and dry seasons over the two years following the study conducted by Georges. Soil microbial biomass, % C and % N were significantly higher under than away from logs felled by both hurricanes (i.e., 1989 and 1998), at all sampling times and at both depths (0–10 and 10–20 cm). Frass from wood boring beetles may contribute to early effects. Root length was greater away from logs during the dry season, and under logs in the wet season. Root length was correlated with microbial biomass C, soil N and soil moisture (R = 0.36, 0.18, and 0.27, respectively; all p values < 0.05). Microbial biomass C varied significantly among seasons but differences between positions (under vs. away) were only suggestive. Microbial C was correlated with soil N (R = 0.35). Surface soil on the upslope side of the logs had significantly more N and microbial biomass, likely from accumulation of leaf litter above the logs on steep slopes. We conclude that decaying wood can provide ephemeral resources that are exploited by tree roots during some seasons.

Keywords: coarse woody debris; nitrogen cycling; soil carbon; soil nitrogen; tree roots; bark beetle frass; fine root proliferation; nutrient hotspots

---

1. Introduction

Coarse woody (CWD) debris can directly and indirectly affect mineral nutrient availability in underlying soil through inputs of nutrients and carbon compounds in leachates, invertebrate frass and particulate decay products [1–7]. CWD as defined by Harmon et al. [3] is fallen or standing dead wood at least 10 cm diameter. Nutrient hotspots that develop in soil beneath decaying CWD may be important for maintaining forest soil fertility and provide important resources for trees [5]. As noted by Johnson et al. (2014) [8], the influence of soil nutrient hotspots and hot moments on plant growth is obscured by pooling samples, averaging values and eliminating outliers from data sets since soil nutrient availability to plants often responds non-linearly to nutrient hotspots and additions of organic matter [9]. Studies in tropical montane [10] and lowland tropical forests [11–14] show that fine roots of trees differentially proliferate in nutrient hotspots (Sayer and Banin 2016) [15]. Two studies in Australian woodlands [16,17] showed significantly higher C, N and P in soil near logs than 80 cm away, but other studies elsewhere showed decreases in soil N associated with CWD [18–21]. A companion study to ours by Zalamea et al. (2016) [22] that was conducted in different plots at Bisley and Río Chiquito located on the opposite side of the Luquillo Mountains from our site, and using somewhat different tree species found no increase in C content in their shallower soil samples (0–5 cm) under decomposing logs than the deeper strata we studied (0–10 and 10–20 cm), though Zalamea et al. (2007) [23] did find more water extractable organic matter beneath logs. Reduced soil N concentrations associated with CWD in North American temperate forests are thought to result from increased leachates with high C:N ratios which depress rates of N mineralization [18–20], while lower extractable N with large hurricane inputs of CWD in a tropical forest was attributed to increased microbial immobilization [21]. Tropical and subtropical forests have fast wood decay rates [7,24,25] and soils that are often dominated by highly weathered iron and aluminum clays that bind tightly to phosphorus and make it less available (P-fixing clays). In such tropical and subtropical sites, increases in soil C from the decay of CWD are predicted by the Century model to increase the amount of readily exchangeable phosphorus beginning several years after wood addition, thereby stimulating forest productivity [26]. CWD can also alter the physical characteristics of the underlying soil, such as pH, moisture and temperature [3,6,16,17,22].

Natural events such as storms, fires and bark beetle outbreaks as well as anthropogenic disturbances such as logging and salvage operations can deposit CWD on forest floors, potentially contributing to spatial heterogeneity of soil carbon and mineral nutrient availability [18]. Increasing sea surface temperatures due to climate change are predicted to produce more frequent high-intensity hurricanes and typhoons [27–29], while increasing winter temperatures and drought associated with climate change favor increased bark beetle outbreaks and forest fires in North America [30,31]. Given the increasing frequency of forest disturbance events, it is important to understand the effects of CWD on soil carbon, soil fertility and forest productivity.

Whether soils under CWD create hotspots of nutrient availability that are exploited by tree roots, and conversely, whether nutrient sinks under logs caused by microbial immobilization of limiting nutrients are avoided by tree roots, have received little attention. However, production of fine roots in Amazonian forest leaf litter was shown to be a response to nutrient hotspots [32], while root proliferation in response to localized fertilization has been found useful for determining which mineral nutrients are limiting to tree growth in Hawaii [13] and the Amazon [11]. These studies indicate that root length can be used as a bioindicator of where limiting resources are available in nutrient hotspots.

The objectives of this study in a tropical forest of Puerto Rico were to determine if soil N and C and tree root lengths were higher under as compared to away from decomposing logs; whether soil microbial biomass and tree root lengths in the same samples varied inversely with each other, thus suggesting competitive exclusion; whether within paired samples, tree roots were more abundant in resource hotspots; and whether there were temporal patterns in soil N, C, microbial biomass C or root length that varied with soil moisture.

2. Materials and Methods

2.1. Study Area

The El Verde Research Area of the Luquillo Experimental Forest (LEF), where this study was conducted, is on the western side of the Luquillo Mountains (18°18′ N; 65°49′ W; Figure 1). The sites ranged from 350–400 m asl. Mean annual rainfall is 3500 mm with a weakly seasonal regime, and mean temperatures are 22–25 °C [32]. The area is classified as subtropical wet forest based on latitude, potential evapotranspiration and low seasonality according to the Holdridge Life Zone system [32], though the moderating effect of the surrounding ocean on temperatures make this comparable to a tropical wet climate despite the higher latitude.

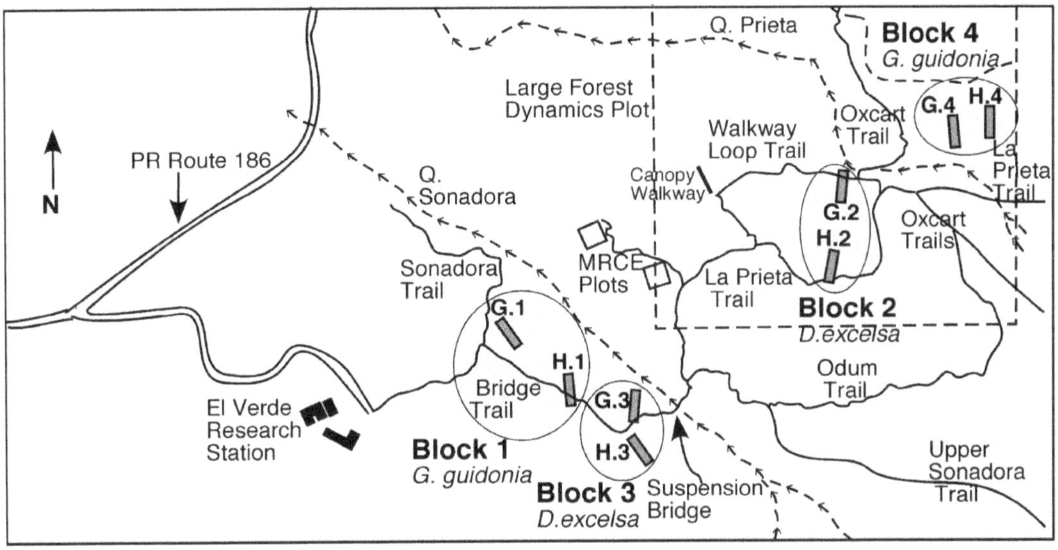

Figure 1. Section of the El Verde Research Area in the Luquillo Experimental Forest of Puerto Rico where this study was conducted, showing the four blocks, each with a pair of logs belonging to two different hurricane cohorts (Hurricane Hugo in 1989 and Hurricane Georges in 1998). The first letter of the log code refers to the hurricane (G for Georges and H for Hugo) while the number refers to the block. Logs in two of the blocks were *Dacryodes excelsa* Vahl. While the logs in the other two blocks were *Guarea guidonia* (L.) Sleumer. Direction of stream flow in the Quebrada Prieta and Quebrada Sonadora is indicated by arrows. The logs are not shown to scale.

Soils in the study area are weathered oxisol clay loams and clays (30%–55% clay, 28%–35% silt and 18%–34% sand) in the Cristal and Coloso series [33], (Table 1). These clays have a high capacity for binding to phosphorus, but the upper horizon has high soil organic matter content and moderately slow permeability. The Coloso clay series has a higher cation exchange capacity than the Cristal clay loams (43 vs. 29 CEC C^{-1}, respectively) [33]. Litter decomposes quickly and humus is usually absent between the forest floor litter and the mineral soil [34], as is typical of mull soils. Leaf litter accumulates on the upslope side of debris dams on slopes >25% [9], resulting in higher soil carbon above than below permanent debris dams [35].

Table 1. Location of study blocks and their soil types in the El Verde Research Area of the Luquillo Experimental Forest.

Location	Block	Latitude, Longitude	Soil Type
Q. Sonadora watershed	1	18°19′24″ N, 65°49′03″ W	Cristal clay-clay loam
Q. Prieta watershed in LFDP cells 02.09 and 07.10	2	18°19′30.7″ N, 65°48′56.6″ W 18°19′31″ N, 65°49′02″ W	Cristal clay-clay loam
Q. Sonadora watershed	3	18°19′24″ N, 65°49′04″ W	Cristal clay-clay loam
Q. Prieta watershed in LFDP cells 10.15 and 10.16	4	18°19′37″ N, 65°49′00″ W 18°19′27″ N, 65°49′00″ W	Coloso clay

2.2. Sampling Design

We examined the effects of decomposing logs of two tree species originating from two major hurricanes, Hugo and Georges, on soil carbon (C) nitrogen (N) content, microbial biomass C and tree root length in the underlying soil. Soil samples collected at two depths beneath the fallen logs were spatially and temporally paired with samples collected 20–50 cm away. This study on the western side of the Luquillo Mountains was conducted when the hurricane-felled logs were 0.6–1.1 and 9.6–11 years after treefall in the two hurricane cohorts, respectively, and was a precursor to a separate study on the eastern side of the Luquillo Mountains by Zalamea et al. [22,23] when their logs had decayed 6 and 15 years. This study differs in that soils were collected from 0–10 and 10–20 cm depths versus 0–5 cm in Zalamea et al. [22,23], and we used the chloroform fumigation-incubation method [36] to estimate soil microbial biomass C rather than substrate induced respiration technique [37] used in Zalamea et al. [22].

Four areas (blocks) were selected for study, each with a pair of logs of the same species that were at least 30 cm diameter, with one felled by Hurricane Hugo and one felled by Hurricane Georges (Figure 1). The two species were *Dacryodes excelsa* Vahl. (Burseraceae) and *Guarea guidonia* (L.) Sleumer (Meliaceae); there were two replicate blocks of each tree species. Fallen trees were >30–80 cm in diameter and at least half of the trunk was in contact with the ground (diameters are given in Table 2). Slopes varied from 20%–40% (Table 2). All logs were decayed by white-rot basidiomycete fungi based on the bleached appearance and spongy-fibrous texture, though rates of decay varied among logs (see Table 2 for details). We used the four-class decay system proposed by Torres (1994) [38] as the decay classification used for logs in temperate forests works poorly in the tropics where the bark can be retained on extensively decayed logs. At the time of this study, the youngest logs (0.6–2 years-old) belonging to the Georges cohort were in decay classes I: no detectable decay, and II: intact bark, sapwood partially soft, and few invading roots present. The older logs belonging to the Hugo cohort (9.5–11 years-old) were in decay II to class III: bark partially lost, sapwood soft, roots invading.

Table 2. Characteristics of logs that were used in this study. The label is comprised of the first letter of the hurricane that felled the tree, followed by the block number. Tree species were *Guarea guidonia* (Meliaceae) and *Dacryodes excelsa* (Burseraceae) but tree species were found to be similar and were not part of the model. Decay class was rated based on description of the logs were selected for study according to the classification of Torres (1994) [38]. Decay rate was classified 18 and 27 years after Hurricane Georges and Hugo, respectively: slow—retaining some bark, diameter loss less than 5 cm; moderate—no bark remaining, diameter loss more than 5 cm but wood remaining; fast—either in the form of a mound of humus invaded by roots or no visible remains). Slope (%) was measured where the logs were located.

Location	Label	Hurricane	Tree Species	Diam. (cm)	Decay Class	Decay Rate	Slope (%)
Q. Sonadora	H.1	Hugo 1989	*G. guidonia*	35–55	2	Slow	28
Q. Sonadora	G.1	Georges 1998	*G. guidonia*	38–80	1	Slow	35
Q. Prieta, LFDP cell 02.09	H.2	Hugo 1989	*D. excelsa*	35–52	3	Fast	35
Q. Prieta, LFDP cell 07.10	G.2	Georges 1998	*D. excelsa*	30–35	2	Fast	40
Q. Sonadora	H.3	Hugo 1989	*D. excelsa*	30–43	2	Moderate	20
Q. Sonadora	G.3	Georges 1998	*D. excelsa*	30–55	1	Slow	40
LFDP cell 10.16	H.4	Hugo 1989	*G. guidonia*	39–53	3	Fast	25
LFDP cell 10.15	G.4	Georges 1998	*G. guidonia*	45–70	2	Fast	25

2.3. Soil Sampling and Analysis

2.3.1. Soil Sampling

Volumetric soil cores (8×8 cm) were extracted to 10 cm depth using a two-part aluminum corer as shown in Figure 2. Soil cores were extracted from under logs by digging an approach trench and driving the rectangular corer in horizontally (Figure 2). Each core from under a log was spatially and temporally paired with a core extracted vertically and located 20–50 cm from the first core and away from the log. These are hence referred to as "under" and "away" positions. On each sample date, one pair of cores was extracted from each log on the uphill side of the hill slope, and a second pair was extracted on the downhill side of the same log to balance the effects of the logs acting as debris dams on steep slopes [9,35] (Figure 2). Samples were not pooled.

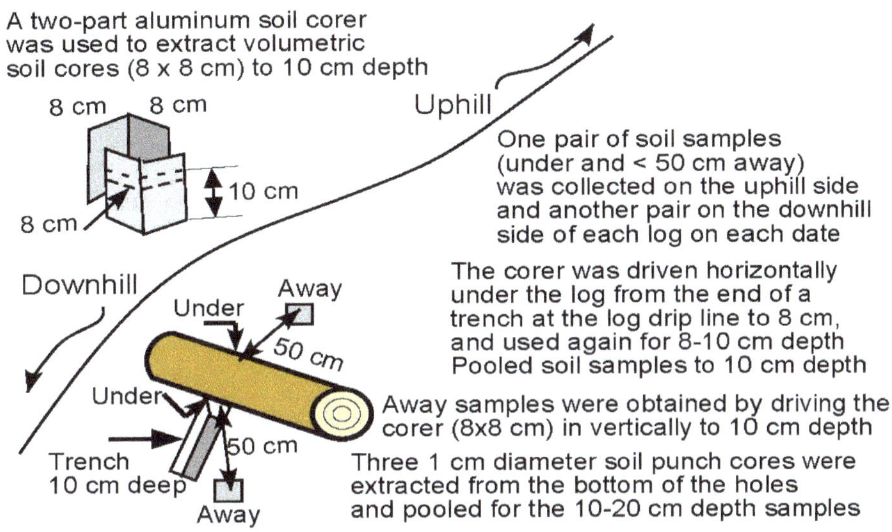

Figure 2. A rectangular, 2-part soil corer was driven into the soil horizontally from the end of an approach trench to obtain volumetric soil samples from 0 to 10 cm depth under logs. Corer halves were used twice (0–8 and 8–10 cm) to obtain soil to 10 cm depth. Paired samples taken away from the logs were extracted vertically and located 20–50 cm away from the under-log core. Two pairs of cores were extracted on each date, one pair on the uphill side and the other on the downhill side of the log. Paired sample locations were randomly selected on each side of the log except for avoiding previously cored areas and large rocks. Soil samples from the 10–20 cm depth were extracted from the bottom of the first core holes using a punch corer.

There were three sampling times: April to early May 1999 (dry season); January 2000 (late wet season) and September 2000 (mid-wet season). These times corresponded to 0.6, 1.4 and 2 years after Hurricane Georges, and 9.5, 10.4 and 11 years after Hurricane Hugo, respectively. Samples in the same block were collected on the same day, and all samples for a particular time were collected within a 12-day period.

2.3.2. Soil Nitrogen and Carbon Content

Soil samples were weighed, hand-sorted to separate soil, roots and rocks; the roots and rocks were then weighed. A 15–20 g subsample was weighed into an aluminum pan, oven dried at 40 °C and re-weighed to determine soil % moisture. Remaining soil was air dried and stored, then oven dried at 65 °C. Total percent soil C and N were determined at the United States Forest Service (USFS) International Institute of Tropical Forestry laboratory using a LECO-2000 CNS analyzer [39] following the procedure of Vitousek and Matson [40]. Two samples of fresh bark beetle frass were also collected from different logs and analyzed for C and N using the CNS analyzer. We were unable to analyze the ammonium and nitrate we had extracted with KCl from fresh soils because of hurricane damage to facilities and infrastructure caused by Hurricane Georges.

2.3.3. Soil Biota

Root length was used as an indicator of resource hotspots. Roots were separated into coarse (>2 mm diameter) and fine (up to 2 mm diameter) from the 0–10 cm depth core samples, weighed and cut into ca. 1 cm long pieces. The decision to estimate root length was made after the first core (G.2) was processed, so those data are missing. Estimates of root lengths were made from three weighed subsamples per sample using the line-intercept method of Newmann [41]. Total root length was estimated by multiplying the mean root length per g of roots by the total weight of the roots. Analyses were made on total root length (fine plus coarse) as coarse roots were negligible.

Soil microbial biomass C was estimated using the chloroform fumigation-incubation method [36]. Soil moisture was adjusted if there was less than 55% field capacity in two 30–40 g subsamples the day after collection. One of the subsamples from each sample was fumigated with chloroform under vacuum, reinoculated with soil and incubated for 10 days in air-tight jars with NaOH for trapping the CO_2 respired; the unfumigated control was also incubated with a CO_2 trap. After 10 days, the amount of CO_2 trapped in the NaOH solution was determined by titration with HCl using phenolphthalein indicator dye. CO_2 in the NaOH stock solution was also determined using titration, and the amount subtracted from the total in the incubation traps. Respired CO_2 in the controls was subtracted from that in the fumigated samples according to the formula in Jenkins and Powlson [36].

2.4. Statistical Analyses

The design was a repeated measures nested randomized complete block. Dependent variables (soil % C, soil % N, microbial biomass C, total root length and soil moisture at 0–10 cm depth) were analyzed separately using a Repeated Measures Analysis; the 10–20 cm depth was only sampled twice for % C and % N, so those data were analyzed separately as a Repeated Measures nested randomized complete block. The independent variables (fixed effects in the models) were hurricane, position (under versus away from logs), hillslope direction (uphill or downhill), and time after Hurricane Georges (0.6, 1.4 and 2 years). Random effects of logs were nested within hurricanes. Block 2 was omitted from the root length analysis because of missing data for log G.2. All statistical analyses were performed using SAS Institute (version 9.3, [42]) software for a generalized linear mixed model using the pseudo-likelihood estimation technique via PROC GLIMMIX. In all analyses, either the normal or lognormal distribution and an identity link function best fit the data. For % N and root length at 0–10 cm, we had to add the groups = hillslope and groups = position option, respectively, to adjust for heterogeneity of variance. In all cases, the correlation between the repeated time was best fit by the unstructured covariance structure. Model selection was based on fit statistics using HQIC (see Table 3 in Results). All models used the Identity Link Function, and all possible interactions were included. No differences were found between tree species, so they were treated as replicates in all the analyses.

Pearson's Product Moment correlations were used to analyze relationships among root length, soil % N, soil % moisture and microbial biomass C. Analyses were performed in Excel. A backwards elimination multiple linear regression was performed on the April 1999 and September 2000 data from the 0–10 cm depth using SAS software [42] after determining that all the variables met assumptions of normality and homoscedasticity of variance. The dependent variable was total root length, and the independent variables were soil microbial C, soil % moisture and soil % N.

Table 3. Summary of the best fit statistical analysis models and results. Probability annotations are as follows: suggestive—sug., $p < 0.1$; * $p < 0.05$; ** $p < 0.01$; *** $p < 0.001$.

Variable (s)	Levels	Probability
0–10 cm soil horizon		
Carbon (total %) Lognormal, Identity Link function, Repeated Measures GLIMMIX, HQIC fit statistics, all possible interactions included in model (shown if significant)		
Position (Under vs. Away)	2	0.0324 *
Hurricane within Block (Hugo vs. Georges)	2	0.3264
Hillslope direction (Upslope vs. Downslope)	2	0.8188
Time (0.6, 1.4 and 2 years post-Georges)	3	0.0374 *
Block (log number, Hurricane nested within Block) true replicates	4	

Table 3. *Cont.*

Variable (s)	Levels	Probability
Nitrogen (total %) Lognormal, Identity Link function, Repeated Measures GLIMMIX, HQIC fit statistics, all possible interactions included in model (shown if significant)		
Position (Under vs. Away)	2	0.0020 **
Hurricane within Block (Hugo vs. Georges)	2	0.2177
Hillslope direction (Upslope vs. Downslope)	2	0.0420 *
Time (0.6, 1.4 and 2 years post-Georges)	3	0.6816
Block (log number, Hurricane nested within Block) true replicates	4	
Microbial biomass C (mg·kg^{-1} soil) Gaussian, Identity Link function, Repeated Measures GLIMMIX, HQIC fit statistics, all possible interactions included (shown if significant)		
Position (Under vs. Away)	2	0.0840 sug.
Hurricane within Block (Hugo vs. Georges)	2	0.9038
Hillslope direction (Upslope vs. Downslope)	2	0.8660
Time (0.6, 1.4 and 2 years post-Georges)	3	< 0.001 ***
Block (log number, Hurricane nested within Block) true replicates	4	
Root length (total) Lognormal, Identity Link function, Repeated Measures GLIMMIX, HQIC fit statistics, all possible interactions included (shown if significant)		
Position (Under vs. Away)	2	0.9049
Hurricane within Block (Hugo vs. Georges)	2	0.0440 *
Hillslope direction (Upslope vs. Downslope)	2	0.0250 *
Time (0.6, 1.4 and 2 years post-Georges)	2	0.0182 *
Block (log number, Hurricane nested within Block) true replicates, deleted one incomplete block	3	
Position by Time interaction	1 DF	0.0368 *
10–20 cm soil horizon		
Carbon (total %) Lognormal, Identity Link function, Repeated Measures GLIMMIX, HQIC fit statistics, all possible interactions included (shown if significant)		
Position (Under vs. Away)	2	0.0273 *
Hurricane within Block (Hugo vs. Georges)	2	0.5899
Hillslope direction (Upslope vs. Downslope)	2	0.7656
Time (0.6, and 1.4 years post-Georges)	2	0.1673
Block (log number, Hurricane nested within Block) true replicates	4	
Hurricane by position interaction		0.0870 sug.
Nitrogen (total %) Lognormal, Identity Link function, Repeated Measures GLIMMIX, HQIC fit statistics, all possible interactions included (shown if significant)		
Position (Under vs. Away)	2	0.0188 *
Hurricane within Block (Hugo vs. Georges)	2	0.7820
Hillslope direction (Upslope vs. Downslope)	2	0.8711
Time (0.6, and 1.4 years post-Georges)	2	0.0579 sug.
Block (log number, Hurricane nested within Block) true replicates	4	

3. Results

3.1. Soil Nitrogen and Carbon

Soil C and N were higher under than away from logs of both hurricanes (the 1989 and 1998 cohorts; Figure 3) and hurricane did not contribute significantly to these models (Table 3). All of the C and N analysis models used a Lognormal distribution and an identity link function based on best fit statistics (Table 3). Significant differences in both C and N concentration (%) were found between under and away from log positions at the 0–10 and 10–20 cm depths (Table 3). Hillslope direction (upslope versus downslope) was not significant except for soil % N at the 0–10 cm depth; N was higher on the upslope side. In most comparisons within logs on a given date, mean C and N (mean of upslope and downslope) were higher under than away from the logs (Figure 3). Percent soil C and N differed significantly among sample dates, but only at the 0–10 cm depth (Table 3).

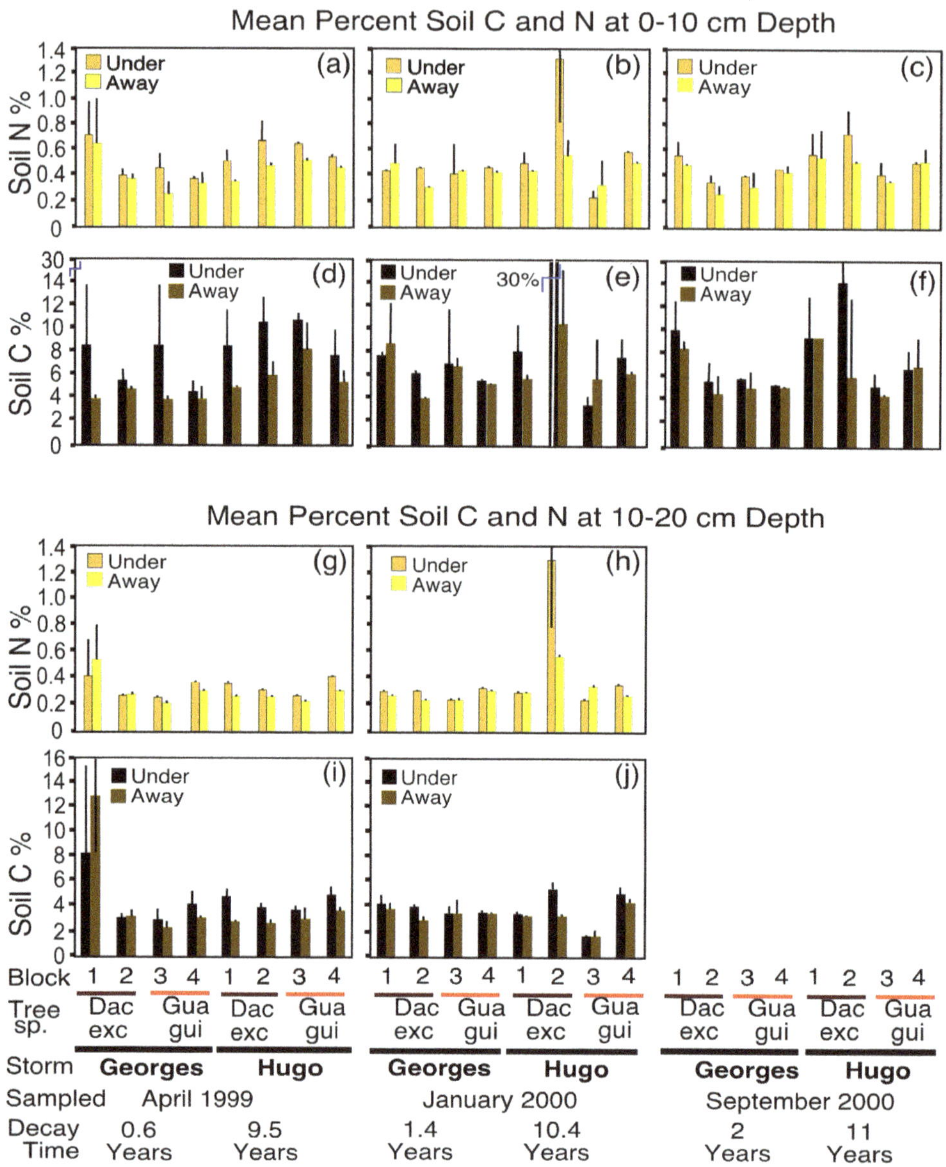

Figure 3. Mean percent soil C and N are presented for paired samples taken under decomposing logs (darker bars) versus samples taken 20–50 cm away and not under the logs (lighter bars) 0.6–2 and 9.5–11 years after Hurricanes Georges and Hugo, respectively. Tree species abbreviations are: Dac exc—*Dacryodes excelsa*, and Gua gui—*Guarea guidonia*. Data are means of two samples, one taken in the upslope direction and the other taken in the downslope direction on the same day (SD bars in black). April 1999 sampling was during the dry season, January 2000 was the end of the wet season, and September 2000 was during the wet season. (**a**–**c**) Mean soil % N at 0–10 cm depth; (**d**–**f**) Mean soil % C at 0–10 cm depth; (**g**,**h**) Mean soil % N at 10–20 cm depth; (**i**,**j**) Mean soil % C at 10–20 cm depth.

Fresh Scoletid beetle frass C and N collected from *Tabebuia heterophylla* (DC.) Britton were 45.0% and 0.4%, respectively, while frass from *Mangifera indica* L. contained 41.8% C and 1.4%. Thus, the C:N ratios were 128.4 for *Tabebuia* and 30.3 for *Mangifera*.

3.2. Soil Biota

3.2.1. Soil Microbial Biomass C

The best fit model for soil microbial biomass C was a Gaussian distribution and an identity link function. Differences in soil microbial biomass C were only suggestive between positions (Table 3). Mean microbial biomass C was generally higher under than away from logs, especially in the April 1999 dry season and January 2000 wet season samples (Figure 4) but there were no significant interactions of position with time (position time $p = 0.634$; position time hurricane $p = 0.268$). Differences in microbial C were significant among sampling dates (Table 3). Microbial biomass C was lower during the April 1999 dry season when soils were drier than in the two wet season samples (January and September 2000; Figure 4).

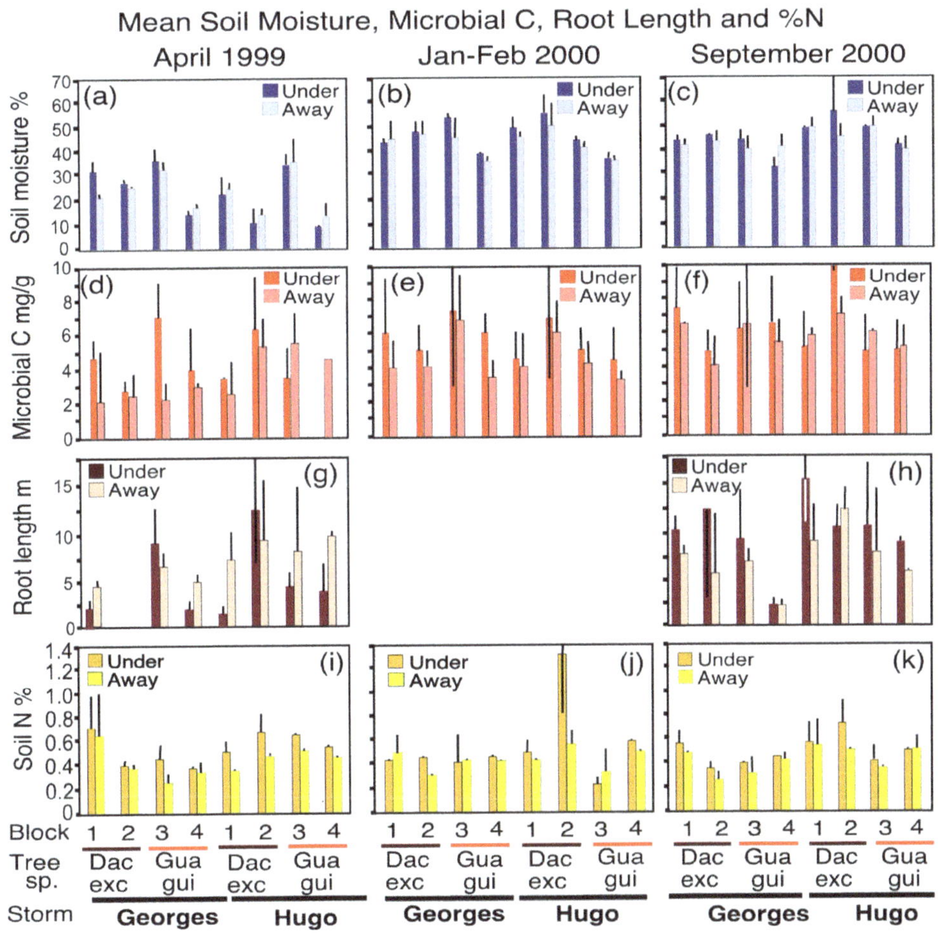

Figure 4. Mean soil % moisture, microbial biomass C, total root length and soil % N are presented for paired (8 × 8 cm) soil cores taken to 10 cm depth under decomposing logs (darker bars) versus samples taken 20–50 cm away and not under the logs (lighter bars). April 1999 sampling was during the dry season, 0.6 and 9.5 years after Hurricane Georges and Hugo, respectively. January 2000 was the end of the wet season, 1.4 and 10.4 years after Hurricanes Georges and Hugo, respectively. September 2000 was during the wet season 2 and 11 years after Hurricanes Georges and Hugo, respectively. Data are means of two samples, one taken in the upslope direction and the other taken in the downslope direction on the same day (SD bars in black). Tree species abbreviations are: Dac exc—*Dacryodes excelsa*, and Gua gui—*Guarea guidonia*. One log from each hurricane (Georges 1998) and Hugo (1989) is within each of the four blocks. (**a–c**): Mean soil % moisture; (**d–f**): Mean microbial biomass C; (**g**,**h**): Mean m root length; (**i–k**): Mean soil % N.

3.2.2. Roots

Most of the tree roots were magnoliod (coarse, little-branched, hairless roots) in which the finest roots ranged from 1 to 2 mm diameter. The best fit model for the data was a Lognormal distribution and an identity link function. There was a significant time by position interaction (Table 3). Root length was greater away from the logs in the April dry season, and greater under the logs in the September wet season samples (Figures 4 and 5). Hurricane was significant (Table 3) as root length was greater for older logs from the Hurricane Hugo cohort than for younger logs from the Hurricane Georges cohort (Figure 4 and comparison of Least Square Means). Although hillslope direction was also significant (Table 3), and a comparison of Least Square Means showed significantly higher root length on the upslope than the downslope side of the logs, the pattern is not readily apparent in Figures 4 and 5 because of the reversal of rooting patterns by position between the dry and rainy seasons.

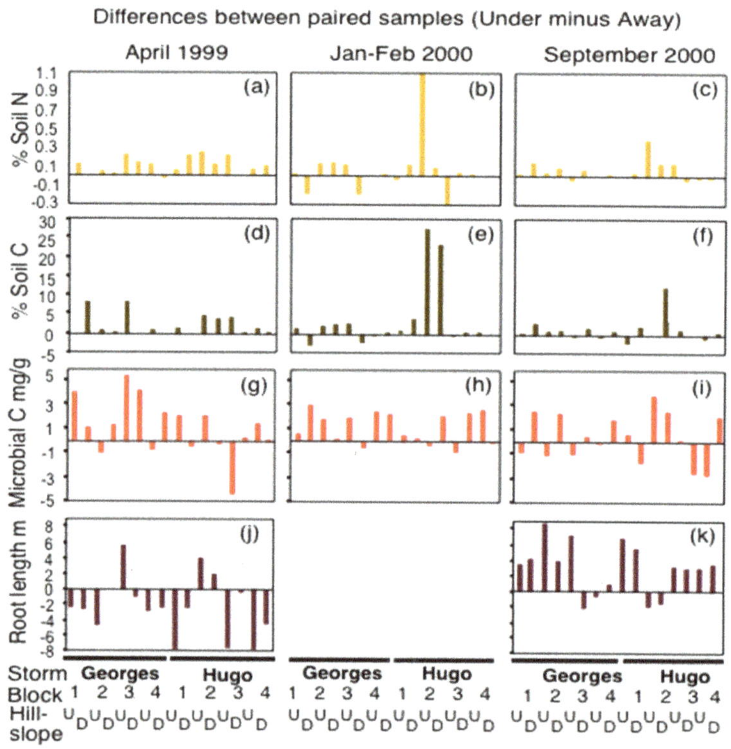

Figure 5. Percent soil N and C, microbial biomass C, and total root length are presented as differences in values between paired (8 × 8) cm soil cores taken to 10 cm depth under decomposing logs minus values for samples taken 20–50 cm away and not under the logs. Thus, bars above the zero line had greater values under than away from the log within a pair while bars below the line had lower values under than away from the log. Bars represent a single pair of samples, one pair taken in the upslope direction (Hillslope U) and the other pair taken in the downslope direction (Hillslope D) on the same day. Logs fell in either Hurricane Hugo (1989) or Hurricane Georges (1998). Two logs, one from each hurricane, were represented in each of the four blocks. April 1999 sampling was during the dry season, 0.6 and 9.5 years after Hurricane Georges and Hugo, respectively. January 2000 was the end of the wet season, 1.4 and 10.4 years after Hurricanes Georges and Hugo, respectively. September 2000 was during the wet season 2 and 11 years after Hurricanes Georges and Hugo, respectively. (**a–c**) Difference (under—away) in soil % N; (**d–f**) Difference (under—away) in soil % C; (**g–i**) Difference (under—away) in microbial biomass C; (**j k**): Difference (under—away) in m total root length.

3.3. Soil Moisture

Soil moisture only differed significantly between seasons ($p < 0.0001$). No differences were found between under versus away positions or for the interaction between season and position. Soils were significantly drier in the April dry season than in the two wet season samples (Figure 4).

3.4. Correlations among Variables

Root length was most strongly correlated with microbial biomass ($R = 0.36$). The relationships between root length, soil % N and microbial biomass C are shown in Figure 4. Root length was significantly positively correlated with soil microbial C ($R = 0.36$), soil moisture ($R = 0.27$) and soil % N ($R = 0.18$), and all the variables were significantly positively correlated with each other ($p < 0.05$). It was therefore surprising that the best multiple linear regression model (highest adjusted R^2 of 0.12; $p = 0.016$) included all three variables, and the coefficient for microbial C changed from positive in the simpler models to negative because of the inverse relationship between root length and microbial C in the April dry season samples (Figures 4 and 5). Microbial biomass C was significantly positively correlated with soil moisture ($R = 0.47$) and % N ($R = 0.27$).

4. Discussion

4.1. Do Tree Roots Proliferate in Resource Hotspots?

Decaying CWD has been found to either augment soil nutrient content or availability through inputs of leachates, insect frass and decayed wood [1,2,4–7,17], or conversely, to diminish available nutrients though responses of soil microbial communities to inputs of decay products with high C:N ratios [18,19,21]. The question of how tree roots respond to nutrient hotspots or local nutrient sinks created by CWD has received less attention [43], though decayed wood itself has been shown to provide an important source of moisture and nutrients for tree roots [3,6,43]. Trees respond readily to patches that contain limiting resources by proliferating fine roots disproportionately in resource hotspots [8,10–14]. Nutrient hotspots associated with CWD may thus play a critical role in maintaining forest productivity [12]. In this study, we used closely paired soil samples taken under and away from decaying logs belonging to two hurricane cohorts to determine if soil nitrogen, carbon, moisture, microbial biomass and tree root lengths were higher or lower under the CWD. Further, we examined whether the patterns of soil N, C and moisture changed through time and season, and whether roots proliferated more in patches with high nitrogen or moisture, and whether roots avoided soil patches with high microbial biomass since microbes can compete with roots for limiting nutrients [21].

4.2. Effects on Soil Nitrogen and Carbon

Unlike the studies by Zalamea et al. [22] and Bantle et al. [1,2], we did not find differences in soil % C or % N between tree species. One of our tree species was shared with the Zalamea et al. study (*D. excelsa*) while our *G. guidonia* is in the same family and is nearly indistinguishable in quality from *S. macrophylla* that was used in Zalamea et al. [22,23]. Fresh wood of *D. excelsa* and *S. macrophylla* (and presumably, *G. guidonia*) has similar N concentrations [22]. Sites and species were confounded in the Zalamea et al. study [22], and while they included soil characteristics as covariates in their ANOVAs to remove effects due to differences between their two sites; it is possible that not all of the site soil differences were accounted for and that differences in clay content may have contributed to what was attributed to differences in tree species. In our study, three of the four blocks were located on the same soil type and all were at the same field site. We found significantly higher % C under than away from logs at both depths (0–10 and 10–20 cm) and at all sample times. However, the subsequent study by Zalamea et al. [22,23] 7 and 15 years after Hurricanes Hugo and Georges did not find any differences between under and away positions in soil % C at 0–5 cm depth in the same forest type located at a different site on the eastern side of the Luquillo Mountains. Clay content is generally well correlated with soil organic C (SOC) [44] and the clay contents of soils in the Zalamea et al. [22] sites (Bisley and Río Chiquito) were lower than at our El Verde site (20%–22% versus 30%–55%). In addition, the soils in the Zalamea et al. study [22,23] were collected only to 5 cm depth; surface soils contain less clay and are also more influenced by inputs of organic matter from leaf litter than are deeper soil horizons. We therefore think that differences in sampling depth and clay content can explain the disparity between our results and those of Zalamea et al. [22,23]. Clay mineralogy is also an important determinant of SOC [44], but all of the highly weathered soils in both the Zalamea et al. [22,23] and this study were high in aluminum and iron [33]. Zinn et al. [45] noted that aluminum and iron cations are the most important interlayer mineral binders of dissolved organic matter (DOM), and Kaiser and Guggenberger [46] showed sorption to the mineral matrix strongly preserved DOM. In contrast, Kayahara et al. [47] did not find greater amounts of organically complexed Fe and Al in soils below forest floors with woody debris versus those without wood in the boreal forest of British Colombia despite finding more total C and greater C in the humic, fulvic and polyphenol fractions of the overlying forest floors containing wood.

While we do not know how stable the C originating from our logs is and thus how much the logs contribute to C sequestration in soils, Kalbitz et al. [48] showed that sorption of recalcitrant compounds was four times greater than labile compounds. Interestingly, one of our logs (H.2) showed high % C (14%–33% at 0–10 cm depth on two of the three sample dates and 4.7% at 10–20 cm depth in January 2000, and % C was also high away from the log in one of the two January 2000 samples. Log H.2 was *D. excelsa*, a species that had historically been used in the El Verde Research Area for making charcoal, so we suspected this log fell onto an old charcoal pit. Soils around the marked log were examined for charcoal, which is known to cause elevated soil C, but none was found. We therefore postulate instead that the paired cores collected in January and September of 2000 from log H.2 coincided with the C imprint of a previously decomposed log whereas the April 1999 core samples did not. The highest % C values in the upper 10 cm of soil both under and away from log H.2 in January 2000 corresponded to our lowest soil bulk density measurements, which suggests an abundance of light fraction soil C. Log G.1 also had remarkably high soil % C and % N, but primarily at the lower, 10–20 cm depth. This suggests that the legacy of the former decaying log near G.1 dated earlier than the legacy of the decayed log associated with H.2, and that light fraction C in the surface soil near G.1 was lost through decomposition and leaching. Our data indicate that 20% of randomly placed soil cores at our site may fall on C and N-rich hotspots that are the legacy of decomposed prior CWD. Consistent with our results, Gutiérrez del Arroyo and Silver [49] found that ten years after a simulated hurricane debris addition experiment at our site, soil C and N were elevated relative to control plots in both surface and deep soils, and that both light fraction organic matter and organic molecules complexed with minerals in the heavy soil fractions were elevated. Both Harmon et al. [3] and Spears et al. [20] noted difficulty in detecting effects of decaying logs on the underlying soil in Oregon in northwestern USA, likely because of overlapping legacies from previously decomposed CWD. The fast rates of CWD decomposition at our wet tropical site (Table 2) and wet tropical forests in general [24] are probably correlated with fast rates of carbon loss belowground, which makes the imprint of prior treefalls on soil C and N in our forest clearer than in the cool temperate forests of Oregon with slower decomposition [3,20].

Our finding of more C under logs only 0.6 years after Hurricane Georges under logs from Georges was surprising as little decomposition was observed. Furthermore, we did not find significant differences in % C in soils beneath logs from two different hurricane cohorts spaced 9 years apart. However, Bantle et al. [1] found concentrations of dissolved organic C (DOC) were 5–10 times higher in runoff from logs than in throughfall even during early decomposition in a temperate beech forest (3 year-old logs). The products of fungal white rot in wood are primarily CO_2 and DOC [34]. Zalamea et al. [23] found significantly more water-soluble organic matter under than away from their decomposing logs, consistent with DOC of Bantle et al. [1], despite finding no differences in total soil C.

We found a significant effect of season on soil % C at 0–10 cm depth, with increases in the dry season (see graphical abstract). Turner et al. [50] found a similar increase in soil C during the dry season in a more strongly seasonal forest on Barro Colorado Island in Panama. However, we did not observe a significant increase in soil N whereas the study in Panama did [50].

Much of the early C and N inputs to soil from fallen logs in Puerto Rico may come from bark beetle frass [21,38,51], and large deposits of frass were noted under our Georges logs beginning 5 months after they had fallen. We also found especially high microbial biomass C in addition to higher C and N under versus away from Georges logs in that sampling (Figure 5). The few studies on the ecological roles of bark beetles, and effects of xylophagous beetle frass on nutrient cycling [4,5,7,51] indicate rapid decay of the inner bark, which contains the highest concentration of nitrogen. Showalter [4] showed rapid loss of inner bark in hardwood logs was related to invertebrate activity and was independent of climate across a broad geographic gradient in North America. Similar to our study, in Australia, Lindsay and Cunningham [16] and Goldin and Hutchinson [17] found significantly higher C, N and P in soil near logs than 80 cm away. Also consistent with our observations, Cobb et al. [51] found in boreal forest of Canada that xylophagous beetles in a conifer forest that had burned followed by salvage logging had significantly increased soil C and N at the bases of dead trees via frass and chewing debris, and a positive and significant relationship between larval abundance and total soil N, presumably due to inputs of frass. They also found that addition of these beetle frass and chewings to soil increase microbial respiration by more than three-fold; furthermore, frass additions decreased germination of two of their three test plant species (as most of the nitrogen returned via frass is immobilized by microbial biomass and is thus not immediately available to plants). There are two mechanisms by which saproxylic organisms can affect N release as decomposition proceeds. They can accelerate release of N immobilized in microbial biomass through indirect interactions with cord-forming fungi [52,53]. They can also promote N_2 fixation by providing conditions conducive to prokaryote activity in their guts as well as in comminuted wood [54]. Results from Asmus [55] suggest that frass maybe a preferred substrate for N_2 fixing bacteria. Thus, our findings contribute to a greater understanding of the potential effects of CWD on soil carbon and nutrient cycling following both natural and anthropogenic forest disturbances.

Further contributing to early impacts of logs on soil nutrient cycling, wood decomposes more quickly in tropical forests [24,25] such as the one at El Verde in Puerto Rico than in temperate forest studied by Bantle et al. [1]. Three of our four logs in blocks 2 and 4 could only be identified by the pink-tipped poles marking their drip lines at both ends of the logs and the fourth log was reduced to a 2 cm deep humus deposit with roots after 18 and 27 years of decomposition for Georges and Hugo cohorts, respectively. These CWD turynover times are consistent with those from elsewhere in the neotropics [24,25]. Lang and Knight [56] proposed a faster wood decomposition in the tropics than in temperate zones because a warm and humid forest floor in the tropics would favor invertebrate activity through the year. In contrast, Torres [38,57] argued that high moisture content, the cool temperatures (caused by cloud cover, rainfall and winds) and a low animal diversity on the logs in the wet forest in Puerto Rico seem to retard wood decay in this habitat. Here, we argue that hurricane conditions can significantly alter both abiotic (changes in light and moisture level conditions) and biotic conditions that can affect the rate of recovery and nutrient cycling in the forest via decomposition processes [58]. Our observations in this study would support the contention that dry exposed wood generated by hurricane disturbance favors saproxylic species [57], which in turn can initiate the decay process. It has been suggested that wood comminution and bark fragmentation caused by these organisms might have increased nutrient loss through leaching [59,60].

Our finding of significantly higher % N in soils at two depths under than away from logs is concordant with the pattern of % C, which is not surprising as soil C and N are usually correlated and C and N mineralization are coupled [61,62]. Bantle et al. [2] found a net release of N from decaying beech logs during early decomposition in temperate forest in Germany, with the largest component being dissolved organic N (DON). Hillslope direction was also significant for the 0–10 cm depth, with % N (but not % C) being higher on the upslope side of the logs. This result is consistent with logs acting as terrestrial debris dams that trap leaf litter on their upslope side [9,35], especially in April 1999 and January 2000 (Figure 5a,b). The dates with high surface soil % N coincide with peaks in leaf litterfall in our site [63]. Tropical forest leaf litter at El Verde in Puerto Rico has high concentrations of N (1% in González et al. and Silver et al. [64,65]; 1.5% in Lodge et al. [66]) relative to temperate coniferous forest litter, as is typical of wet lowland and low elevation montane tropical forests, and net mineralization usually begins immediately and proceeds steadily rather than being preceded by immobilization phase in the litter during decomposition [67,68]. It is thus not surprising that soil % N was higher in the upper soil horizon on the upslope side of our logs.

4.3. *Effects on Biota*

4.3.1. Root Responses

Root length was greater away from the logs during the April dry season, but the pattern reversed in the September wet season leading to a significant time by position interaction. Although the shifting of roots away from soil under logs in the dry season corresponds to when soil microbial biomass C was highest under logs, this might be interpreted as competitive exclusion of roots for nutrients by soil microbial biomass, the correlation between roots and soil microbial biomass C was modest (R = 0.36). The correlation between root length and soil moisture was higher (R = 0.47) while the correlation with soil % N was smaller (0.18); the correlation of microbial C with soil % moisture (R = 0.47) was higher than with soil % N (R = 0.27). The change in the microbial C coefficient from positive to negative in the best-fit multiple linear regression model that included all independent variables was informative and consistent with the inverse relationship between microbial C and root length in the 0.6 year dry season samples from Georges logs. These patterns are consistent with root proliferation where nutrients were more available. On the other hand, mean root length was higher under than away from logs whereas mean soil microbial biomass C did not differ between positions in September (wet season), which is not consistent with the competitive exclusion hypothesis. However, inputs of bark beetle frass were greatest during the first sampling in the dry season, so the inverse correlation between roots and microbial C in the dry season may be related to effects of early frass deposition. Root lengths were also significantly higher on the upslope sides of logs, concordant with significantly higher % N on the upslope side in the upper 10 cm of soil, suggesting that roots may proliferate where N is more available. Availability of other nutrients, such as P, may have contributed to the rooting patterns since % N is correlated with % C, and C is known to increase P-availability in weathered tropical clay soils [26]. Soil moisture is another resource that can increase root proliferation [3,43]. Although soil moisture was slightly higher under than away from Georges cohort logs in the dry season and the January 2000 wet season sampling (Figures 4 and 5), the differences were slight. One of the lowest soil moistures was recorded for a Hugo log in Block 2 (Figure 4a), which corresponded to some of the highest root lengths; this pattern is not consistent with moisture being a limiting resource that determines root proliferation. Soil microbial C and soil % N were also high in those samples, so nutrient availability also may not have been limiting in the wet season under H.2. Competitive displacement can only occur where critical resources are limited.

4.3.2. Soil Microbial Biomass Responses

We found the strongest differences in soil microbial C between seasons, with values being significantly lower during the April dry than in the two wet season samplings. The soil was so dry at the beginning of the April sampling that one soil sample did not lose weight on oven drying, and some of the others probably lost water of hydration rather than free water. Fungal biomass [68] and the microbial community [69,70] in wet tropical forest of Puerto Rico were shown to be highly sensitive to drying, though community effects were most evident in the very driest soils.

Soil microbial biomass was often (suggestively) higher under than away from logs at the 0–10 cm depth, especially for Georges logs 0.6 years after the hurricane in the April dry season samples. This coincided with lower root lengths under than away from the Georges logs in the April dry season samples. Zalamea et al. [22] did not find differences in microbial biomass between positions, but they did not sample during the early stages of decomposition. Two possible mechanisms seem plausible for explaining this pattern: competition between microbial biomass and plant roots for limiting nutrients and allelophathic effects of beetle frass on roots. As noted above in the discussion of soil % C, the input of Scoletid beetle frass may have stimulated the soil microbial biomass immobilization of N that was observed by Zimmerman et al. [21] beginning 4–6 months after Hurricane Hugo. Cobb et al. [35] similarly found higher microbial activity with input of xylophagous beetle frass and chewings in temperate conifer forest. The C:N ratio of the beetle frass and chewings in the Cobb et al. [55] study was 84–55 while the Scoletid beetle frass we analyzed had C:N ratios of 30–128; all higher than ratios >12–24 that are associated with N immobilization [62,71]. Frass extracts of bark beetle larvae have been attributed to potent allelochemicals that might inhibit germination and seedling growth [72]. Thus, the results of our study would suggest a conservative mechanism of recovery and nutrient retention following a hurricane in this tropical forest that is similar to that observed in the boreal forest of Canada after post fire disturbance. The observation of greater root length away than underneath the decaying logs could potentially be explained by allochemical inhibition by secondary compounds contained in bark beetle frass on roots.

In addition to DOC contributions to soil C from bark and sapwood leachates, frass from bark beetles in the Scoletideae may have contributed to early increases in soil C below our logs. The fourth author recorded copious amounts of frass from Scoletid bark beetles underneath our recently fallen logs five months after Hurricane Georges. Our analyses of two representative samples of Scoletid beetle frass had 42%–45% C. Because Scoletid beetle frass has a fine texture resembling sawdust due to comminution by the beetles and processing in their guts and a high C:N ratio, it is likely to be decomposed quickly by fungi. Our observations of higher soil microbial biomass under than away from Georges logs 7 months after falling (Figure 5) are consistent with the N immobilization by soil microbial biomass and the slowing of forest canopy cover beginning 4–6 months after Hurricane Hugo found by Zimmerman et al. [21]. In a canopy trimming experiment conducted at our field site in El Verde, Silver (personal communication, 11 March 2014) found increases in soil respiration for several years after forest plots were subjected to a simulated hurricane as compared to control plots. Further, Xianbin Liu (personal communication, 12 January 2016) found more labile C in soils of the hurricane simulation plots for at least 1.5 years after the second application of trimming. Thus, it is likely that at least some of the C originating from coarse woody debris from disturbances such as tropical cyclones is labile and readily decomposed once it reaches the soil and is not sequestered. This labile carbon, however, can contribute to microbial immobilization of nutrients, leading to temporarily lower nutrient availability for tree roots.

5. Conclusions

Tropical cyclonic storms, bark beetle outbreaks, fires and anthropogenic disturbances deposit CWD on forest floors, which significantly alters the soil C and N dynamics, and could potentially alter soil fertility and forest productivity. Decomposing logs from two hurricanes spaced 9 years apart left a significant signature on the underlying soil, even as early as 0.6 years after the trees fell. Percent soil C and N were significantly higher under than near logs at both 0–10 and 10–20 cm depths at all sample times. Legacies of previously decomposed logs in the soil C and N were also detected. These signatures of decomposing logs on soil C and nutrients contribute to high spatial heterogeneity in soil characteristics, and lead to nutrient hotspots as well as sinks. The close spatial pairing (<50 cm apart) of soil samples taken under and away from decomposing logs in this study was effective in reducing background variation sufficiently to detect the effects of the logs on soil C, N, microbial biomass C and roots, and was also comparable to root 'choice experiments'. Early increases in soil microbial C under logs felled by Hurricane Georges were detected 0.6 years after the hurricane, and may have been stimulated by inputs of insect frass with high C:N ratios from Scoletid bark beetles in addition to leachates from bark and sapwood. The soils beneath the logs in these early, dry season samples may have been a nutrient sink caused by microbial nutrient immobilization. Correspondingly, roots were less abundant under the logs than away from the logs at that time. Soils on the upslope side of the logs differed only from the downslope side for soil % N and total root length in the upper 0–10 cm, likely because of N fluxes from leaf litter accumulations on the upslope sides of logs on steep slopes and corresponding proliferation of fine roots where N was more available. This also indicates that root proliferation was tracking nutrient availability in transient hotspots. The longevity of the C in soils originating from the decomposing logs was not determined, but spatial anomalies (hotspots) in some our samples suggest that higher soil C may persist after any remains of a decomposed log have disappeared. Evidence from other studies in the same area also suggest, however, that part of the soil C originating from coarse woody debris is labile and is respired quickly by soil microbial biomass as CO_2. The proliferation of roots under decaying logs in the wet season but not the dry season highlights the importance of maintaining spatial heterogeneity in management of coarse woody debris following natural disturbances and logging plus salvage operations in order to allow tree roots to track and exploit nutrient and resource hotspots as they change through time.

Acknowledgments: This research was performed under grants DEB-0218039 and 1239764 from the National Science Foundation to the Institute of Tropical Ecosystem Studies, University of Puerto Rico, and the United States Department of Agriculture, Forest Service, International Institute of Tropical Forestry as part of the Long-Term Ecological Research Program in the Luquillo Experimental Forest. Additional support for G. González was provided by the Luquillo LCZO grant (EAR-1331841). We are thankful to Maria M. Rivera, Mary Jeane Sánchez, Maysaá Ittayem, Edwin Lopéz, Carmen Marrero, Marybelís Santiago and other staff at the International Institute of Tropical Forestry—Soil Laboratory who helped process soil samples; and Xiaoming Zou for advice on measuring soil microbial biomass. We thank the statistician for the United States Department of Agriculture, Forest Service, Northern Research Station, John Stanovick, for statistical analyses and review, and Walter Shortle and Tana Wood for pre-reviews that improved the manuscript. Additional support was provided by the University of Puerto Rico, Rio Piedras.

Author Contributions: D. Jean Lodge and Naomi Clum conceived designed the experiment, and Naomi Clum located the logs for this study and conducted the April 1999 sampling with help from D. Jean Lodge; D. Jean Lodge conducted the January 2000 sampling and Dirk Winter conducted the September 2000 sampling with assistance from D. Jean Lodge. Dirk Winter corrected the data sets and coordinated soil analyses. D. Jean Lodge, Dirk Winter and Grizelle González wrote the paper.

References

1. Bantle, A.; Borken, W.; Ellerbrock, R.H.; Schulze, E.D.; Weisser, W.W.; Matzner, E. Quantity and quality of dissolved organic carbon released from coarse woody debris of different tree species in the early phase of decomposition. *For. Ecol. Manag.* **2014**, *329*, 287–294. [CrossRef]
2. Bantle, A.; Borken, W.; Matzner, E. Dissolved nitrogen release from coarse woody debris of different tree species in the early phase of decomposition. *For. Ecol. Manag.* **2014**, *334*, 277–283. [CrossRef]
3. Harmon, M.E.; Franklin, J.F.; Swanson, F.J.; Sollins, P.; Gregory, S.V.; Lattin, J.D.; Anderson, N.H.; Cline, S.P.; Aumen, N.G.; Sedell, J.R.; et al. Ecology of coarse woody debris in temperate ecosystems. *Adv. Ecol. Res.* **1986**, *15*, 133–302.
4. Edmonds, R.L.; Eglitis, A. The role of Douglas-fir beetle and wood borers in the decomposition of and nutrient release from Douglas-fir logs. *Can. J. For. Res.* **1989**, *19*, 853–859. [CrossRef]
5. Schowalter, T.D. Heterogeneity and nutrient dynamics of oak (*Quercus*) logs during the first 2 years of decomposition. *Can. J. For. Res.* **1992**, *22*, 161–166. [CrossRef]
6. Stevens, V. *The Ecological Role of Coarse Woody Debris: An Overview of the Ecological Importance CWD in BC Forests*; Work Paper 30; British Columbia Ministry of Forests: Victoria, BC, Canada, 1997.
7. Ricker, M.C.; Graeme Lockaby, B.; Blosser, G.D.; Conner, W.H. Rapid wood decay and nutrient mineraliztion in an old-growth bottomland hardwood forest. *Biogeochemistry* **2016**, *127*, 323–338. [CrossRef]
8. Johnson, D.W.; Woodward, C.; Meadows, M.W. A three-dimensional view of nutrient hotspots in a Sierra Nevada forest soil. *Soil Sci. Soc. Am. J.* **2014**, *78*, S225–S236. [CrossRef]
9. Lodge, D.J.; McDowell, W.H.; Macy, J.; Ward, S.K.; Leisso, R.; Claudio Campos, K.; Kuhnert, K. Distribution and role of mat-forming saprobic basidiomycetes in a tropical forest. In *Ecology of Saprobic Basidiomycetes*; Boddy, L., Frankland, J.C., Eds.; Academic Press, Elsevier Ltd.: Amsterdam, The Netherlands, 2008; pp. 195–208.
10. Stewart, C.G. A test of nutrient limitation in two tropical montane forests using root ingrowth cores. *Biotropica* **2000**, *32*, 369–373. [CrossRef]

11. Cuevas, E.; Medina, E. Nutrient dynamics within Amazonian forests. II. Fine root growth, nutrient availability, and leaf litter decomposition. *Oecologia* **1988**, *76*, 222–235. [CrossRef]
12. Haines, B. Impact of leaf-cutting ants on vegetation development at Barro Colorado Island. In *Tropical Ecological Systems: Trends in Terrestrial and Aquatic Research*; Golley, F.G., Medina, E., Eds.; Springer: New York, NY, USA, 1975; pp. 99–101.
13. Raich, J.W.; Riley, R.H.; Vitousek, P.M. Use of root-ingrowth cores to assess nutrient limitations in forest ecosystems. *Can. J. For. Res.* **1994**, *24*, 2135–2138. [CrossRef]
14. St. John, T.V.; Coleman, D.C.; Reid, C.P. Growth and spatial distribution of nutrient-absorbing organs: Selective exploitation of soil heterogeneity. *Plant Soil* **1983**, *71*, 487–493. [CrossRef]
15. Sayer, E.J.; Banin, L.F. Tree nutrient cycling in tropical forest—Lessons from fertilization experiments. In *Tropical Tree Physiology: Adaptation and Responses in a Changing Environment*; Goldstein, G., Santiago, L.S., Eds.; Tree Physiology Volume 6; Springer: New York, NY, USA, 2016; pp. 275–297.
16. Goldin, S.R.; Hutchinson, M.F. Coarse woody debris modifies surface soils of degraded temperate eucalypt woodlands. *Plant Soil* **2013**, *370*, 461–469. [CrossRef]
17. Lindsay, A.E.; Cunningham, S.A. Native grass establishment in grassy woodlands with nutrient-enriched soil and exotic grass invasion. *Restor. Ecol.* **2011**, *19*, 131–140. [CrossRef]
18. Hafner, S.D.; Groffman, P.M. Soil nitrogen cycling under litter and coarse woody debris in a mixed forest in New York State. *Soil Biol. Biochem.* **2005**, *37*, 2159–2162. [CrossRef]
19. Kwak, J.H.; Chang, S.X.; Naeth, M.A.; Schaaf, W. Coarse woody debris extract decreases nitrogen availability in two reclaimed oil sands soils in Canada. *Ecol. Eng.* **2015**, *84*, 13–21. [CrossRef]
20. Spears, J.D.H.; Holub, S.M.; Harmon, M.E.; Lajtha, K. The influence of decomposing logs on soil biology and nutrient cycling in an old-growth mixed coniferous forest in Oregon, USA. *Can. J. For. Res.* **2003**, *33*, 2193–2201. [CrossRef]
21. Zimmerman, J.K.; Pulliam, W.M.; Lodge, D.J.; Quiñones-Orfila, V.; Fetcher, N.; Guzman-Grajáles, S.; Parrotta, J.A.; Asbury, C.E.; Walker, L.R.; Waide, R.B. Nitrogen immobilization by decomposing woody debris and the recovery of tropical wet forest from hurricane damage. *Oikos* **1995**, *72*, 314–322. [CrossRef]
22. Zalamea, M.; González, G.; Lodge, D.J. Physical, chemical and biological properties of soil under decaying wood in a tropical wet forest in Puerto Rico. *Forests* **2016**, *7*, 168. [CrossRef]
23. Zalamea, M.; González, G.; Ping, C.L.; Michaelson, G. Soil organic matter dynamics under decaying Wood in a subtropical wet Forest: Effect of tree species and decay stage. *Plant Soil* **2007**, *296*, 173–185. [CrossRef]
24. Clark, D.B.; Clark, D.A.; Brown, S.; Oberbauer, S.F.; Veldkamp, E. Stocks and flows of coarse woody debris across a tropical rain forest nutrient and topography gradient. *For. Ecol. Manag.* **2002**, *164*, 237–248. [CrossRef]
25. Delaney, M.; Brown, S.; Lugo, A.E.; Torres-Lezama, A.; Quintero, N.B. The quantity and turnover of dead wood in permanent forest plots in six life zones of Venezuela. *Biotropica* **1998**, *30*, 2–11. [CrossRef]
26. Sanford, R.L., Jr.; Parton, W.J.; Ojima, D.S.; Lodge, D.J. Hurricane effects on soil organic matter dynamics and forest production in the Luquillo Experimental Forest, Puerto Rico: Results of simulation modeling. *Biotropica* **1991**, *23*, 364–372. [CrossRef]
27. Emanuel, K.A. Increasing destructiveness of tropical cyclones over the past 30 years. *Nature* **2005**, *436*, 686–688. [CrossRef] [PubMed]
28. Knutson, T.R.; McBride, J.L.; Chan, J.; Emanuel, K.; Holland, G.; Landsea, C.; Held, I.; Kossin, J.P.; Srivastava, A.K.; Sugi, M. Tropical cyclones and climate change. *Nat. Geosci.* **2010**, *3*, 157–163. [CrossRef]
29. Mei, W.; Xie, S.P. Intensification of landfalling typhoons over the northwest Pacific since the late 1970s. *Nat. Geosci.* **2016**, *9*, 753–757. [CrossRef]
30. Bentz, B. Climate Change and bark beetles of the western United States and Canada: Direct and indirect effects. *Bioscience* **2010**, *60*, 602–613. [CrossRef]
31. Dale, V.H.; Joyce, L.A.; McNulty, S.; Neilson, R.P.; Ayres, M.P.; Flannigan, M.D.; Hanson, P.J.; Irland, L.C.; Lugo, A.E.; Peterson, C.J.; et al. Climate change and forest disturbance. *Bioscience* **2001**, *51*, 723–734. [CrossRef]
32. Brown, S.; Lugo, A.E.; Silander, S.; Liegel, L. *Research History and Opportunities in the Luquillo Experimental Forest*; General Technical Report, SO-44; United States Forest Service: Washington, DC, USA, 1983.
33. Soil Survey Staff. *Order 1 Soil Survey of the Luquillo Long-Term Ecological Research Grid, Puerto Rico*; United States Department of Agriculture, National Resources Conservation Service: Lincoln, NE, USA, 1995.

34. Lodge, D.J.; Cantrell, S.A.; González, G. Effects of canopy opening and debris deposition on fungal connectivity, phosphorus movement between litter cohorts and mass loss. *For. Ecol. Manag.* **2014**, *332*, 11–21. [CrossRef]
35. Shanley, J.B.; Lodge, D.J.; Krabbenhoft, D.P.; Olson, M.L.; McDowell, W.H. New and old mercury fluxes from mercury amendments to Puerto Rico soil columns. In Presented at the American Geophysical Union Fall 2008 Meetings, San Francisco, CA, USA, 15–19 December 2008; Available online: http://adsabs.harvard.edu/abs/2008AGUFM.B13C0457S (accessed on 17 August 2009).
36. Jenkinson, D.S.; Powlson, D.S. The effects of biocidal treatments on metabolism in soil—V: A method for measuring soil biomass. *Soil Biol. Biochem.* **1976**, *8*, 209–213. [CrossRef]
37. Lin, Q.; Brookes, P.C. An evaluation of the substrate-induced respiration method. *Soil Biol. Biochem.* **1999**, *31*, 1969–1983. [CrossRef]
38. Torres, J.A. Wood decomposition of *Cyrilla racemiflora* in a tropical montane forest. *Biotropica* **1994**, *26*, 124–140. [CrossRef]
39. Tabatabai, M.A.; Bremmer, J.M. Automated instruments for determination of total carbon, nitrogen, and sulfur in soils by combustion techniques. In *Soil Analysis, Modern Instruments Techniques*; Marcel Dekker, Inc.: New York, NY, USA, 1991; pp. 261–286.
40. Vitousek, P.M.; Matson, P.A. Mechanisms of nitrogen retention in forest ecosystems: A field experiment. *Science* **1984**, *225*, 51–52. [CrossRef] [PubMed]
41. Newmann, E.I. A method of estimating the total length of root in a sample. *J. App. Ecol.* **1966**, *3*, 139–145. [CrossRef]
42. SAS Institute Inc. *SAS/STAT® 9.3 User's Guide*; SAS Institute Inc.: Cary, NC, USA, 2011.
43. Harvey, A.E.; Jurgensen, M.F.; Larssen, M.J. *Effects of Soil Organic Matter on Regeneration in Northern Rocky Mountain Forests*; GTR-PNW-163; USDA Forest Service, Pacific Northwest Research Station: Portland, OR, USA, 1983; pp. 239–242.
44. Price, S.P.; Bradford, M.A.; Ashton, M.S. Characterizing organic carbon stocks and flows in forest soils. In *Managing Forest Carbon in a Changing Environment*; Ashton, M.S., Tyrrell, M.L., Spalding, D., Gentry, B., Eds.; Springer Science and Business Media: New York, NY, USA, 2012; pp. 7–30.
45. Zinn, Y.L.; Lal, R.; Bigham, J.M.; Resck, D.V.S. Edaphic controls on soil organic carbon retention in the Brazilian Cerrado: Texture and mineralogy. *Soil Sci. Soc. Am. J.* **2007**, *71*, 1204–1214. [CrossRef]
46. Kaiser, K.; Guggenberger, G. The role of DOM sorption to mineral surfaces in the preservation of organic matter in soils. *Org. Geochem.* **2000**, *31*, 711–725. [CrossRef]
47. Kayahara, G.J.; Klinka, K.; Lavkulich, L.M. Effects of decaying wood on eluviation, podzolization, acidification and nutrition in soils with different moisture regimes. *Environ. Monit. Assess.* **1991**, *39*, 485–492. [CrossRef] [PubMed]
48. Kalbitz, K.; Schwesig, D.; Rethemeyer, J.; Matzner, E. Stabilization of dissolved organic matter by sorption to the mineral soil. *Soil Bio. Biochem.* **2005**, *37*, 1319–1331. [CrossRef]
49. Gutiérrez del Arroyo, O.; Silver, W.L. How deep does disturbance go? The legacy of hurricanes on tropical forest soil biogeochemistry. In Proceedings of the American Geophysical Union Meeting, San Francisco, CA, USA, 12–16 December 2016; Available online: https://agu.confex.com/agu/fm16/meetingapp.cgi/Paper/168890 (accessed on 21 October 2016).
50. Turner, B.L.; Yavitt, J.B.; Harms, K.E.; Garcia, M.N.; Wright, S.J. Seasonal changes in soil organic matter after a decade of nutrient addition in a lowland tropical forest. *Biogeochemistry* **2015**, *123*, 221–235. [CrossRef]
51. Cobb, T.P.; Hannam, K.D.; Kishchuk, B.E.; Langor, D.W.; Quideau, S.A.; Spence, J.R. Wood-feeding beetles and soil nutrient cycling in burned forests: Implications of post-fire salvage logging. *Agric. For. Entomol.* **2010**, *12*, 9–17. [CrossRef]
52. Boddy, L.; Watkinson, S.C. Wood decomposition, higher fungi, and their role in nutrient redistribution. *Can. J. Bot.* **1995**, *73*, S1377–S1383. [CrossRef]
53. Boddy, L. Saprotrophic cord-forming fungi: Meeting the challenge of heterogeneous environments. *Mycologia* **1999**, *91*, 13–32. [CrossRef]
54. Ulyshen, M.D. Insect-mediated nitrogen dynamics in decomposing wood. *Ecol. Entomol.* **2015**, *40* (Suppl. 1), 97–112. [CrossRef]
55. Ausmus, B.S. Regulation of wood decomposition rates by arthropod and annelid populations. *Ecol. Bull.* **1977**, *25*, 180–192.

56. Lang, G.E.; Knight, D.H. Decay rates for boles of tropical trees in Panama. *Biotropica* **1979**, *11*, 316–317. [CrossRef]
57. Torres, J.A.; González, G. Wood Decomposition of *Cyrilla racemiflora* (Cyrillaceae) in Puerto Rican Dry and Wet Forests: A 13-year Case Study. *Biotropica* **2005**, *37*, 452–456. [CrossRef]
58. Shiels, A.; González, G.; Lodge, D.J.; Willig, M.R.; Zimmerman, J.K. Cascading effects of canopy opening and debris deposition from a large-scale hurricane experiment in a tropical rainforest. *Bioscience* **2015**, *65*, 871–881. [CrossRef]
59. Bouget, C.; Duelli, P. The effects of windthrow on forest insect communities: A literature review. *Biol. Conserv.* **2004**, *118*, 281–299. [CrossRef]
60. Swift, M.J.; Boddy, L. Animal-microbial interactions in wood decomposition. In *Invertebrate-Microbial Interactions*; Anderson, J.M., Rayner, A.D.M., Walton, D.W.H., Eds.; Cambridge University Press: Cambridge, UK, 1984; pp. 89–131.
61. Winsor, G.W.; Pollard, A.G. Carbon-nitrogen relationships in soil. IV. Mineralization of carbon and nitrogen. *J. Sci. Food Agric.* **1956**, *7*, 618–624. [CrossRef]
62. Cleveland, C.C.; Liptzin, D. C:N:P stoichiometry in soil: Is there a 'Redfield' ratio for the microbial biomass? *Biogeochemistry* **2007**, *85*, 235–252. [CrossRef]
63. Zalamea, M.; González, G. Leaf fall phenology in a subtropical wet forest in Puerto Rico: From species to community patterns. *Biotropica* **2008**, *40*, 295–304. [CrossRef]
64. González, G.; Lodge, D.J.; Richardson, B.A.; Richardson, M.J. A canopy trimming experiment in Puerto Rico: The response of litter decomposition and nutrient release to canopy opening and debris deposition in a subtropical wet forest. *For. Ecol. Manag.* **2014**, *332*, 32–46. [CrossRef]
65. Silver, W.L.; Hall, S.J.; González, G. Differential effects of canopy trimming and litter deposition on litterfall and nutrient dynamics in a wet subtropical forest. *For. Ecol. Manag.* **2014**, *332*, 47–55. [CrossRef]
66. Lodge, D.J.; Scatena, F.N.; Asbury, C.E.; Sánchez, M.J. Fine litterfall and related nutrient inputs resulting from Hurricane Hugo in subtropical wet and lower montane tropical forest of Puerto Rico. *Biotropica* **1991**, *23*, 336–342. [CrossRef]
67. Zou, X.; Zucca, C.P.; Waide, R.B.; McDowell, W.H. Long-term influence of deforestation on tree species composition and litter dynamics of a tropical rain forest in Puerto Rico. *For. Ecol. Manag.* **1995**, *78*, 147–157. [CrossRef]
68. Lodge, D.J. Nutrient cycling by fungi in wet tropical forests. In *Aspects of Tropical Mycology*; BMS Symposium Series 19; Isaac, S., Frankland, J.C., Watling, R., Whalley, A.J.S., Eds.; Cambridge University Press: Cambridge, UK, 1993; pp. 37–57.
69. Bouskill, N.J.; Lim, H.C.; Borglin, S.; Salve, R.; Wood, T.E.; Silver, W.L. Pre-exposure to drought increases the resistance of tropical forest soil bacterial communities to extended drought. *ISME J.* **2013**, *7*, 384–394. [CrossRef] [PubMed]
70. Bouskill, N.J.; Wood, T.E.; Baran, R.; Ye, Z.; Bowen, B.P.; Lim, H.C.; Zhou, J.; Van Norstrand, J.D.; Nico, P.; Northern, T.R.; et al. Belowground response to drought in a tropical forest soil. I. Changes in microbial functional potential and metabolism. *Front. Microbiol.* **2016**, *7*, 525. [CrossRef] [PubMed]
71. Paul, E.A.; Clark, E.E. *Soil Microbiology and Biochemistry*; Academic Press: San Diego, CA, USA, 1996.
72. Khan, D.; Sahito, Z.A.; Dawar, S.; Zaki, M.J. Frass of saproxylic-cerambycid larvae from dead twigs of *Acacia stenophylla* A. Cunn. EX. Benth. and its effects on germination and seedling growth of *Lactuca sativa* L. var. grand rapids. *Int. J. Biol. Biotechnol.* **2016**, *13*, 461–470.

13

Nutrient Resorption and Phenolics Concentration Associated with Leaf Senescence of the Subtropical Mangrove *Aegiceras corniculatum*: Implications for Nutrient Conservation

Hui Chen [1,2,†], Benbo Xu [1,†], Shudong Wei [1,2,*], Lihua Zhang [3,4,*], Haichao Zhou [2] and Yiming Lin [2]

1 College of Life Science, Yangtze University, Jingzhou 434025, China; chenhui_17@126.com (H.C.); benboxu@yangtzeu.edu.cn (B.X.)
2 Key Laboratory of the Ministry of Education for Coastal and Wetland Ecosystems, Xiamen University, Xiamen 361005, China; seapass2004@163.com (H.Z.); linym@xmu.edu.cn (Y.L.)
3 Key Laboratory of Coastal Environmental Processes and Ecological Remediation, Yantai Institute of Coastal Zone Research (YIC), Chinese Academy of Sciences (CAS), Yantai 264003, China
4 Shandong Provincial Key Laboratory of Coastal Zone Environmental Processes, YICCAS, Yantai 264003, China
* Correspondence: weishudong2005@126.com (S.W.); lhzhang@yic.ac.cn (L.Z.)
† These authors contributed equally to this work.

Academic Editors: Scott X. Chang and Xiangyang Sun

Abstract: *Aegiceras corniculatum* (L.) Blanco, a mangrove shrub species in the Myrsine family, often grows at the seaward edge of the mangrove zone in China. In the present study, seasonal dynamics of nutrient resorption and phenolics concentration associated with leaf senescence of *A. corniculatum* were investigated in order to evaluate its possible nutrient conservation strategies in the subtropical Zhangjiang river estuary. It was found that the nitrogen (N) and phosphorus (P) concentrations in mature leaves showed similar seasonal changes with the highest concentrations in winter and the lowest in summer, and were significantly higher than those in senescent leaves. The N:P ratios of mature leaves through the year were found to be less than 14, indicating that the *A. corniculatum* forest was N-limited. The nitrogen resorption efficiency (NRE) was higher than phosphorus resorption efficiency (PRE), and N resorption was complete. In addition, *A. corniculatum* leaves contained high total phenolics (TPs) and total condensed tannin (TCT) levels (both above 20%). TPs concentrations in mature and senescent leaves were all inversely related to their N or P concentrations. TPs:N and TCT:N ratios in senescent leaves were significantly higher than those in mature leaves. The obtained results suggested that high NRE during leaf senescence and high TPs:N and TCT:N ratios in senescent leaves might be important nutrient conservation strategies for the mangrove shrub *A. corniculatum* forest growing in N-limited conditions.

Keywords: *Aegiceras corniculatum*; nutrient resorption; phenolics; leaf senescence

---

1. Introduction

Mangrove species that flourish in low-nutrient environments have very efficient mechanisms for retaining and recycling nutrients [1]. *Aegiceras corniculatum* (L.) Blanco is a cryptoviviparous mangrove species, often grows at the seaward edge of the mangrove zone in China. This species seems most characteristic of the seaward mangal fringe and occurs typically as an isolated low shrub [2]. Compared with other mangrove species, *A. corniculatum* shrubs have high annual litter fall production in subtropical China [3]. The resulting litter from the leaf fall must be decomposed and the nutrients contained in that litter must be remineralized to become available again for plant uptake. However, most of this leaf fall is washed away by tides [4]. Such a high leaf fall rate will inevitably bring about a great loss of nutrients. Therefore, mangrove soils are generally low in nutrient concentrations [5], which can lead to the slow growth of mangrove species. However, mangroves are characterized by their high primary productivity [6]. To sustain the high levels of productivity, a substantial amount of nutrients is required. How mangrove plants, including *A. corniculatum*, cope with these conditions has provoked little attention [7].

Many previous studies showed that the concentrations of nutrients in the mangrove leaves varied with season and species [8,9]. In particular, seasonal changes in foliar nutrients occur in response to resorption before senescence [6]. Nutrient transfer from senescing leaves is the ecophysiological process by which plants withdraw nutrients from these leaves, making those nutrients available for later investment in new structures [10]. According to the reports by Lin et al. [8], Wei et al. [9], and Lin and Sternberg [11], the process of retranslocation is closely associated with leaf senescence and conservation of nutrients, and thus it is considered to be an important mechanism enabling plants to maintain growth in nutrient-poor sites. Nutrients may be used more efficiently at nutrient-poor sites, and this efficient nutrient use could reduce plant dependence on soil nutrient supply [12]. Feller [13], Reef et al. [14], and McKee et al. [15] suggested that nitrogen (N) and phosphorus (P) availability are important factors responsible for mangrove growth. Plant N:P ratio is a useful variable for consideration in ecological research because it reflects the gradual and dynamic nature of nutrient limitation rather than the fixed characteristics, such as N-limited versus P-limited [16].

Phenolic compounds, including tannins, are a significant component of plant secondary metabolites. In mangrove species, tannins are an abundant component, constituting as high as 20% dry weight [17]. Due to tannins being complex and energetically costly molecules to synthesize, their widespread occurrence and abundance indicate that tannins play an important role in plant function and evolution [18]. High tannin concentrations in plants are often associated with infertile site conditions. It has been suggested that there is an evolutionary advantage to higher tannin production. By reducing decomposition rates and decreasing N leaching potential, tannins may provide a nutrient conservation mechanism. In comparison to foliar nutrients, the detailed changes of leaf tannins between seasons and during leaf senescence of the mangrove species has been less intensively studied [8,9,19,20]. According to the reports by Serrano [21] and Northup et al. [22], this variability may determine not only the susceptibility of plants to herbivore attack, but also important aspects of nutrient cycling in terrestrial and aquatic ecosystems.

Therefore, nutrient resorption and high tannin production may be important for *A. corniculatum* to conserve nutrients in coastal environments. The objective of the present study was to evaluate possible nutrient conservation strategies of the mangrove shrub species *A. corniculatum* under low nutrient conditions. The questions asked here were regarding whether (1) leaf N and P concentrations follow a similar seasonal pattern, and decrease during leaf senescence; (2) a strong nutrient limitation will show higher resorption of the limiting nutrients; and (3) the production of phenolics will increase during leaf senescence under nutrient limited conditions.

2. Materials and Methods

2.1. Study Area

The experiments were carried out in Zhangjiangkou National Mangrove Nature Reserve (23°55′ N, 117°24′ E), which is located in the gulf of Gulei, Yunxiao, Fujian Province, in southern China. This nature reserve occupies 2358 hectares, and consists mostly of mangroves and salt marshes along the coastline of the Zhangjiang estuary. This nature reserve is the northernmost national nature mangrove in China, and it was included into the Ramsar List in 2008. The climate of the region is characteristic of a southern subtropical maritime monsoon climate. Based on Yunxiao Meteorological Administration, during 2001–2008 the mean air temperature was 21.15 ± 0.68 °C, 28.13 ± 0.28 °C, 24.07 ± 0.54 °C, and 15.34 ± 0.60 °C for spring, summer, autumn, and winter, respectively. The mean annual precipitation and evaporation were 187.21 mm and 1718.4 mm, respectively. The soil is clay, and the concentration of organic matter is 39.60 ± 1.90 mg/g. N and P concentrations in the soil are 2.20 ± 0.10 mg/g and 0.40 ± 0.00 mg/g, respectively [23]. The tide is irregular semidiurnal, the soil can be flooded for 1–2 h per day during spring tides, but the soil can be exposed for 3–7 days during the neap tide. The salinity of the soil is above 10‰, and the salinity of tidal water salinity ranges between 1‰ and 22‰ [24]. In the study site, *A. corniculatum* was the dominant mangrove species with a few *Kandelia candel*. The mean height of *A. corniculatum* plants was 2.8 m and the canopy density was 0.90.

2.2. Sample Collection

Leaf samples of *A. corniculatum* were collected from Zhangjiangkou National Mangrove Nature Reserve, Yunxiao, Fujian Province, China in 15 July 2009 (summer), 13 October 2009 (autumn), 15 December 2009 (winter), and 17 March 2010 (spring). Thirty trees with similar height and growth conditions were selected and labeled. The thirty trees were divided into five groups (six trees in one group) as five replications. The developmental stages of leaves were demarcated into two stages, green mature leaf and yellow senescent leaf. The green mature leaf was the third pair of developmentally matured leaves and did not show any sign of senescence. The yellow senescent leaf was collected by gently tapping the petiole. Only those senescent leaves that could be detached by this method were collected. Leaves damaged by insects and disease or mechanical factors were avoided. All samples were taken to the laboratory immediately after sampling, cleaned with distilled water, and then freeze dried using a desktop freeze-dryer at −56 °C for 72 h. The freeze-dried leaves were ground finely and stored at −20 °C prior to analysis.

2.3. Measurement of N and P Concentrations

The freeze-dried leaf samples were digested with sulfuric acid and hydrogen peroxide. The N concentration was determined based on the micro-Kjeldahl method described by Yoshida et al. [25], and the P concentration was evaluated according to the ascorbic acid-antimony reducing phosphate colorimetric method [26].

2.4. Determination of Total Phenolics (TPs), Extractable Condensed Tannins (ECT), Protein-Bound Condensed Tannins (PBCT), and Fiber-Bound Condensed Tannins (FBCT)

Established procedures demonstrated by Lin et al. [20] were used. TPs were measured with the Prussian blue method [27], and ECT, PBCT, and FBCT were determined by the butanol-HCl method [28] using purified tannins from *A. corniculatum* leaves as the standard. The concentration of total condensed tannins (TCT) was calculated by adding the respective quantities of ECT, PBCT, and FBCT [28].

2.5. Calculations

Resorption efficiency (RE) is the percentage of N or P recovered from the senescing leaves [29,30]. RE was calculated using the following equation: RE (%) = $(A_1 - A_2)/A_1 \times 100\%$. Where A_1 is N or P concentration in mature leaves, and A_2 is N or P concentration in senescent leaves.

Nutrient resorption proficiency referred to the minimum level to which a plant can reduce an element in senescing leaves. N and P resorption proficiency values were determined as absolute nutrient concentrations in senescent leaves [30]. Lower final nutrient concentrations correspond to higher proficiencies, which are usually expressed as percentages. This index's proxy seems to be more closely correlated than RE to the nutrient status of a population [31].

2.6. Statistical Analysis

All measurements were replicated five times. A one-way analysis of variance (ANOVA) was performed with season as the treatment factor. The Student-Newman-Keuls multiple comparison method was used to test significant differences between any two seasons. All analyses were performed by SPSS13.0 for Windows (SPSS Inc., Chicago, IL, USA).

3. Results

3.1. Seasonal Changes of N and P Concentrations, N:P Ratios, and Nutrient Resorption in Mature and Senescent Leaves of *A. corniculatum*

The N and P concentrations in mature leaves showed similar changes, with the highest concentrations in winter (17.94 ± 0.48 mg/g and 1.56 ± 0.05 mg/g) and the lowest in summer (9.27 ± 0.54 mg/g and 0.85 ± 0.04 mg/g), respectively. The N concentrations in senescent leaves ranged from 3.83 ± 0.15 mg/g to 4.77 ± 0.10 mg/g, with the lowest in spring and no significant differences among other three seasons. The P concentrations in senescent leaves were higher in summer (0.60 ± 0.04 mg/g) and autumn (0.59 ± 0.05 mg/g) than in winter (0.52 ± 0.03 mg/g) and spring (0.49 ± 0.03 mg/g) (Figure 1). Statistical results showed that the concentrations of N and P in mature leaves were significantly higher than those in senescent leaves in all seasons.

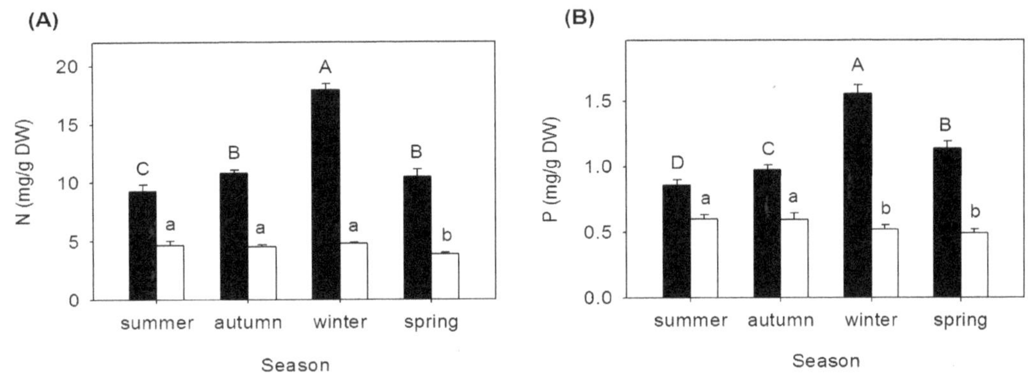

Figure 1. Seasonal changes in the (**A**) nitrogen (N) and (**B**) phosphorus (P) concentrations in leaves of *A. corniculatum*. Symbols are black bars for mature leaves and white bars for senescent leaves. Different capital letters indicate significant differences among seasons for N and P concentrations in mature leaves at $p < 0.05$. Different lowercase letters indicate significant differences among seasons for N and P concentrations in senescent leaves at $p < 0.05$.

The N:P ratios in mature leaves ranged from 9.23 ± 0.92 to 11.55 ± 0.55 with the lowest in spring. The nitrogen resorption efficiency (NRE) was significantly higher than phosphorus resorption efficiency (PRE) during leaf senescence in every season. The NRE and PRE followed the identical pattern, reaching the highest in winter (73.39% ± 1.06% and 66.51% ± 2.37%) and the lowest in summer (49.87% ± 4.39% and 30.01% ± 4.43%), respectively (Table 1).

The N concentrations in senescent leaves were all below 0.48%, and the P concentrations in senescent leaves were all above 0.05%, respectively (Table 1).

Table 1. Seasonal changes in the nitrogen and phosphorus ratios (N:P), nitrogen resorption efficiency (NRE), phosphorus resorption efficiency (PRE), nitrogen resorption proficiency (NRP), and phosphorus resorption proficiency (PRP) of *A. corniculatum* leaves.

Season	N_m:P_m	NRE (%)	PRE (%)	NRP (%)	PRP (%)
Summer	10.90 ± 1.13 a	49.87 ± 4.39 d	30.01 ± 4.43 d	0.46 ± 0.04 a	0.06 ± 0.00 a
Autumn	11.09 ± 0.41 a	58.18 ± 2.33 c	38.84 ± 5.94 c	0.45 ± 0.02 a	0.06 ± 0.00 a
Winter	11.55 ± 0.55 a	73.39 ± 1.06 a	66.51 ± 2.37 a	0.48 ± 0.01 a	0.05 ± 0.00 b
Spring	9.23 ± 0.92 b	63.36 ± 2.03 b	57.08 ± 3.48 b	0.38 ± 0.02 b	0.05 ± 0.00 b

N_m, P_m correspond to the N and P concentrations in mature leaves, respectively; Different letters in the same column indicate significant differences among seasons at $p < 0.05$.

There were significantly positive correlations between N and P concentrations in mature and senescent leaves. N concentrations in mature leaves and P concentrations in mature and senescent leaves were correlated with NRE and PRE, respectively. Meanwhile, the correlation between NRE and PRE was also highly significant (Table 2).

Table 2. Correlations between nitrogen (N) and phosphorus (P) concentrations, N and nitrogen resorption efficiency (NRE), P and phosphorus resorption efficiency (PRE), and NRE and PRE of *A. corniculatum* leaves.

Correlation	*F*	*R*	*p*
N_m-P_m	108.070	0.926	**<0.001**
N_s-P_s	5.765	0.493	**0.027**
N_m-NRE	55.731	0.869	**<0.001**
N_s-NRE	0.117	−0.080	0.737
P_m-PRE	91.214	0.914	**<0.001**
P_s-PRE	32.805	−0.804	**<0.001**
NRE-PRE	82.361	0.906	**<0.001**

N_m, P_m correspond to the N and P concentrations in mature leaves, respectively; N_s, P_s correspond to the N and P concentrations in senescent leaves, respectively. *p* values less than 0.05 are marked as bold numbers.

3.2. Seasonal Changes of Tannins Concentrations in Mature and Senescent Leaves of *A. corniculatum*

TPs concentrations in senescent leaves were higher than those in mature leaves. TPs concentrations reached the lowest in winter (210.19 ± 5.67 mg/g) for mature leaves. However, TPs concentrations in senescent leaves were the highest in spring (318.72 ± 12.66 mg/g) (Figure 2). TPs concentrations in mature and senescent leaves were all inversely correlated to N or P concentrations. A significant negative correlation between TPs concentrations in senescent leaves and N:P ratios in mature leaves was also observed (Table 3).

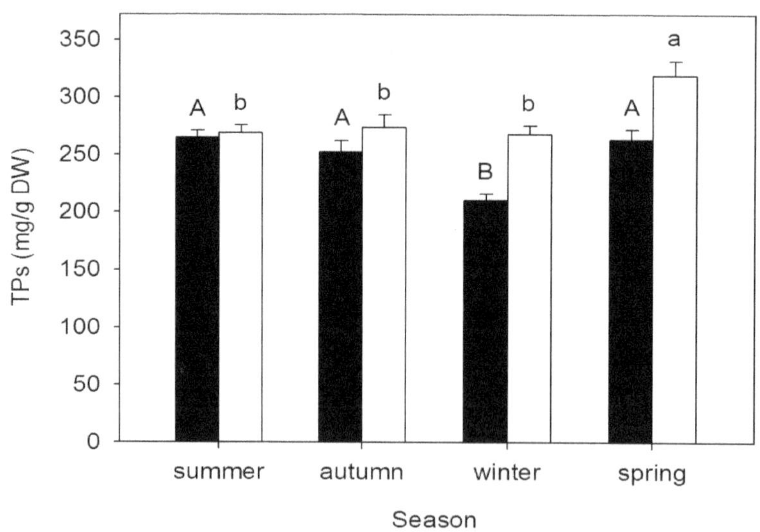

Figure 2. Seasonal changes in the total phenolics (TPs) concentrations in leaves of *A. corniculatum*. Symbols are black bars for mature leaves and white bars for senescent leaves. Different capital letters indicate significant differences among seasons for TPs concentrations in mature leaves at $p < 0.05$. Different lowercase letters indicate significant differences among seasons for TPs concentrations in senescent leaves at $p < 0.05$.

Table 3. Correlations between total phenolics (TPs) and nutrient concentrations (N and P), and N:P ratios of *A. corniculatum* leaves.

Correlation	*F*	*R*	*p*
TPs_m-N_m	122.338	−0.934	**<0.001**
TPs_s-N_s	34.399	−0.810	**<0.001**
TPs_m-P_m	56.098	−0.870	**<0.001**
TPs_s-P_s	11.631	−0.627	**0.003**
TPs_m-N_m:P_m	3.940	−0.424	0.063
TPs_s-N_m:P_m	12.143	−0.635	**0.003**

N_m, P_m correspond to the N and P concentrations in mature leaves, respectively; N_s, P_s correspond to the N and P concentrations in senescent leaves, respectively; TPs_m, TPs_s correspond to the TPs concentrations in mature and senescent leaves, respectively. *p* values less than 0.05 are marked as bold numbers.

ECT concentrations in mature leaves ranged from 190.77 ± 17.21 mg/g to 230.18 ± 33.10 mg/g, and remained relatively stable between the seasons. ECT concentrations in senescent leaves were significantly higher than those in mature leaves in winter and spring (Figure 3A).

PBCT in mature and senescent leaves were both higher in summer and winter than those in autumn and spring (Figure 3B). However, FBCT concentrations did not significantly change during leaf senescence, except in spring, and FBCT in mature leaves was the highest (1.56 ± 0.17 mg/g) in autumn (Figure 3C).

TCT concentrations had a similar changing trend to that of ECT concentrations during leaf senescence through the seasons (Figure 3D).

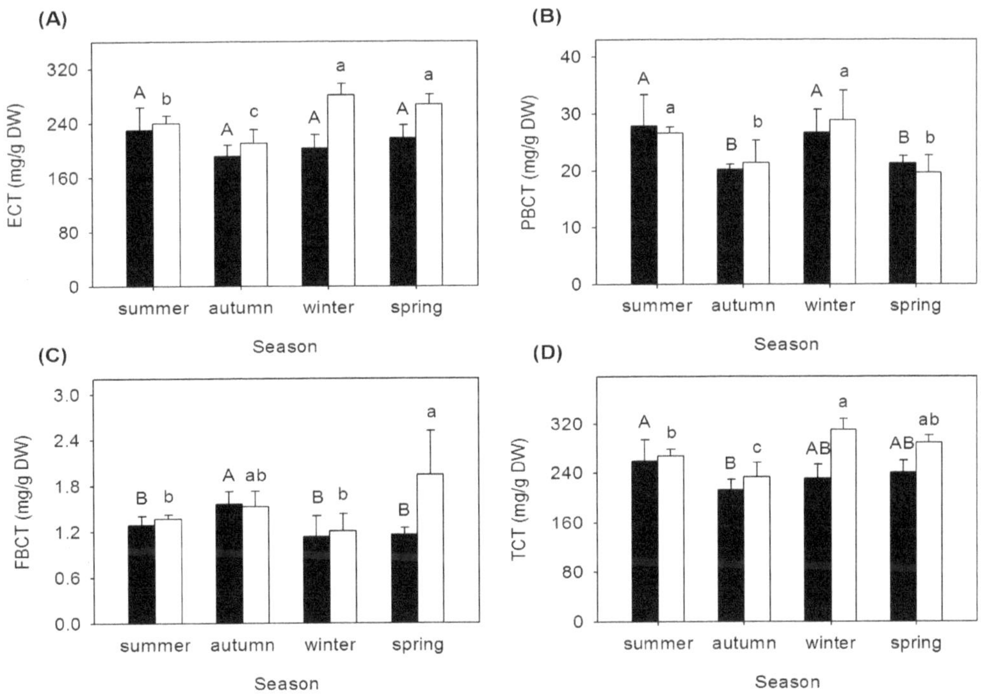

Figure 3. Seasonal changes in the concentrations of (**A**) extractable condensed tannins (ECT); (**B**) protein-bound condensed tannins (PBCT); (**C**) fiber-bound condensed tannins (FBCT); and (**D**) total condensed tannins (TCT) in leaves of *A. corniculatum*. Symbols are black bars for mature leaves and white bars for senescent leaves. Different capital letters indicate significant differences among seasons for ECT, PBCT, FBCT, and TCT concentrations in mature leaves at $p < 0.05$. Different lowercase letters indicate significant differences among seasons for ECT, PBCT, FBCT and TCT concentrations in senescent leaves at $p < 0.05$.

3.3. *Seasonal Changes of TPs:N and TCT:N Ratios in Mature and Senescent Leaves of A. corniculatum*

Seasonal changes of TPs:N and TCT:N ratios during leaf senescence are shown in Figure 4. TPs:N and TCT:N ratios in senescent leaves were significantly higher than those in mature leaves for all seasons. TPs:N and TCT:N ratios in mature leaves reached the highest in summer and the lowest in winter, while in senescent leaves they were higher in spring than in the other three seasons.

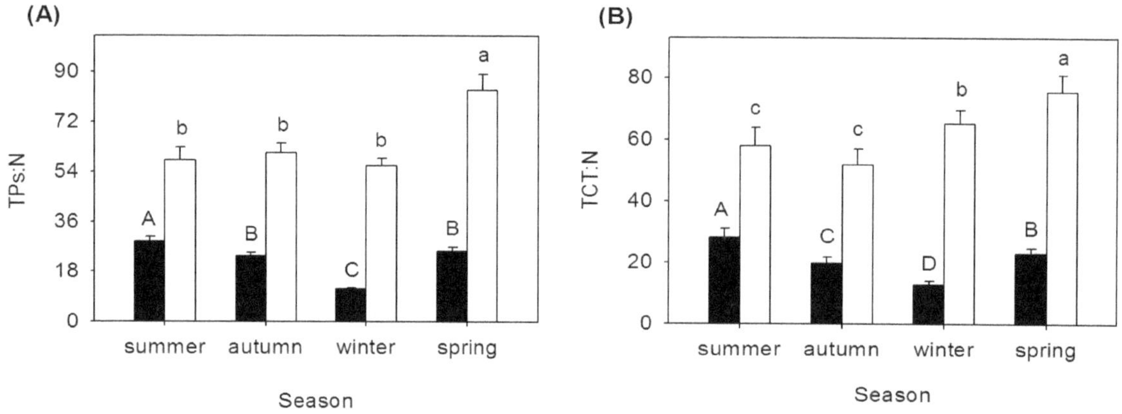

Figure 4. Seasonal changes of (**A**) total phenolics and nitrogen ratios (TPs:N) and (**B**) total condensed tannins and N ratios (TCT:N) in leaves of *A. corniculatum*. Symbols are black bars for mature leaves and white bars for senescent leaves. Different capital letters indicate significant differences among seasons for TPs:N and TCT:N ratios in mature leaves at $p < 0.05$. Different lowercase letters indicate significant differences among seasons for TPs:N and TCT:N ratios in senescent leaves at $p < 0.05$.

4. Discussion

N and P are the most common nutrients limiting plant growth [32]. The N and P concentrations in mature leaves of *A. corniculatum* followed a similar seasonal pattern, with the highest concentrations in winter and the lowest in summer (Figure 1). This was in agreement with the observations made in a study by Wang et al. [6], which showed higher N and P concentrations in cold seasons and lower concentrations in warm seasons for mature leaves. Similarly, Aerts et al. [33] also suggested that summer warming reduced N concentrations of mature and senescent leaves in *Rubus*. First, a portion of N and P were allocated to other plant parts (e.g., roots and fruits). Second, N and P concentrations were diluted by leaf mass accumulation during summer when *A. corniculatum* grew rapidly.

Nutrient resorption during leaf senescence serves to reduce the plant's dependence on current nutrient supply, and has been recognized as one of the most important strategies used by plants to conserve nutrients [8,9,34,35]. It has been estimated that approximately 50% of the nutrients (N and P) are resorbed during leaf senescence [29,36]. In the present study, the averaged NRE and PRE values were 61.20% $\pm$ 9.10% and 48.11% $\pm$ 15.31%, respectively (Table 1). These high N and P resorption values indicated that internal cycling of N and P can supply a significant fraction of the required nutrients for *A. corniculatum* growth. In addition, both NRE and PRE values were the highest in winter and the lowest in summer. This may indicate that *A. corniculatum* mainly resorbed N and P from senescent leaves in winter, but absorbed N and P from the soil in summer to adapt to low nutrient environments.

The N:P ratios of mature leaves are in response to natural variations in N and P supply, reflecting variation in N and P concentrations or both together [16]. A review of 40 fertilization studies revealed that an N:P ratio > 16 indicates P limitation on a community level, while an N:P ratio < 14 is indicative of N limitation [37]. As critical N:P ratios are successfully used to establish the nature of nutrient limitation in aquatic and agricultural ecosystems, the general concept may be applicable to a much wider range of ecosystems. N:P ratios in mature leaves of *A. corniculatum* in the Zhangjiang river estuary ranged from 9.23 to 11.55, and the lowest N:P ratio occurred in spring (Table 1). These results, both less than 14, suggested the presence of strong community level N limitation in this forest. Accordingly, N-limited *A. corniculatum* forest had significantly higher NRE than PRE. This result supported the hypothesis that a strong nutrient limitation will show higher resorption of the limiting nutrient.

Killingbeck [30] modified the concepts of nutrient resorption efficiency and proposed the use of "resorption proficiency", which is the level to which nutrient concentrations are reduced in senescent leaves. One of the important features of knowing the levels to which N and P can be reduced in senescent leaves is that these values offer an objective gauge by which resorption can be measured [30]. The potential or ultimate resorption proficiency, the lowest levels to which a nutrient is reduced in senescent leaves, is a reflection of biochemical limits to resorption and, unlike efficiency, is not subject to the temporal variations in nutrient concentrations present in green leaves and the timing of sampling [30,38]. For these reasons, resorption proficiency has been argued to be the more definitive measure of resorption success than efficiency [39]. By analyzing N and P levels in the senescent leaves of 89 species of deciduous and evergreen woody perennials, Killingbeck [30] proposed that in evergreen species, nutrient concentrations reduced below 0.7% N and 0.04% P or above 1.0% N and 0.05% P in senescent leaves are believed to represent complete or incomplete N and P resorption, respectively. In the present study, N concentrations in senescent leaves ranging from 0.38% to 0.48% were indicative of complete resorption of N; P concentrations in senescent leaves were all above 0.05%, reflecting that P resorption was incomplete (Table 1). Alongi et al. [5] argued that mangrove soils are generally low in nutrient concentrations, especially N. Our results suggested that there is greater need to resorb N from senescent leaves in N-limited mangrove forests. All the same, NRP and PRP values were lowest in spring when *A. corniculatum* grew rapidly. This is in agreement with the findings by Wei et al. [9], which suggested that lower N and P concentrations of senescent leaves in growing seasons in comparison to other seasons make it difficult to resorb additional nutrients.

N and P concentrations correlated with each other in mature and senescent leaves (Table 2). Chapin and Kedrowski [40] found a direct correlation between proportional nutrient retranslocation from the leaves during senescence and nutrient concentration in tree leaves. The significant positive correlation between N concentrations in mature leaves and NRE was consistent with the findings of Chapin and Kedrowski [40]. The significant negative correlation between P concentrations in senescent leaves and PRE was consistent with results reported in a previous study [8]. NRE had a significant positive correlation with PRE. This correlation suggested that N resorption may be controlled by biochemical processes similar to those that control P resorption, in accordance with the results of studies of wood species in tropical forests [41,42] and wetland graminoids [43]. However, other studies have found no significant correlations between the N and P resorption efficiencies of certain plants [40].

Traditionally, justification for the high metabolic cost associated with the production of tannins is attributed to improved herbivore defense [44]. In mangrove species, tannin is an abundant component as high as 20% dry weight which prevents damage from herbivory [17]. *A. corniculatum* leaves had high total phenolics and total condensed tannin levels (both above 20%), which increased during leaf senescence (Figures 2 and 3). The observed changes were in agreement with the findings reported for another mangrove species, *Rhizophora stylosa* [8]. As polyphenols are water-soluble and susceptible to leaching [45], leaching of polyphenols or tannins from green leaves by sporadic rain might be a cause for the net enrichment in senescent leaves [46]. However, Mafongoya et al. [47] indicated that much of the soluble carbon compounds, including polyphenols, are expected to be translocated from leaves during senescence. Constantinides and Fownes [48] found both an increase and a decrease in concentrations of polyphenols with the development and senescence of leaves, while Kuhajek et al. [49] observed no significant effects of leaf age on both condensed tannins and total phenolics. As for bound condensed tannins, only the FBCT concentrations in spring were significantly higher in senescent leaves than in mature leaves. These results were not exactly the same as the findings reported by Zhang et al. [34], which indicated that PBCT concentrations are significantly higher in senescent leaves than in mature leaves in all seasons.

The TPs concentrations in mature leaves were lowest in winter, and in senescent leaves concentrations were highest in spring (Figure 2). The TPs concentrations in mature and senescent leaves were all inversely related to N or P concentrations (Table 3). It is common to find a negative correlation between N and secondary compound concentrations, such as phenolics and tannins [50]. This pattern lends support to source-sink hypotheses, such as the carbon-nutrient balance (CNB) hypothesis and the growth-differentiation (GDB) hypothesis that predict increased C allocation to secondary C compounds under low nutrient conditions. Under nutrient limited circumstances, the diversion of C to secondary compounds may be essentially cost-free with respect to growth or reproduction [51]. Kandil et al. [52] suggested that the production and accumulation of the polyphenolics is favored in the high light, high temperature, high salinity, and low nutrient environment in which the mangroves occur, particularly when carbon used for growth or reproduction is precluded by lack of nutrient resources.

TPs:N and TCT:N ratios in senescent leaves were significantly higher than those in mature leaves (Figure 4). Condensed tannins are considered to have an important impact for N immobilization in soils, in particular during fall when fresh litter materials are introduced into soils [53]. Through both litter deposition and foliar leaching, soils in tannin rich plant communities receive appreciable tannin inputs. Because high tannin concentrations in plants are often associated with infertile site conditions [22], it has been suggested that there is an evolutionary advantage to higher tannin production. By reducing decomposition rates and decreasing N leaching potential, tannins may provide a nutrient conservation mechanism [8,54]. Therefore, parameters such as TPs:N and TCT:N ratios may be the best predictors of litter quality [55]. In the present study, higher TPs:N and TCT:N ratios in senescent leaves may slow down the rate of litter decomposition, thereby increasing the nutrient use efficiency and may be, in part, an adaptation to nutrient limited conditions.

In conclusion, N:P ratios of mature leaves are a good indicator for nutrient limitation in mangrove ecosystems. The results confirmed that resorption efficiency during leaf senescence depends on the type of nutrient limitation, and NRE was much higher than PRE in N-limited habitat. The resorption proficiency of N and P indicated that N resorption was complete and P resorption was incomplete. *A. corniculatum* leaves contained high total phenolics, extractable condensed tannin, and total condensed tannin levels. TPs:N and TCT:N ratios in senescent leaves were significantly higher than those in mature leaves. Our results indicated that high NRE during leaf senescence and high TPs:N and TCT:N ratios in senescent leaves might be important nutrient conservation strategies for the mangrove shrub *A. corniculatum* forest growing under the N-limited conditions. Furthermore, it has also been previously reported that mangroves are highly plastic in their responses to multiple abiotic and biotic factors such as destructive weather, herbivory, etc. As such, further work is required for understanding a more thorough temporal variation of nutrient resorption and phenolics concentration for *A. corniculatum* in long-term observations.

Acknowledgments: This work was supported by the National Natural Science Foundation of China (41201293), the WEL Visiting Fellowship Program of Xiamen University (WEL201504), and the Yangtze Youth Fund of Yangtze University (2015cqn60).

Author Contributions: Hui Chen, Benbo Xu, Shudong Wei, and Haichao Zhou designed and performed the experiments; Hui Chen, Shudong Wei and Lihua Zhang analyzed the data; Benbo Xu and Yiming Lin contributed reagents/materials/analysis tools; Hui Chen wrote the paper; Shudong Wei and Lihua Zhang revised the paper and modified the language.

References

1. Twilley, R.W.; Lugo, A.E.; Patterson-Zucca, C. Litter production and turnover in basin mangrove forests in southwest Florida. *Ecology* **1986**, *67*, 670–683. [CrossRef]
2. Tomlinson, P.B. *The Botany of Mangroves*; Cambridge University Press: New York, NY, USA, 1986.
3. Li, M. Nutrient dynamics of a Futian mangrove forest in Shenzhen, South China. *Estuar. Coast. Shelf. Sci.* **1997**, *45*, 463–472. [CrossRef]

4. Odum, W.E.; Heald, E.J. The detritus-based food web of an estuarine mangrove community. In *Estuarine Research*; Cronin, L.E., Ed.; Academic Press: New York, NY, USA, 1975; pp. 265–286.
5. Alongi, D.M.; Boto, K.G.; Robertson, A.I. Nitrogen and phosphorus cycles. In *Tropical Mangrove Ecosystems*; Robertson, A.I., Alongi, D.M., Eds.; American Geographysical Union: Washington, DC, USA, 1992; pp. 251–292.
6. Wang, W.Q.; Wang, M.; Lin, P. Seasonal changes in element contents in mangrove element retranslocation during leaf senescence. *Plant Soil* **2003**, *252*, 187–193. [CrossRef]
7. Cai, G. Salt content and regulation of mangroves. *Ecol. Sci.* **1990**, 2, 123–127.
8. Lin, Y.M.; Liu, X.W.; Zhang, H.; Fan, H.Q.; Lin, G.H. Nutrient conservation strategies of a mangrove species *Rhizophora stylosa* under nutrient limitation. *Plant Soil* **2010**, *326*, 469–479. [CrossRef]
9. Wei, S.D.; Liu, X.W.; Zhang, L.H.; Chen, H.; Zhang, H.; Zhou, H.C.; Lin, Y.M. Seasonal changes of nutrient levels and nutrient resorption in *Avicennia marina* leaves in Yingluo Bay, China. *South. For.* **2015**, *77*, 237–242.
10. Millard, P.; Neilsen, G.H. The influence of nitrogen supply on the uptake and remobilization of stored N for the seasonal growth of apple trees. *Ann. Bot.* **1989**, *63*, 301–309.
11. Lin, Y.; Sternberg, L.D.L. Nitrogen and phosphorus dynamics and nutrient resorption of *Rhizophora mangle* leaves in south Florida, USA. *Bull. Mar. Sci.* **2007**, *80*, 159–169.
12. Pugnaire, F.I.; Chapin, F.S. Controls over nutrient resorption from leaves of evergreen Mediterranean species. *Ecology* **1993**, *74*, 124–129. [CrossRef]
13. Feller, I.C. Effects of nutrient enrichment on growth and herbivory of dwarf red mangrove (*Rhizophora mangle*). *Ecol. Monogr.* **1995**, *65*, 477–505. [CrossRef]
14. Reef, R.; Feller, I.C.; Lovelock, C.E. Nutrition of mangroves. *Tree Physiol.* **2010**, *30*, 1148–1160. [CrossRef] [PubMed]
15. McKee, K.L.; Feller, I.C.; Popp, M.; Wanek, W. Mangrove isotopic (δ^{15}N and δ^{13}C) fractionation across a nitrogen vs. phosphorus limitation gradient. *Ecology* **2002**, *83*, 1065–1075. [CrossRef]
16. Güsewell, S. N:P ratios in terrestrial plants: Variation and functional significance. *New Phytol.* **2004**, *164*, 243–266. [CrossRef]
17. Hernes, P.J.; Benner, R.; Cowie, G.L.; Goni, M.A.; Bergamaschi, B.A.; Hedges, J.I. Tannin diagenesis in mangrove leaves from a tropical estuary: A novel molecular approach. *Geochim. Cosmochim. Acta* **2001**, *65*, 3109–3122. [CrossRef]
18. Zucker, W.V. Tannins: Does structure determine function? An ecological perspective. *Am. Nat.* **1983**, *121*, 335–365. [CrossRef]
19. Lin, Y.M.; Liu, J.W.; Xiang, P.; Lin, P.; Ding, Z.H.; Sternberg, L.D.L. Tannins and nitrogen dynamics in mangrove leaves at different age and decay stages (Jiulong River Estuary, China). *Hydrobiologia* **2007**, *583*, 285–295. [CrossRef]
20. Lin, Y.M.; Liu, J.W.; Xiang, P.; Lin, P.; Ye, G.; Sternberg, L.D.L. Tannin dynamics of propagules and leaves of *Kandelia candel* and *Bruguiera gymnorrhiza* in the Jiulong River Estuary, Fujian, China. *Biogeochemistry* **2006**, *78*, 343–359. [CrossRef]
21. Serrano, L. Leaching from vegetation of soluble polyphenolic compounds, and their abundance in temporary ponds in the Doñana National Park (SW Spain). *Hydrobiologia* **1992**, *229*, 43–50. [CrossRef]
22. Northup, R.R.; Yu, Z.; Dahlgren, R.A.; Vogt, K.A. Polyphenol control of nitrogen release from pine litter. *Nature* **1995**, *377*, 227–229. [CrossRef]
23. Huang, G.M. Growth characteristics of *Spartina alterniflora* and its relative competitive ability with *Kandelia candle* in mangrove areas of the Zhangjiang estuary. Master's Thesis, Xiamen University, Xiamen, China, 2009.
24. Chen, H.; Lu, W.; Yan, G.; Yang, S.; Lin, G. Typhoons exert signifificant but differential impacts on net ecosystem carbon exchange of subtropical mangrove forests in China. *Biogeosciences* **2014**, *11*, 5323–5333. [CrossRef]
25. Yoshida, S.; Forno, D.A.; Cock, J.H.; Gomez, K.A. *Laboratory Manual for Physiological Studies of Rice*, 2nd ed.; The International Rice Research Institute: Manila, Philippines, 1972; pp. 7–9, 36–38.
26. Nanjing Institute of Soil Science. *Physical and Chemical Analysis of Soils*; Science and Technology Press: Shanghai, China, 1978; pp. 1–320.
27. Graham, H.D. Stabilization of the Prussian blue color in the determination of polyphenols. *J. Agric. Food Chem.* **1992**, *40*, 801–805. [CrossRef]

28. Terrill, T.H.; Rowan, A.M.; Douglas, G.B.; Barry, T.N. Determination of extractable and bound condensed tannin concentrations in forage plants, protein concentrate meals and cereal grains. *J. Sci. Food Agric.* **1992**, *58*, 321–329. [CrossRef]
29. Aerts, R. Nutrient resorption from senescing leaves of perennials: Are there general patterns? *J. Ecol.* **1996**, *84*, 597–608. [CrossRef]
30. Killingbeck, K.T. Nutrients in senesced leaves: Keys to the search for potential resorption and resorption proficiency. *Ecology* **1996**, *77*, 1716–1727. [CrossRef]
31. Rejmánková, E. Nutrient resorption in wetland macrophytes: Comparison across several regions of different nutrient status. *New Phytol.* **2005**, *167*, 471–482. [CrossRef] [PubMed]
32. Lambers, H.; Chapin, F.S.; Pons, T.L. *Plant Physiological Ecology*; Springer: New York, NY, USA, 1998.
33. Aerts, R.J.; Barry, T.N.; McNabb, W.C. Polyphenols and agriculture: Beneficial effects of proanthocyanidins forages. *Agr. Ecosyst. Environ.* **1999**, *75*, 1–12. [CrossRef]
34. Zhang, L.H.; Ye, G.F.; Lin, Y.M.; Zhou, H.C.; Zeng, Q. Seasonal changes in tannin and nitrogen contents of *Casuarina equisetifolia* branchlets. *J. Zhejiang Uni. Sci. B* **2009**, *10*, 103–111. [CrossRef] [PubMed]
35. Ye, G.F.; Zhang, S.J.; Zhang, L.H.; Lin, Y.M.; Wei, S.D.; Liao, M.M.; Lin, G.H. Age-related changes in nutrient resorption patterns and tannin concentration of *Casuarina equisetifolia* plantations. *J. Trop. For. Sci.* **2012**, *24*, 546–556.
36. Aerts, R.; Chapin, F.S. The mineral nutrition of wild plants revisited: A re-evaluation of processes and patterns. *Adv. Ecol. Res.* **2000**, *30*, 1–67.
37. Koerselman, W.; Meuleman, A.F.M. The vegetation N:P ratio: A new tool to detect the nature of nutrient limitation. *J. Appl. Ecol.* **1996**, *33*, 1441–1450. [CrossRef]
38. Kobe, R.K.; Lepczyk, C.A.; Iyer, M. Resorption efficiency decreases with increasing green leaf nutrients in a global data set. *Ecology* **2005**, *86*, 2780–2792. [CrossRef]
39. Wright, I.J.; Westoby, M. Nutrient concentration, resorption and lifespan: Leaf traits of Australian sclerophyll species. *Funct. Ecol.* **2003**, *17*, 10–19. [CrossRef]
40. Chapin, F.S.; Kedrowski, R.A. Seasonal changes in nitrogen and phosphorus fractions and autumn retranslocation in evergreen and deciduous taiga trees. *Ecology* **1983**, *64*, 376–391. [CrossRef]
41. Lal, C.B.; Annapurna, C.; Raghubanshi, A.S.; Singh, J.S. Effect of leaf habit and soil type on nutrient resorption and conservation in woody species of a dry tropical environment. *Can. J. Bot.* **2001**, *79*, 1066–1075.
42. Cai, Z.; Bongers, F. Contrasting nitrogen and phosphorus resorption efficiencies in trees and lianas from a tropical montane rain forest in Xishuangbanna, south-west China. *J. Trop. Ecol.* **2007**, *23*, 115–118. [CrossRef]
43. Güsewell, S. Nutrient resorption of wetland graminoids is related to the type of nutrient limitation. *Funct. Ecol.* **2005**, *19*, 344–354. [CrossRef]
44. Feeny, P. Seasonal changes in oak leaf tannins and nutrients as a cause of spring feeding by winter moth caterpillars. *Ecology* **1970**, *51*, 565–581. [CrossRef]
45. Hättenschwiler, S.; Vitousek, P.M. The role of polyphenols in terrestrial ecosystem nutrient cycling. *Trends Ecol. Evol.* **2000**, *15*, 238–243. [CrossRef]
46. Teklay, T. Seasonal dynamics in the concentrations of macronutrients and organic constituents in green and senesced leaves of three agroforestry species in southern Ethiopia. *Plant Soil* **2004**, *267*, 297–307. [CrossRef]
47. Mafongoya, P.L.; Giller, K.E.; Palm, C.A. Decomposition and nitrogen release patterns of tree prunings and litter. *Agrofor. Syst.* **1998**, *38*, 77–97. [CrossRef]
48. Constantinides, M.; Fownes, J.H. Nitrogen mineralization from leaves and litter of tropical plants: Relationship to nitrogen, lignin and soluble polyphenol concentrations. *Soil Biol. Biochem.* **1994**, *26*, 49–55. [CrossRef]
49. Kuhajek, J.M.; Payton, I.J.; Monks, A. The impact of defoliation on the foliar chemistry of southern rata (*Metrosideros umbellata*). *N. Z. J. Ecol.* **2006**, *30*, 237–249.
50. Mansfield, J.L.; Curtis, P.S.; Zak, D.R.; Pregitzer, K.S. Genotypic variation for condensed tannin production in trembling aspen (*Populus tremuloides*, Salicaceae) under elevated CO_2 and in high-and low-fertility soil. *Am. J. Bot.* **1999**, *86*, 1154–1159. [CrossRef] [PubMed]
51. Bryant, J.P.; Chapin, F.S., III; Klein, D.R. Carbon/nutrient balance of boreal plants in relation to vertebrate herbivory. *Oikos* **1983**, *40*, 357–368. [CrossRef]
52. Kandil, F.E.; Grace, M.H.; Seigler, D.S.; Cheeseman, J.M. Polyphenolics in *Rhizophora mangle* L. leaves and their changes during leaf development and senescence. *Trees* **2004**, *18*, 518–528. [CrossRef]

53. Maie, N.; Behrens, A.; Knicker, H.; Kogel-Knabner, I. Changes in the structure and protein binding ability of condensed tannins during decomposition of fresh needles and leaves. *Soil Biol. Biochem.* **2003**, *35*, 577–589. [CrossRef]
54. Kraus, T.E.C.; Dahlgren, R.A.; Zasoski, R.J. Tannins in nutrient dynamics of forest ecosystems—A review. *Plant Soil* **2003**, *256*, 41–66. [CrossRef]
55. Kraus, T.E.C.; Zasoski, R.J.; Dahlgren, R.A. Fertility and pH effects on polyphenol and condensed tannin concentrations in foliage and roots. *Plant Soil* **2004**, *262*, 95–109. [CrossRef]

14

Importance of the Local Environment on Nutrient Cycling and Litter Decomposition in a Tall Eucalypt Forest

Jessie C. Buettel *, Elise M. Ringwaldt, Mark J. Hovenden and Barry W. Brook

Discipline of Biological Sciences, School of Natural Sciences, University of Tasmania, Hobart 7000, Australia; Elise.Ringwaldt@utas.edu.au (E.M.R.); Mark.Hovenden@utas.edu.au (M.J.H.); Barry.Brook@utas.edu.au (B.W.B.)
* Correspondence: jessie.buettel@utas.edu.au

Abstract: The relative abundance of nitrogen-fixing species has been hypothesised to influence tree biomass, decomposition, and nitrogen availability in eucalypt forests. This prediction has been demonstrated in experimental settings (two-species mixtures) but is yet to be observed in the field with more realistically complex communities. We used a combination of (a) field measurements of tree-community composition, (b) sampling of soil from a subset of these sites (i.e., the local environment), and (c) a decomposition experiment of forest litter to examine whether there is a local-scale effect of the nitrogen-fixing *Acacia dealbata* Link (presence and abundance) on nitrogen availability, and whether increases in this essential nutrient led to greater biomass of the canopy tree species, *Eucalyptus obliqua* L'Hér. Average *A. dealbata* tree size was a significant predictor of forest basal area in 24 plots (12% deviance explained) and, when combined with average distance between trees, explained 29.1% variance in *E. obliqua* biomass. However, static patterns of local nitrogen concentration were unrelated to the presence or size of *A. dealbata*, despite our experiments showing that *A. dealbata* leaf litter controls decomposition rates in the soil (due to three times higher N). Such results are important for forest management in the context of understanding the timing and turnover of shorter-lived species like acacias, where higher N (through either litter or soil) might be better detected early in community establishment (when growth is faster and intraspecific competition more intense) but with that early signal subsequently dissipated.

Keywords: forests; nutrients; disturbance; management; diversity; biomass

---

1. Introduction

Understanding the mechanisms that control the biomass and density of trees in forests is important for ecological, economic, and conservation motivations [1]. For instance, growing forests sequester substantial amounts of carbon [2], and as global atmospheric carbon dioxide (CO_2) continues to rise, understanding tree growth and the mechanisms that drive carbon accumulation and storage has become a pressing issue. Abiotic factors like climate, soil type, and disturbance frequency are known to be among the major determinants of the productivity of plant communities [3]. However, in recent decades, interest has shifted towards how community properties (e.g., number and relative abundance of species), and functional traits (e.g., [4]), influence productivity [5]. This is particularly important in studies of community structure, because these features are amenable to experimentation and as a management tool. Additionally, since different plant species vary in their nutrient requirements [6], it is possible that nutrient cycling is a fundamental feature contributing to the observed link between diversity and productivity.

The availability of nitrogen (N) limits productivity in many terrestrial ecosystems, and its increase typically promotes plant-community productivity [7,8]. This effect is further facilitated through complementarity, because differences in N-use efficiency lead to positive feedbacks in N-cycling rates [9]. Nitrogen cycling is heavily dependent on the input and output of various forms of N [10,11], which is often supplied in various quantities by different plant species (and their litter). The plant litter can comprise a mixture of fine detritus (e.g., leaf litter, bark, twigs) and coarse material (branches and whole stems of fallen trees), which are then decomposed and transformed into available nutrients, such as nitrogen [12], via ammonification, nitrification, and N mineralisation. Nitrification is the dominant process in most soils, where nitrifying organisms are often constrained by, and are dependent on, the rate of ammonification [13]. Because of the scarcity of N in soils, soil-N transformations have a strong influence on the potential productivity of an ecosystem [14,15].

It is well known that decomposition rates can vary due to differences in leaf quality (C:N ratio) among plant species, and often increase when litter of different species is combined [16,17] due to complementary stimulation among microbial groups [18]. For example, sclerophyllous eucalypts produce litter that has low nutrient concentrations and low rates of decomposition [19] whereas, by contrast, the litter of N-fixing acacias has comparatively high N concentrations and decomposition rates [20,21]. This has led to speculation that Australian forest communities containing species with more readily decomposable and nutrient-rich litter (such as *Acacia* spp.) might have higher litter decomposition rates than those with *Eucalyptus* litter alone [22,23]. However, this idea has yet to be explored in natural forest systems or linked to overall tree size and biomass.

Here we use field and laboratory data to determine the local-scale effect of an N-fixing tree species (*Acacia dealbata* Link) on the size and total biomass of the dominant eucalypt species (*Eucalyptus obliqua* L'Hér) in a tall hardwood forest community in Tasmania, southern Australia. We used a combination of:

1. Field measurements on the standing pattern (size structure, density, and species composition) of mature trees, used as a proxy for productivity;
2. Sampling and analysis of the soil-nutrient composition in the local environment (i.e., across field sites) and assessment of its relationship to site biomass; and
3. Evaluation of the association between forest-litter quality (different species mixes) and quantity, and the rate of decomposition, as a measure of turnover.

2. Materials and Methods

2.1. Study Site

The field sites were located in a tall eucalypt forest at an altitudinal range of 310–440 m above sea level, on the foothills of Mount Wellington, in south-eastern Tasmania (42.895° S–147.268° E). Vegetation in the region is dominated by *E. obliqua*, a widespread and abundant tall tree found across south-eastern Australia, with the occasional occurrence of *Eucalyptus regnans* F.Muell and *Eucalyptus delegatensis* R.T.Baker. The community also contains two acacia species as sub-dominants: *A. dealbata* and *Acacia melanoxylon* R.Br. The understorey species included *Bedfordia salicina* D.C., *Coprosma quadrifida* (Labill.) B.L.Rob., *Pittosporum bicolor* Hook., and *Pomaderris apetala* Labill. The site is predominately covered by grey-brown podzolic soils on mudstone, classified as Kurosol under the Australian Soil Classification system [24]. Kurosol soil is known as being acidic and nutrient poor [25] with vegetation structure strongly dependent on rainfall [24]. The climate has warm, dry summers and cool, moist winters, with a mean annual precipitation of 720 mm and a mean annual temperature of 17.5 °C. We used the basal area of trees per unit area as a proxy for long-term integrated 'productivity' (>four decades of canopy-tree-biomass accumulation). The forest stand is even-aged, being a natural regeneration following a 1967 stand-replacing fire. The few trees that had obviously lived through the fire (e.g., diameter at breast height >100 cm were avoided during field measurements or excluded from analysis). For this study, we were interested in local heterogeneity across space, rather than temporal comparisons across different site histories. Landscape features were homogeneous across sites (e.g., little variation in topography, canopy, understory, and rock cover).

2.2. Site Selection and Survey Technique

To collect data on community structure and to examine the relationship between tree size and total basal area we surveyed 24 sites using a modified point-quarter method [26], where each site was chosen using a random bearing and distance greater than 50 m from a randomly selected starting point located within the forest (Table S2). The tree closest in proximity to this distance and bearing was designated the starting tree. Each starting tree was identified to species level, and diameter at breast height (DBH in cm) was measured. From the starting tree, the nearest neighbour in each of four compass-bearing quadrants (northeast, northwest, southeast and southwest) was selected. These four trees were also identified to species level and their DBH was also measured, with distance from the starting tree determined using a laser meter. This nearest-neighbour selection method was repeated for each of the four initial trees, giving a total of 17 trees (including the starting tree) at each of the 24 sites.

2.3. Soil Sampling and Processing

Soil samples were collected during early autumn from within the study community. Nine of the 24 sites were selected to represent variation in the proportion of *Acacia* (nitrogen fixers) present, relative to total basal area of all trees (Table S1). At each of these *Acacia* sites, five sampling locations were selected using a random bearing (in NW, SE, SW, NE directions) and a random distance from the starting tree. Five soil cores (2×15 cm) were randomly placed in a 2×2 m quadrant within each location for a total of 25 soil core samples per site. A site's soil samples were then mixed and homogenised by passing through a 2 mm sieve, which removed large roots, litter, and rocks. To determine relative soil water content, a 10 g subsample from each site was dried in an oven at 110 °C for 48 h. A 5 g soil subsample from each site was also ground in a ball mill (MM200 Mixer Mill, Retsch GmbH, Haan, Germany) for total C and N determination via a controlled combustion process in a PerkinElmer 2400 Series II Elemental Analyser (PerkinElmer, Australia [27]).

2.4. Assessing Field Nutrient Availability

To determine nutrient availability in the forest community, a 10 g subsample of each sieved soil sample was extracted with 40 mL of 2 M KCl to determine available NH_4^+ and NO_3^- concentrations (Table S3). The tubes containing the soil and solution were shaken for 1 h, then centrifuged for 3 min at 4000 rpm. A total of 10 mL of supernatant was removed with a syringe and filtered through a Whatman No. 42, 2 μm syringe filter. The filtered extracts were analysed for NH_4^+ and NO_3^- concentrations in a SmartChem Discrete Analyser (Westco Scientific Instruments, Brookfield, CT, USA). Additionally, a portion of the soil was also assessed to determine phosphorus (P) concentrations in the soil at each site using the extraction methods described in [28]: 2 g of soil was mixed with 40 mL 0.005 M $NaCl_2$ solution and shaken for 17 h, then the mixed solution was extracted and analysed using the syringe method described above, with available P concentration determined using a SmartChem Discrete Analyser.

2.5. Determining Nitrogen Mineralisation Rate

The N mineralisation rate of the soil from each of the nine subsampled sites was determined using published protocols [29]. Briefly, 20 g of fresh soil from each composite sample was incubated in a sealed 500 mL glass jar placed in a darkened incubator at 25 °C for 34 days. Jars were ventilated regularly to prevent soils becoming anoxic. Net N mineralisation, nitrification, and ammonification were calculated as the difference between initial total available N, NO_3^-, and NH_4^+ concentrations and those measured following the incubation period (Table S4). The rates of nitrogen mineralisation (mg N g^{-1} soil) in relation to the proportion of *A. dealbata* in the community was analysed from incubation measurements from soil taken from nine sites, as the sum of NH_4^+ and NO_3^-. Each of the nine sites had a different proportion of *A. dealbata*, and this natural cross-site variation was used to determine whether: (a) There were local-scale differences between nitrogen mineralisation rates and (b) there was a relationship between the proportion of *A. dealbata* and mineralisation rates of nitrogen.

Results were analysed with a linear regression, using the resampling technique described below in Section 2.6.

2.6. Statistical Methods

All statistical analyses were done in Program R v3.3.2 (R Core Development Team 2016, Vienna, Austria). A likelihood-based generalised linear modelling framework was used to assess experimental predictors and dependencies. For each data set, an *a priori* model set was created, and strength of support was evaluated by using Akaike's information criterion, adjusted for small sample sizes (AIC_c). Model weights were calculated using the MuMIn package [30]. Akaike weights (w_i) were used to assess model support; these represent the relative likelihood of a given model compared to others in the evaluation set (which can be of varying complexity and non-nested), by adjusting the likelihood for bias associated with fitting the parameters [31,32]. The relative support from the data for competing models was evaluated using the evidence ratio (ER), the ratio of the respective model's w_i. The percentage deviance (%DE) explained by each model relative to the null (intercept only) was used to assess structural goodness of fit.

In cases where multiple measurements of the dependent variable were taken at each level of a predictor (see Section 3.4), resampling was used (with one outcome selected randomly for any given predictor level and repeated 10,000 times) to avoid pseudoreplication, with the models fitted repeatedly to each subsample. The median prediction from the set of models fitted to the resampled data was used to determine the slope of the predictor and 95% confidence intervals. As a check, these were also compared to models fitted to the medians of the raw data (i.e., ignoring within-sample heterogeneity).

The 24 plots were analysed to assess the influence of *Acacia* sp. on *Eucalyptus* sp., and total site productivity. In this case, the *a priori* models represented additive combinations of the following predictors (of varying complexity): Proportion of *A. dealbata* biomass, the average size of *A. dealbata*, the number of species, or the variance of the pairwise distances between trees. Dependent variables were total basal area of a site or, alternatively, the average size of the eucalypts (DBH cm).

To test whether the availability of soil nutrients (C and N) had an influence on the average size of eucalypts or the total basal area (see Section 3.3), fitted linear models were restricted to either one or two terms due to sample-size limitations (nine total sites). Average size of *A. dealbata* was determined to be the most important factor influencing site productivity during the 24-plot analysis (Section 3.1 and see Results), and therefore was included as a control term in all *a priori* models. The other predictor variable assessed was either: C, N, or pre-total mineralisation of N. Phosphorus was also considered but there were negligible differences across sites.

2.7. Laboratory Litter Experiment

The impact of litter quality and quantity on decomposition rate was determined experimentally by manipulating the relative contribution of litter from the three co-occurring species in the field sites (*E. obliqua*, *A. dealbata*, and *P. apetala*). Newly senesced leaves of the three tree species were collected from each of the 24 sites. The leaf litter was combined to give a representative sample of each of the three species across the entire field site and subsequently dried in an oven at 110 °C for 24 h. Leaves were shredded in an adapted coffee grinder and then ground to powder in a ball mill (MM200 Mixer Mill, Retsch, Haan Germany).

We developed seven treatments ranging from 100% contribution of a single species to equal proportions (i.e., an even mixture; Table 1). The total weight of litter was controlled at 2 g, with 10% of the total weight as sieved pasteurised sand (20 g), which eliminated the effect of existing soil organic matter interfering with the effect of litter amendment on the incubations. Three replicates were prepared for each treatment, each with a different dominant species contributing the largest percentage of litter to the mix, yielding a total of 63 replicates, with three 'no litter' controls. Litter was apportioned into appropriate weights based on the percentage contribution of the species (from 0% to 100% contribution: Table 1) and placed in a 60 mL specimen container in a 500 mL glass jar containing the pasteurised sand. A 2 mL soil slurry solution was added to each jar, being prepared by mixing 50 g (fresh weight) of soil taken from the study site (Section 2.1), with 100 mL of deionised water. The soil was added to the incubation containers with a minimum of additional soil organic matter such that there was a consistency across each treatment. Additionally, samples were supplemented with 2.6 mL of deionised water to bring them to 60% field capacity, which is within the range of soil water content (50%–70%) that has been shown to encourage optimal microbial activity [33]. Field capacity was determined by flooding a 3 g subsample of soil and allowing it to stand for 30 min before draining the free water on a porous ceramic plate and then weighing to obtain water content at field capacity [34]. The subsample was then oven-dried for 24 h at 105 °C to obtain soil dry weight. To control for the respiration rate of the organic matter present in the soil slurry, inoculum blanks consisting of sand and the soil slurry (but no added litter) were also prepared. The sealed jars were incubated in the dark at 25 °C. Gas samples were obtained after 1, 2, 4, 7, 10, 14, 18, 23, 30, and 35 days of incubation.

Table 1. Relative proportion of litter across the seven treatments (ranging from an even mixture of 1, to one dominated by a single species (0). The three species (*Eucalyptus obliqua* L'Hér, *Acacia dealbata* Link, and *Bedfordia salicina* D.C.) are the most common found across the site.

Evenness	*E. obliqua*	*A. dealbata*	*B. salicina*
0	100% 0% 0%	0% 100% 0%	0% 0% 100%
0.2	95% 2.5% 2.5%	2.5% 95% 2.5%	2.5% 2.5% 95%
0.4	88% 6% 6%	6% 88% 6%	6% 6% 88%
0.6	79% 10.5% 10.5%	10.5% 79% 10.5%	10.5% 10.5% 79%
0.8	65% 17.5% 17.5%	17.5% 65% 17.5%	17.5% 17.5% 65%
0.9	55% 22.5% 22.5%	22.5% 55% 22.5%	22.5% 22.5% 55%
1	33.3% 33.3% 33.3%	33.3% 33.3% 33.3%	33.3% 33.3% 33.3%

Daily C mineralisation results for the litter experiment were analysed to determine the amount of potentially mineralisable C using a nonlinear differential model (two-pool model with one pool constant), as follows:

$$\frac{dC}{dt} = C_a.k_a e^{(-k_{at})} + r, \tag{1}$$

where $\frac{dC}{dt}$ is the cumulative amount of C mineralised at time t (per day in this case), C_a is the amount of C in the fast or labile C pool, k_{at} is the respiration rate constant of CO_2 evolution from the active pool, and r is the non-labile soil C respiration rate, which is effectively a constant over timescales less than a year (e.g., [33]). We assumed the labile C pool to have come directly from recent plant inputs (e.g., labile C compounds in litter [35]). Decomposition rates were corrected for CO_2 release from the inoculum by subtracting the daily mean CO_2 release of inoculum blanks from each experimental unit.

We then explored the differences in the rate of decomposition of the labile carbon per day and the proportion of *Acacia* in the litter over time. A linear regression was fitted to the resampled nitrogen composition taken at each proportion of *A. dealbata* (see next section). To determine the differences in carbon mineralisation of litter over time, each of three species: *E. obliqua*, *A. dealbata*, and *Bedfordia* sp. were measured for cumulative CO_2 evolution during litter incubation over a 40 day period and analysed using the Michaelis–Menten equation (a three-term nonlinear saturating model).

Although the Michaelis–Menten equation (MME) is commonly used as a simple representation of enzyme kinetics, it can also serve as a useful two-parameter characterisation of any system that exhibits some type of reaction velocity with a non-linear saturation (via estimation of a maximal rate and half-saturation parameter, with a start at the origin). We agree that soils, being complex mixtures, are unlikely to follow any single, simple reaction dynamics. We instead assume that the overall response will be MME-like but fit the model as a phenomenological rather than overtly mechanistic descriptor of the aggregate system response. The fact that it fits extremely well (See results, R^2 ranging from 90%–94%) gives confidence in this approach.

3. Results

3.1. Relationship between Size of Trees and Total Basal Area across All 24 Plots

The average size of *A. dealbata* (biomass) was the best predictor for total basal area across the 24 plots (df = 3, $w\text{AIC}_c$ = 0.32), explaining 12% of the total deviance. There was little information-theoretic support for the saturated model (containing all three variables), nor for the two predictor variables (i) variance of distances between trees (df = 3, $w\text{AIC}_c$ = 0.13) and (ii) proportion of *A. dealbata* (biomass)(df = 3, $w\text{AIC}_c$ = 0.10); these were no better supported than the null expectation.

Average tree size of *A. dealbata* (biomass) in additive combination with the variance of distances between trees was the best model for predicting average size of eucalypts (df = 4, $w\text{AIC}_c$ = 0.46, %DE = 29.1), followed by the saturated model (df = 5, $w\text{AIC}_c$ = 0.35, %DE = 36.5). A post hoc analysis suggested that of the two best predictors for average size of eucalypts, variance of distances between trees was 1.6 times better supported than the average size of *A. dealbata* (biomass). See Supplementary Materials for all raw data associated with the linear models and associated analyses.

3.2. Local Soil Nutrient Availability

There was a positive relationship between soil mineral N availability and soil C content (%) (R^2 = 0.51, slope = 1.680, 95% CI = 0.4–2.90; Figure 1a), indicating that nitrogen availability was dependent upon the amount of organic matter in the soil. However, neither soil total N (%) content nor available mineral nitrogen concentrations were related to the basal area of *A. dealbata* across the nine sites sampled (R^2 = 0.03, slope = −0.01, 95% CI = −0.03–0.02 and R^2 = 0.04, slope = −0.29, 95% CI = −1.40–0.81, respectively; Figure 1).

Conversely, *A. dealbata* average size did show a strong negative relationship to soil mineral N availability (R^2 = 0.47, slope = −0.61, 95% CI = −1.10–0.12), but not to soil N content (%) (R^2 = 0.19, slope = −0.01, 95% CI = −0.03–0.01). This result shows that none of the predictors for nutrient availability in the soil were predictive of total basal area of *A. dealbata*.

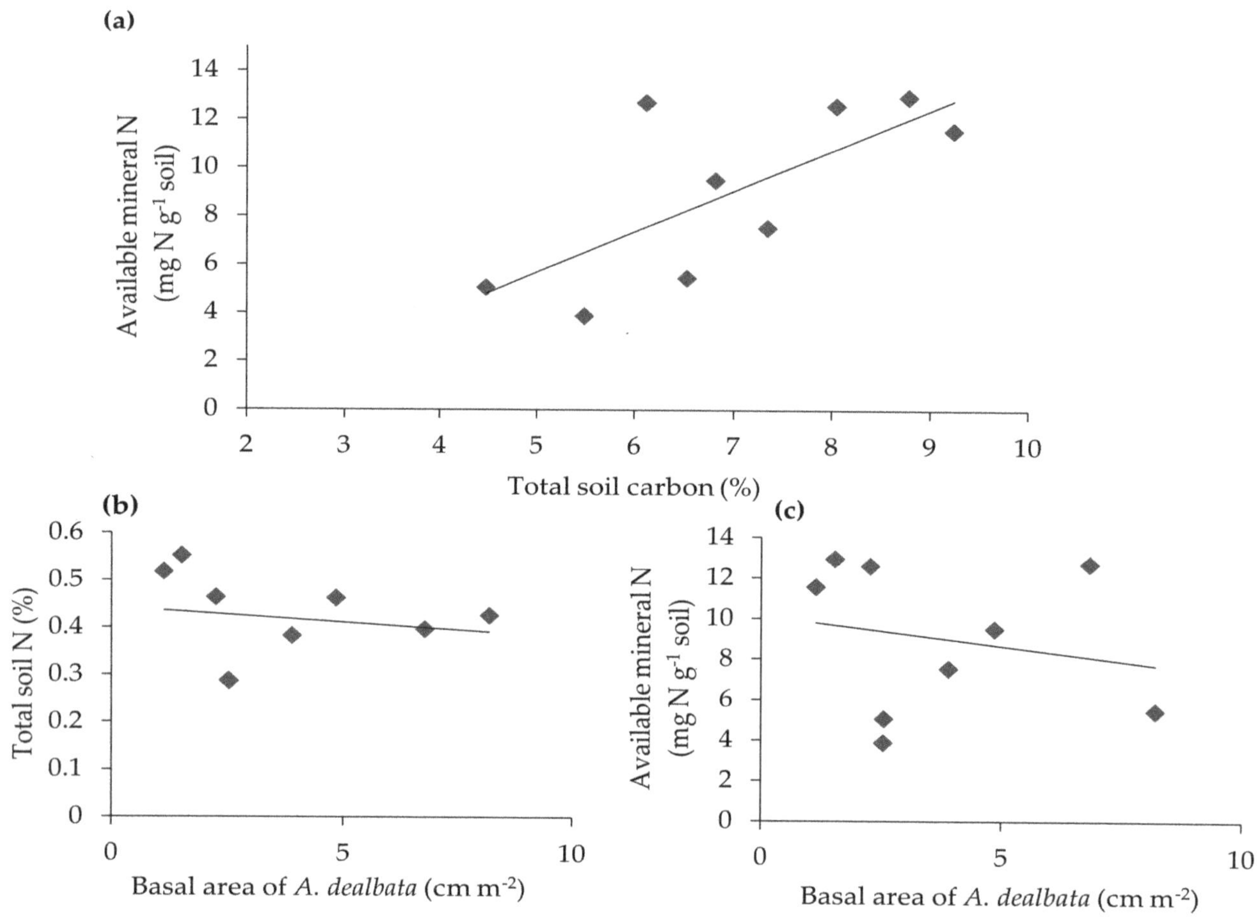

Figure 1. Relationship between total basal area of *A. dealbata* and the amount of nitrogen at each soil site in two different forms: (**a**) Relationship between total soil carbon (%) and available mineral nitrogen (N). Concentrations are determined from soil collected from each of the nine sites in a eucalypt forest. (**b**) Total soil organic matter quantity in terms of total soil nitrogen (%) and the total basal area at each soil site and (**c**) Available mineral N present at each site and the relationship to total basal area of *A. dealbata*.

3.3. Local C and N Dynamics

Since C mineralisation is an enzymatic process, its rate is dependent upon the availability of substrate; as expected, there was a clear relationship in our analysis between C mineralisation rate and soil organic matter content, whether expressed as total soil C or N (R^2 = 0.47, slope = 1371, CI % = 275–2467 and R^2 = 0.65, slope = 96.0, CI % = 42.4–149.6, Figure 2). Thus, the rate of soil organic matter decomposition was dependent upon the quantity of organic matter present.

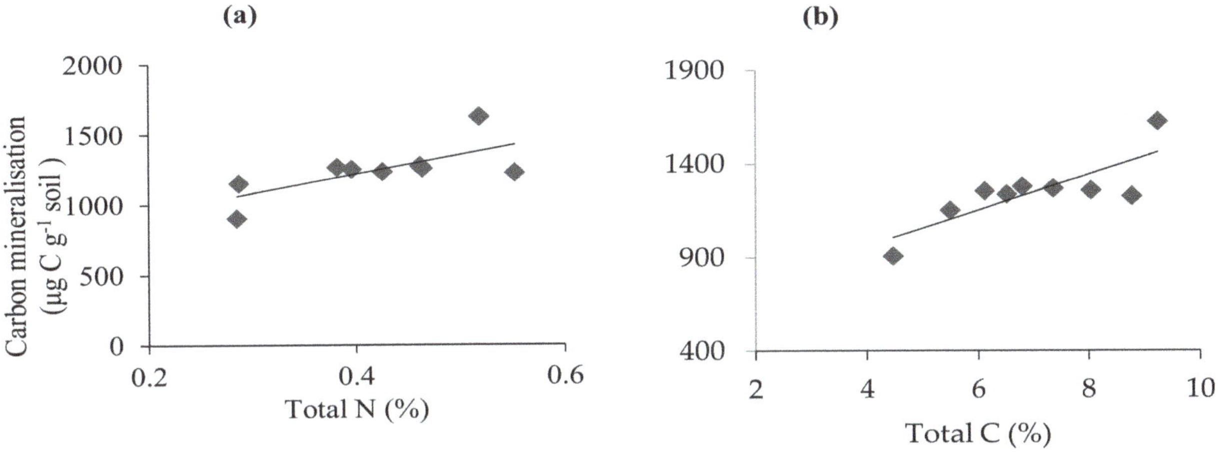

Figure 2. Relationship between total carbon mineralised (µg C g^{-1} soil) and organic matter quality demonstrated as two different forms. (**a**) Total soil nitrogen (N%) $R^2 = 0.47$ and (**b**) total soil carbon (C%) $R^2 = 0.65$. Data were determined from soil collected at nine different sites in a eucalypt forest.

There was a strong positive correlation between field-mineral-N availability and measured nitrification rate in the laboratory ($r = 0.79$). Therefore, the measurement of soil nitrogen availability of field samples corresponded well with measured rates of nitrogen transformation. There was also a strong relationship between rates of N transformation, ammonification, and nitrification ($r = 0.64$), demonstrating that the rate of nitrification was dependent on the rate of net ammonification, and vice versa.

However, there was only a weak correlation between nitrification and total N mineralisation ($r = 0.33$). Indeed, nitrification rates were relatively stable across the range of sites in comparison to the wide range of ammonification rates (Figure 3). There was by contrast, a strong positive relationship between net ammonification and net N mineralisation rates ($r = 0.94$, Figure 3), indicating that the supply of mineral N in this system was related to the conversion of organic N to ammonium.

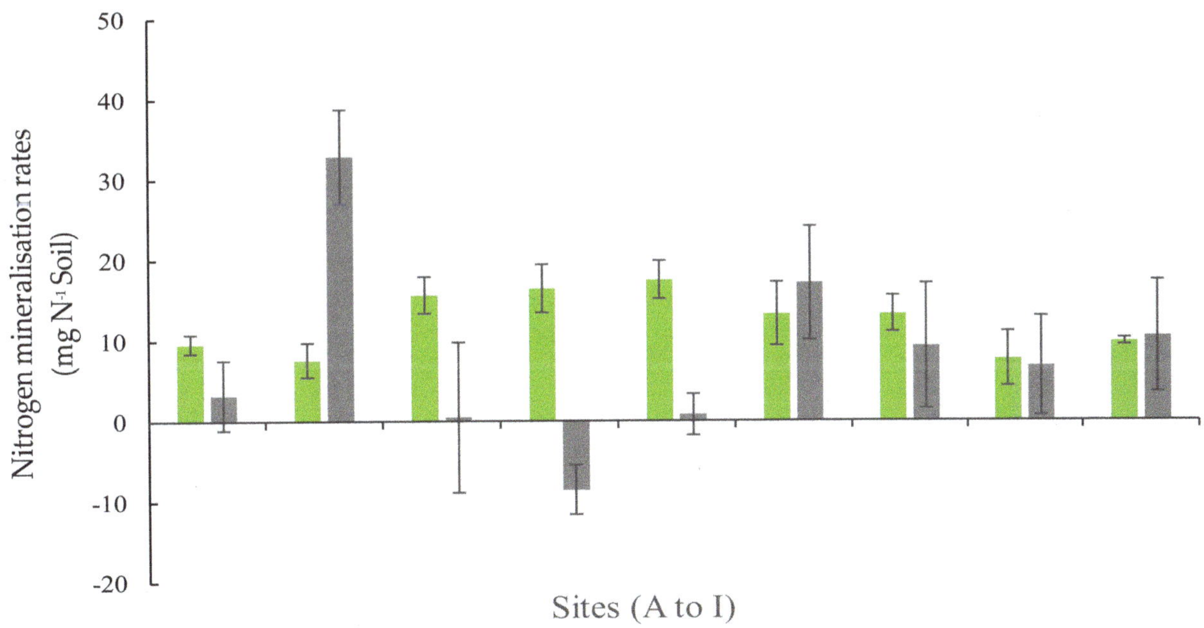

(**a**)

Figure 3. *Cont.*

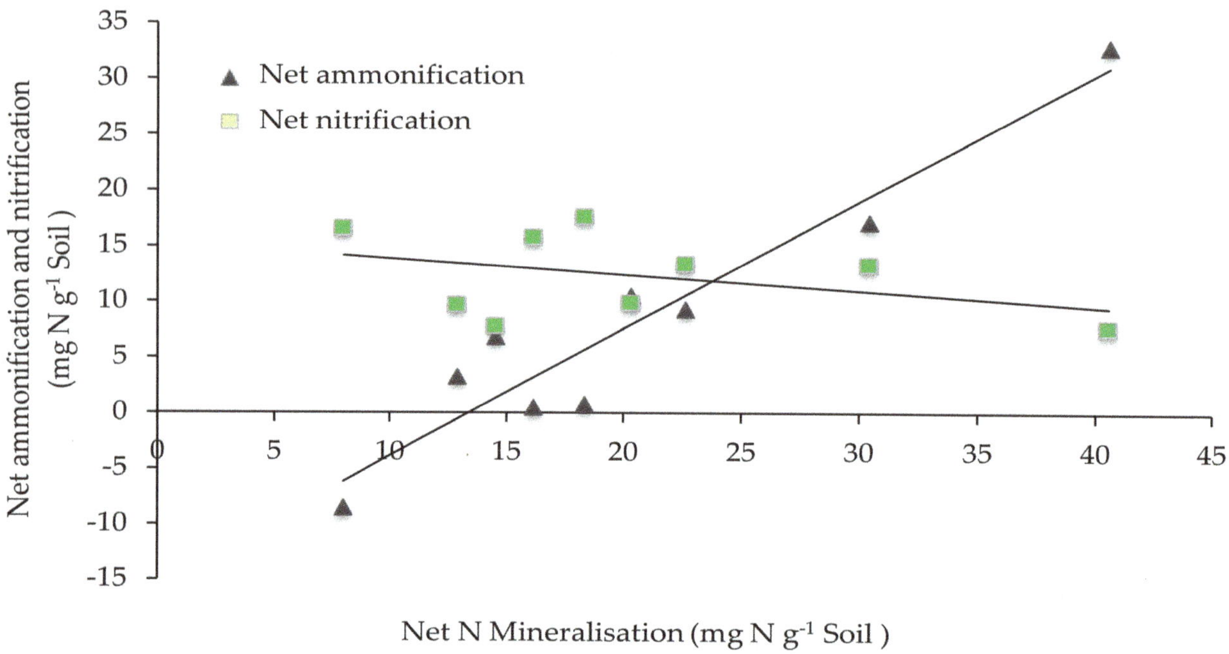

(b)

Figure 3. Rates of nitrogen (N) transformation obtained from incubation measurements from soil taken from nine sites. (**a**) Rates of ammonification (mg N g^{-1} soil); as the conversion of C and N (%) into NH_4^+; and the rates of nitrification (mg N g^{-1} soil); as the conversion of NH_4^+ into NO_3^-. One standard error (+ and −) is shown for each site measurement. (**b**) Relationship between net N mineralisation rates and net ammonification and net nitrification rates from soil collected from the nine sites.

3.4. Relationship between Litter Quality, Quantity, and Litter Decomposition

Field samples from communities with *A. dealbata* had almost three times more nitrogen in the litter (2.57% ± 0.11%) compared to *E. obliqua* (0.53% ± 0.03%) and *Bedfordia* sp. (0.80% ± 0.05%), and thus a lower C:N ratio. A higher proportion of *A. dealbata* in the litter (%) subsequently led to a higher release of labile carbon during the litter-decomposition experiment (Table 2; Figure 4). Both approaches to fitting this relationship statistically: (i) The median of the resampled points and (ii) the initial linear model to the median points, yielded nearly identical R^2 (indicating a negligible effect of pseudoreplication).

Table 2. Litter decomposition table showing the intercept, slope, and corresponding 95% confidence intervals for each, the initial model (fitted to the raw data) and the median of model fits to resampled data points; goodness of fit is indicated by the R^2.

Model (Median)	Intercept (α)	95% CI	Slope (β)	95% CI	Sigma	R^2%
Initial	2838	1818–3859	177	155–199	2829	81.8
Resampled	2877	290–5509	169	36–267	2192	81.7

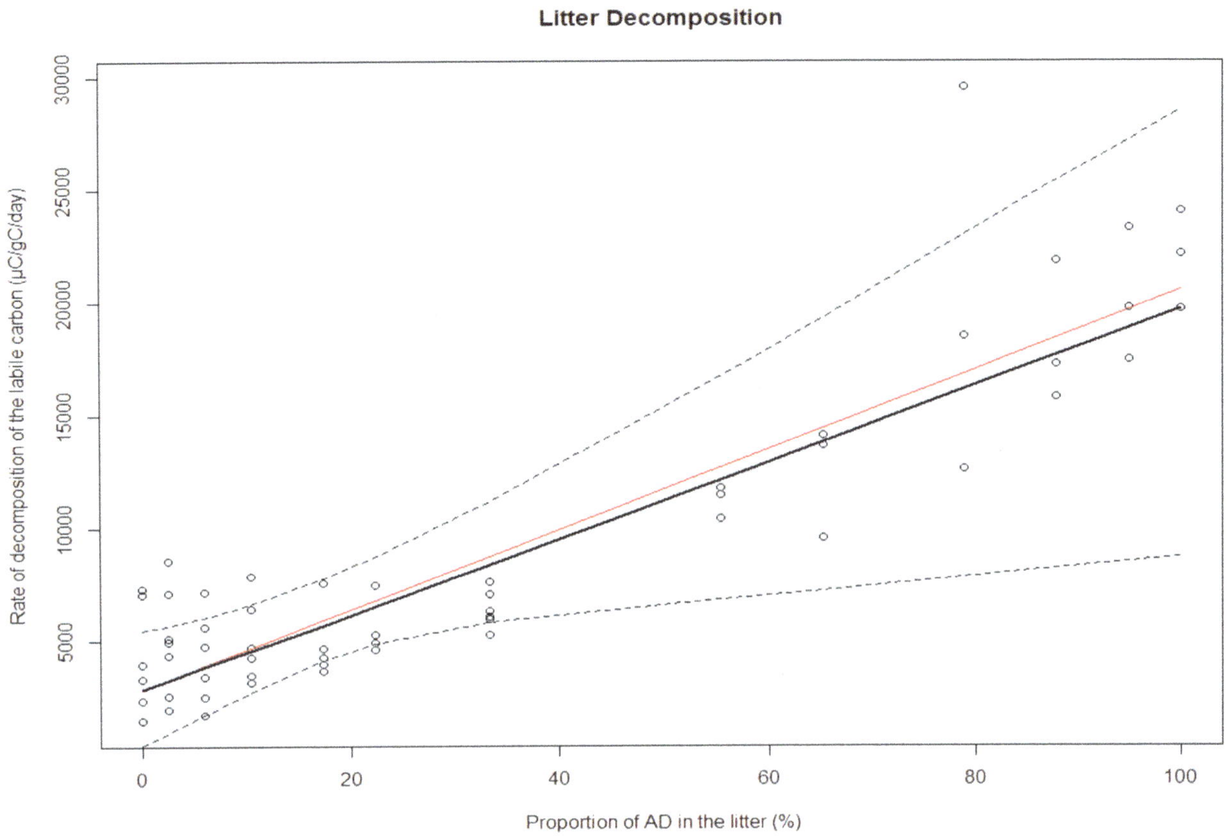

Figure 4. The relationship between the rate of labile carbon decomposition and of the nitrogen fixed *A. dealbata* in the litter. The red line represents the initial linear regression model to the median points (R^2: 81.8%), and the black line shows the median of the resampled points (R^2: 81.7%). Dotted lines represent the confidence bounds for the resampled data.

The resampled fits of the Michaelis–Menten saturating equation (MM) for the cumulative CO_2 evolved from the decomposition of the litter was similar for each of the three species (Table 3; Figure 5). Although both median and resampled fits yielded a similar R^2, the resampled model did reveal subtle deviations in the data, but had little to no influence on the R^2, with *A. dealbata* showing the largest variation of R^2 between models (1.9% difference). *A. dealbata* had the lowest half-saturation point for the cumulative carbon mineralisation, reaching it faster than *E. obliqua* and *Bedfordia*. However, all species released a comparable total amount of CO_2 per gram of litter, as indicated by the similar asymptotes of the MM equation (Table 3).

Table 3. The model fit (R^2) for each species for the cumulative CO_2 evolved from the decomposition of the litter: Initial Michaelis–Menten equation fit to the median data points, and the resampled data (median model prediction). Shown is the asymptote (α), half saturation point (β), and corresponding 95% confidence intervals, for each species.

Model (Median)	Asymptote (α)	95% CI	Half Saturation Point (β)	95% CI	Sigma	R^2 (%)
			Eucalyptus obliqua			
Initial	92,189	84,612–101,407	9.9	7.8–12.7	2210	93.3
Resampled	90,526	72,756–116,217	9.8	5.7–18.0	8355	93.7
			Acacia dealbata			
Initial	89,561	81,141–100,055	5.6	3.9–8.0	3769	90.9
Resampled	89,563	74,410–105,699	5.7	3.5–9.1	6729	92.8
			***Bedfordia* sp.**			
Initial	92,476	76,899–117,482	10.1	5.8–18.0	4637	89.1
Resampled	84,947	67,487–113,376	7.9	4.4–7.9	8859	90.2

a) *Eucalyptus obliqua*

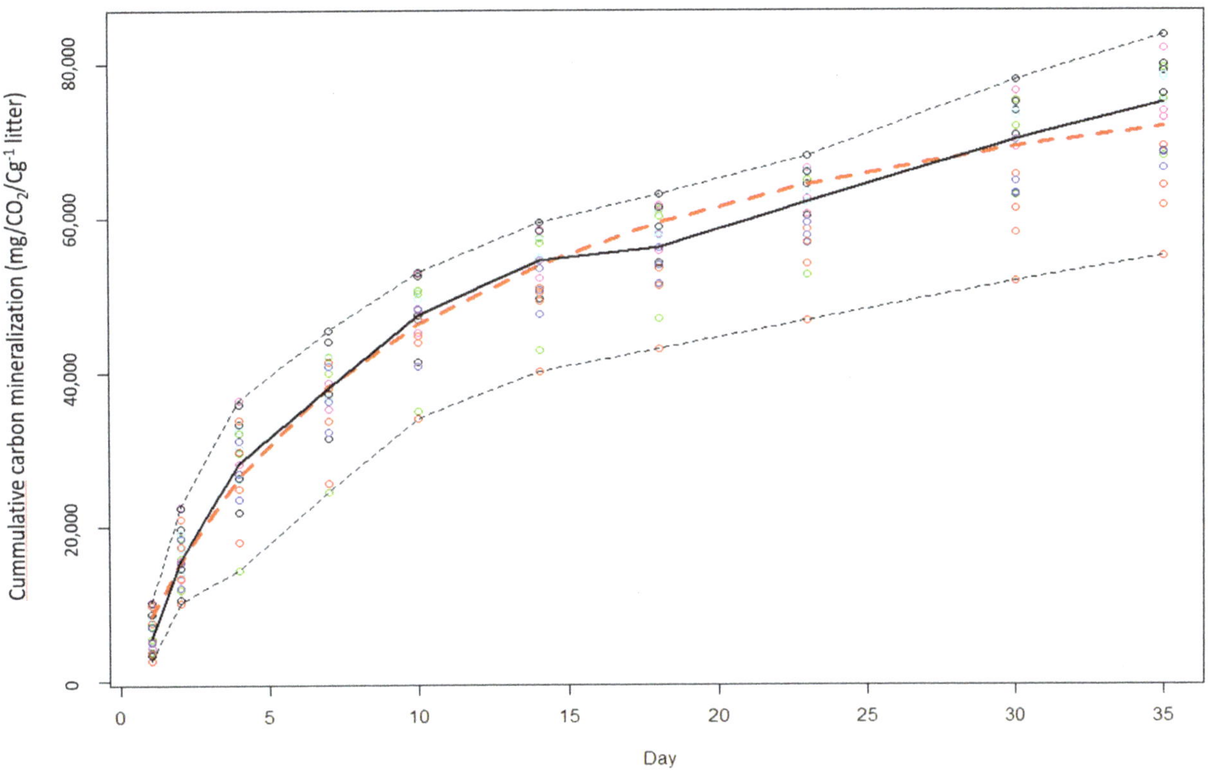

b) *Acacia dealbata*

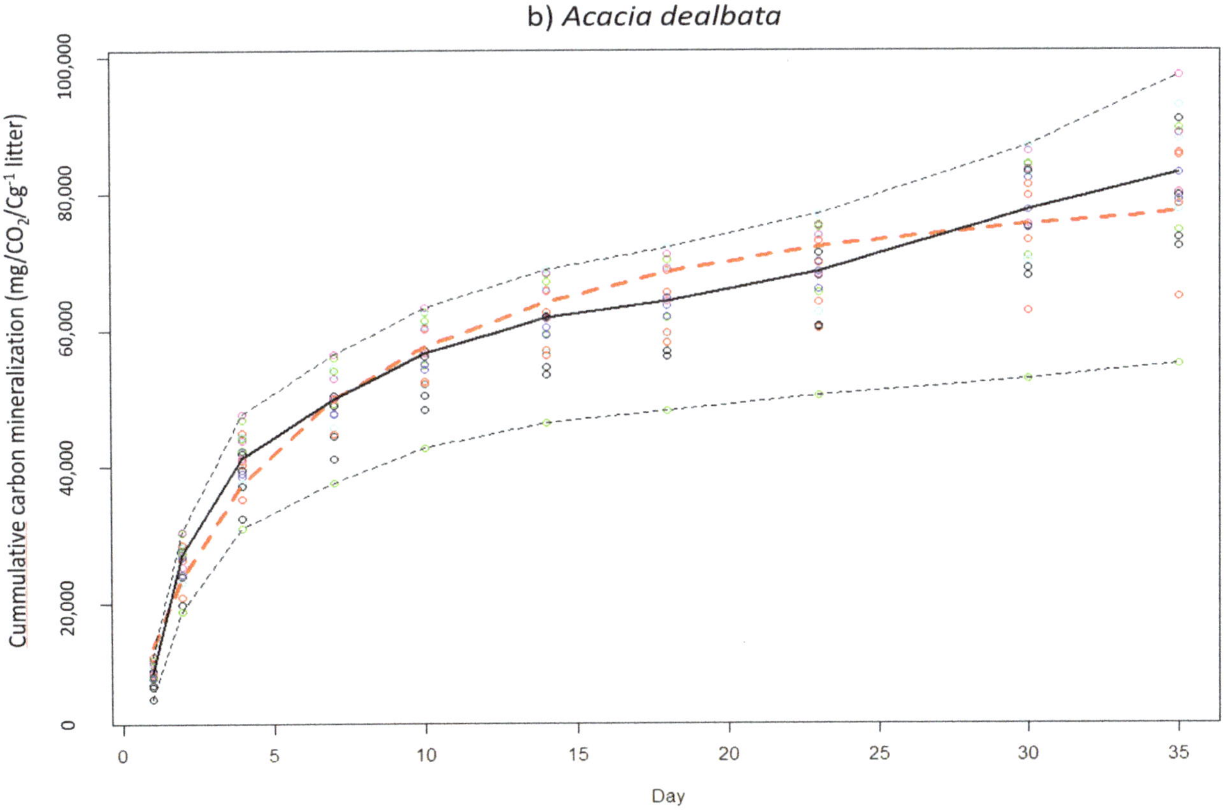

Figure 5. *Cont.*

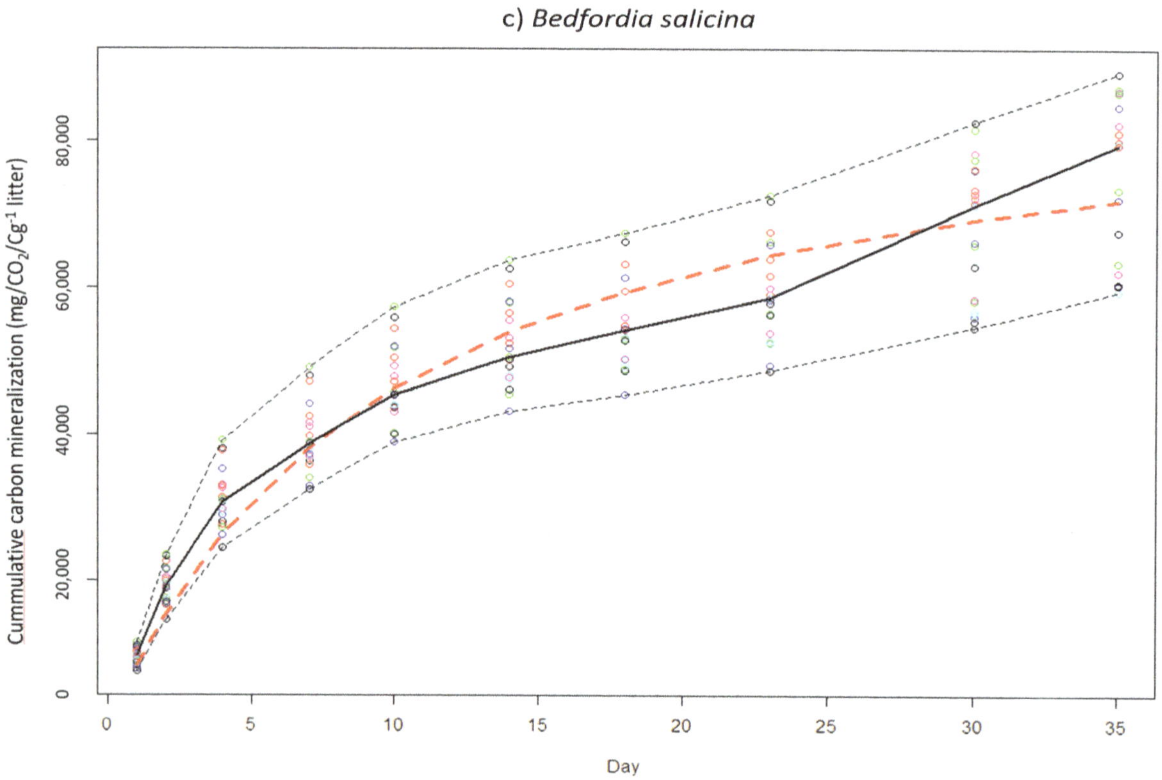

Bedfordia sp. (**c**)

Figure 5. Cumulative carbon mineralisation rate over time (40 days) against all data points for each species. The red dotted line represents the Michaelis–Menten equation (MM) fit to the median data points. The black line is the median of the resampled data fits using the MM equation, with corresponding black dotted 95% confidence bounds: (**a**) *E. obliqua* initial MM fit (R^2: 93.3) and resampled median (R^2: 93.7%); (**b**) *A. dealbata* initial MM fit (R^2: 90.9%) and resampled median (R^2: 92.8%); (**c**) *Bedfordia* sp. initial MM fit (R^2: 89.1%) and resampled median (R^2: 90.2%).

4. Discussion

We have shown that experimental mixtures of plant species from Australian eucalypt forests that contain a higher proportion of nitrogen-fixing *A. dealbata* litter have higher decomposition rates (Figures 4 and 5). This was driven primarily by the three-fold increase in nitrogen (N) content in *A. dealbata* litter compared to, for example, *B. salicina* (another understorey tree) and the canopy-forming *E. obliqua*. This result is supported mechanistically by previous studies (e.g., [36,37]) showing that decomposition rates of litter mixtures are driven predominantly by substrate quality, particularly the amount of C and N in the litter mixtures. Plant functional traits, such as nitrogen fixation, are key drivers of higher-quality substrates in forest ecosystems, with the potential to mediate and control plant abundance, biomass, and composition via inherent differences between species in quality (C:N, lignin:N) and quantity (%C, %N) of plant litter [38]. Indeed, the quantity of carbon (C) in the soil at our Tasmanian forest sites was the best predictor of average eucalypt tree size, followed by the average size of *A. dealbata* (Figure 1). Studies that have examined this phenomenon in other forests have demonstrated that the identity and functional traits of litter-creating species can surpass species diversity as the main driver of decomposition [39–42].

Given that *A. dealbata* presence and abundance indirectly controls litter decomposition via nitrogen availability and was consequently a useful predictor of average *E. obliqua* tree size, this research supports the hypothesis that N availability and cycling would be higher in sites with a greater *A. dealbata* biomass. Our results align with those of [23], who demonstrated that the presence of *Acacia mearnsii* (another N-fixer) influenced the quantity and rates of N and P cycling in *Eucalyptus globulus* mixed forests compared to monocultures, in an 11-year experimental trial. Similar relationships have also been found in boreal forests, grasslands, and tropical forests, through field N fertilisation trials [42–44] and in laboratory experiments [45,46]. However, in our study, there was no observable relationship between *A. dealbata* biomass and the total amount of organic matter (C and N) or the concentrations of NH_4^+, NO_3^- or with total mineral N and P in the soil, despite high heterogeneity among sites in N availability and cycling (Figure 3). Further, we found no correlation between total available soil-mineral N and average *A. dealbata* tree size. As *A. dealbata* was the sub-dominant canopy species in our forest system, and there was 30% more biomass of *A. dealbata* relative to other sites, the potential for N-rich soil from N-fixers with high N litter should have been realised. As such, this null result demands an explanation.

There are several plausible reasons as to why a higher N content was not observed at sites with high *A. dealbata* biomass (Figure 1), with these inferences potentially useful for forest management during early succession and development. Two leading candidates are: (1) *A. dealbata* is a nitrogen fixer and therefore is not reliant on high soil N to grow, whereas this might be limiting for other species (such as *E. obliqua*); and (2) due to the relatively short lifespan of *A. dealbata* (~40 years) compared to *E. obliqua* (>100 years) [47], a higher N litter might have been present during the early stages of forest succession, when growth was faster and intraspecific competition more intense, but that early signal has since dissipated (e.g., [48,49]). The development of a forest is characterised by both changes in structure and succession, with concomitant consequences for productivity of standing forests and regrowth [50]. For example, restoration of forest ecosystems with multiple forest species helps rebuild multiple functions, such as species structure, litter composition, and soil nutrients of an area [9,51]. Selective logging of natural forests is a form of disturbance, and management plans require some older trees to be left in a logged area for animal habitat and seedling regeneration. Our results suggest that complimentary tree species, such as *A. dealbata* and *E. obliqua*, should be left in such areas, to maintain the overall functionality, especially during early successional development [52]. However, our experimental results only show that complementarity and nitrogen facilitation can occur, whereas determining this relationship explicitly within a natural setting (such as [53,54] is challenging to demonstrate, especially in the face of inevitable inter-site heterogeneity).

In terms of confounding effects, there are also many factors that can influence environmental and topographic heterogeneity within eucalypt forests, which might act to mask any relationship and thus cause the field results to deviate from the expectations of the experimental incubations. For example, the realised distribution of *A. dealbata* litter at the field sites might be influenced by localised erosion or mineral dilution of litter and topographic variation, causing the build-up of litter in soil depressions or exclusions from mounds, but which were not reflected in controlled soil collections. The few studies that have examined patterns of litter fall dispersal and collection in field conditions support the notion of high heterogeneity in litter-sample collection, finding large disparities between the location of the litter source (local tree canopy) and the actual litterfall sample that was collected [55]. Similarly, [56] found that under-representation of certain components of vegetation, such as litter, can have important implications for the estimation of forest productivity and nutrient cycling, leading to a scale mismatch between sample points and the character of the soil across the forest.

While this study focussed in detail on N transformations and cycling within a Tasmanian eucalypt forest system, N losses could not be determined explicitly or assessed in the field. Leaching of N from the soil at field sites might contribute to the low N availability; this suggests that future studies should focus on determining how much N is below the 15 cm of topsoil that was collected in our sampling protocol. Stable isotope tracing is one method that could be used to track N pathways through the soil, providing information of N availability and movement at various depth stratifications. If leaching does occur, then it is likely that N will accumulate further down the soil profile. This effect might have been exacerbated by the overwhelming dominance of the deep-rooted *E. obliqua* (numerically and in terms of biomass), whose leaves, due to variations in litter characteristics between the two species such as litter size, mass, decomposability and chemical composition [57,58], dominate the surface litter composition, and yet it is able to access historically deposited nutrients that are now only found deep in the soil profile.

5. Conclusions

In conclusion, we have presented an experimental and field-observational snapshot of nutrient dynamics within a typical eucalypt forest community from southern Australia. We have demonstrated the importance of understanding litter decomposition, but also the need to quantify how different sources of N can impact N availability and cycling. Repeated sampling across a larger range of seasons in Tasmanian eucalypt forests (including wet and dry periods) would be important to provide a temporal component to a spatially complex process. Our results also highlight the difficulty in defining ecosystem processes and function using any single mechanism in forests—even one so apparently clear cut as N availability—because of the many processes that ultimately interact to determine the size of trees and relative abundance of species.

Author Contributions: J.C.B. and M.J.H. conceived and designed the experiments; J.C.B. performed the experiments; B.W.B., J.C.B, and E.M.R. analysed the data; and all authors contributed to the manuscript drafts and writing and gave final approval for publication.

References

1. Becker, K.; Wulfmeyer, V.; Berger, T.; Gebel, J.; Münch, W. Carbon farming in hot, dry coastal areas: An option for climate change mitigation. *Earth Syst. Dyn.* **2013**, *4*, 237–251. [CrossRef]
2. Law, B.E.; Hudiburg, T.W.; Berner, L.T.; Kent, J.J.; Buotte, P.C.; Harmon, M.E. Land use strategies to mitigate climate change in carbon dense temperate forests. *Proc. Natl. Acad. Sci. USA* **2018**, *115*, 3663–3668. [CrossRef] [PubMed]
3. Johnson, K.H.; Vogt, K.A.; Clark, H.J.; Schmitz, O.J.; Vogt, D.J. Biodiversity and the productivity and stability of ecosystems. *Trends Ecol. Evol.* **1996**, *11*, 372–377. [CrossRef]
4. Wright, I.J.; Westoby, M. Differences in seedling growth behaviour among species: Trait correlations across species, and trait shifts along nutrient compared to rainfall gradients. *J. Ecol.* **1999**, *87*, 85–97. [CrossRef]
5. Stirling, G.; Wilsey, B. Empirical relationships between species richness, evenness, and proportional diversity. *Am. Nat.* **2001**, *158*, 286–299. [CrossRef] [PubMed]
6. Krebs, C.J. *Ecology*, 5th ed.; Benjamin Cummings: Boston, MA, USA, 2001.
7. Vitousek, P.M.; Howarth, R.W. Nitrogen limitation on land and in the sea: How can it occur? *Biogeochemistry* **1991**, *13*, 87–115. [CrossRef]
8. Wardle, D.A.; Bardgett, R.D.; Klironomos, J.N.; Setälä, H.; Van Der Putten, W.H.; Wall, D.H. Ecological linkages between aboveground and belowground biota. *Science* **2004**, *304*, 1629–1633. [CrossRef]
9. Forrester, D.I.; Bauhus, J. A Review of Processes Behind Diversity—productivity relationships in forests. *Curr. For. Rep.* **2016**, *2*, 45–61. [CrossRef]
10. Harrison, K.A.; Bol, R.; Bardgett, R.D. Preferences for different nitrogen forms by coexisting plant species and soil microbes. *Ecology* **2007**, *88*, 989–999. [CrossRef]
11. Parton, W.; Silver, W.L.; Burke, I.C.; Grassens, L.; Harmon, M.E.; Currie, W.S.; King, J.Y.; Adair, E.C.; Brandt, L.A.; Hart, S.C. Global-scale similarities in nitrogen release patterns during long-term decomposition. *Science* **2007**, *315*, 361–364. [CrossRef]
12. Sharma, J.; Sharma, Y. Nutrient cycling in forest ecosystems—A review. *Agric. Rev.* **2004**, *25*, 157–172.
13. Rennenberg, H.; Dannenmann, M.; Gessler, A.; Kreuzwieser, J.; Simon, J.; Papen, H. Nitrogen balance in forest soils: Nutritional limitation of plants under climate change stresses. *Plant Biol.* **2009**, *11*, 4–23. [CrossRef]
14. Reich, P.B.; Grigal, D.F.; Aber, J.D.; Gower, S.T. Nitrogen mineralization and productivity in 50 hardwood and conifer stands on diverse soils. *Ecology* **1997**, *78*, 335–347. [CrossRef]
15. McGuire, A.D.; Melillo, J.M.; Joyce, L.; Kicklighter, D.W.; Grace, A.; Moore, B., III; Vorosmarty, C.J. Interactions between carbon and nitrogen dynamics in estimating net primary productivity for potential vegetation in North America. *Glob. Biogeochem. Cycles* **1992**, *6*, 101–124. [CrossRef]

16. Conn, C.; Dighton, J. Litter quality influences on decomposition, ectomycorrhizal community structure and mycorrhizal root surface acid phosphatase activity. *Soil Biol. Biochem.* **2000**, *32*, 489–496. [CrossRef]
17. King, R.F.; Dromph, K.M.; Bardgett, R.D. Changes in species evenness of litter have no effect on decomposition processes. *Soil Biol. Biochem.* **2002**, *34*, 1959–1963. [CrossRef]
18. Chapman, S.K.; Newman, G.S. Biodiversity at the plant–soil interface: Microbial abundance and community structure respond to litter mixing. *Oecologia* **2010**, *162*, 763–769. [CrossRef] [PubMed]
19. Forrester, D.I.; Bauhus, J.; Cowie, A.L. Nutrient cycling in a mixed-species plantation of *Eucalyptus globulus* and *Acacia mearnsii*. *Can. J. For. Res.* **2005**, *35*, 2942–2950. [CrossRef]
20. Parrotta, J.A. Productivity, nutrient cycling, and succession in single-and mixed-species plantations of *Casuarina equisetifolia*, *Eucalyptus robusta*, and *Leucaena leucocephala* in Puerto Rico. *For. Ecol. Manag.* **1999**, *124*, 45–77. [CrossRef]
21. Attiwill, P.M.; Leeper, G.W. *Forest Soils and Nutrient Cycles*; Melbourne University Press: Carlton, VIC, Australia, 1987.
22. Adams, M.; Attiwill, P. Nutrient cycling and nitrogen mineralization in eucalypt forests of south-eastern Australia. *Plant Soil* **1986**, *92*, 341–362. [CrossRef]
23. Forrester, D.I.; Bauhus, J.; Cowie, A.L.; Vanclay, J.K. Mixed-species plantations of Eucalyptus with nitrogen-fixing trees: A review. *For. Ecol. Manag.* **2006**, *233*, 211–230. [CrossRef]
24. Isbell, R. *The Australian Soil Classification*; CSIRO Publishing: Collingwood, Australia, 2016.
25. Warren, R.J.; Giladi, I.; Bradford, M.A. Competition as a mechanism structuring mutualisms. *J. Ecol.* **2014**, *102*, 486–495. [CrossRef]
26. Mitchell, K. Quantitative analysis by the point-centered quarter method. *arXiv* **2010**, arXiv:1010.3303.
27. Jimenez, R.R.; Ladha, J.K. Automated elemental analysis: A rapid and reliable but expensive measurement of total carbon and nitrogen in plant and soil samples. *Commun. Soil Sci. Plant Anal.* **1993**, *24*, 1897–1924. [CrossRef]
28. Mendham, D.S.; Smethurst, P.J.; Holz, G.K.; Menary, R.C.; Grove, T.S.; Weston, C.; Baker, T. Soil analyses as indicators of phosphorus response in young eucalypt plantations. *Soil Sci. Soc. Am. J.* **2002**, *66*, 959–968. [CrossRef]
29. Osanai, Y.; Bougoure, D.S.; Hayden, H.L.; Hovenden, M.J. Co-occurring grass species differ in their associated microbial community composition in a temperate native grassland. *Plant Soil* **2013**, *368*, 419–431. [CrossRef]
30. Barton, K.; Barton, M.K. Mu-MIn: Multi-model inference. R Package Version 1.43.6. 2019. Available online: https://cran.r-project.org/web/packages/MuMIn/MuMIn.pdf (accessed on 18 January 2019).
31. Anderson, D.R. *Model Based Inference in the Life Sciences: A Primer on Evidence*; Springer Science & Business Media: New York, NY, USA, 2007.
32. Burnham, K.P.; Anderson, D.R. *Model Selection and Multimodel Inference: A Practical Information—Theoretic Approach*; Springer Science & Business Media: New York, NY, USA, 2003.
33. Pendall, E.; Osanai, Y.U.I.; Williams, A.L.; Hovenden, M.J. Soil carbon storage under simulated climate change is mediated by plant functional type. *Glob. Chang. Biol.* **2011**, *17*, 505–514. [CrossRef]
34. Osanai, Y.; Flittner, A.; Janes, J.K.; Theobald, P.; Pendall, E.; Newton, P.C.; Hovenden, M.J. Decomposition and nitrogen transformation rates in a temperate grassland vary among co-occurring plant species. *Plant Soil* **2012**, *350*, 365–378. [CrossRef]
35. Dijkstra, F.A.; Hobbie, S.E.; Reich, P.B. Soil processes affected by sixteen grassland species grown under different environmental conditions. *Soil Sci. Soc. Am. J.* **2006**, *70*, 770–777. [CrossRef]
36. Freschet, G.T.; Cornwell, W.K.; Wardle, D.A.; Elumeeva, T.G.; Liu, W.; Jackson, B.G.; Onipchenko, V.G.; Soudzilovskaia, N.A.; Tao, J.; Cornelissen, J.H. Linking litter decomposition of above-and below-ground organs to plant–soil feedbacks worldwide. *J. Ecol.* **2013**, *101*, 943–952. [CrossRef]
37. García-Palacios, P.; Maestre, F.T.; Kattge, J.; Wall, D.H. Climate and litter quality differently modulate the effects of soil fauna on litter decomposition across biomes. *Ecol. Lett.* **2013**, *16*, 1045–1053. [CrossRef] [PubMed]
38. Makkonen, M.; Berg, M.P.; Handa, I.T.; Hättenschwiler, S.; Van Ruijven, J.; Van Bodegom, P.M.; Aerts, R. Highly consistent effects of plant litter identity and functional traits on decomposition across a latitudinal gradient. *Ecol. Lett.* **2012**, *15*, 1033–1041. [CrossRef] [PubMed]
39. Hobbie, S.E. Effects of plant species on nutrient cycling. *Trends Ecol. Evol.* **1992**, *7*, 336–339. [CrossRef]
40. Wardle, D.; Bonner, K.; Nicholson, K. Biodiversity and plant litter: Experimental evidence which does not

support the view that enhanced species richness improves ecosystem function. *Oikos* **1997**, *79*, 247–258. [CrossRef]

41. Jiang, L.; Wan, S.; Li, L. Species diversity and productivity: Why do results of diversity-manipulation experiments differ from natural patterns? *J. Ecol.* **2009**, *97*, 603–608. [CrossRef]
42. Hyvönen, R.; Ågren, G.I.; Linder, S.; Persson, T.; Cotrufo, M.F.; Ekblad, A.; Freeman, M.; Grelle, A.; Janssens, I.A.; Jarvis, P.G.; et al. The likely impact of elevated [CO_2], nitrogen deposition, increased temperature and management on carbon sequestration in temperate and boreal forest ecosystems: A literature review. *New Phytol.* **2007**, *173*, 463–480. [CrossRef] [PubMed]
43. Nordin, A.; Näsholm, T.; Ericson, L. Effects of simulated N deposition on understorey vegetation of a boreal coniferous forest. *Funct. Ecol.* **1998**, *12*, 691–699. [CrossRef]
44. Matson, P.; Lohse, K.A.; Hall, S.J. The globalization of nitrogen deposition: Consequences for terrestrial ecosystems. *Ambio* **2002**, *31*, 113–119. [CrossRef] [PubMed]
45. Wedin, D.A.; Tilman, D. Species effects on nitrogen cycling: A test with perennial grasses. *Oecologia* **1990**, *84*, 433–441. [CrossRef] [PubMed]
46. Temperton, V.M.; Mwangi, P.N.; Scherer-Lorenzen, M.; Schmid, B.; Buchmann, N. Positive interactions between nitrogen-fixing legumes and four different neighbouring species in a biodiversity experiment. *Oecologia* **2007**, *151*, 190–205. [CrossRef] [PubMed]
47. Koch, A.J.; Driscoll, D.A.; Kirkpatrick, J. Estimating the accuracy of tree ageing methods in mature *Eucalyptus obliqua* forest, Tasmania. *Aust. For.* **2008**, *71*, 147–159. [CrossRef]
48. Bouillet, J.-P.; Laclau, J.-P.; Gonçalves, J.L.d.M.; Voigtlaender, M.; Gava, J.L.; Leite, F.P.; Hakamada, R.; Mareschal, L.; Mabiala, A.; Tardy, F.; et al. Eucalyptus and Acacia tree growth over entire rotation in single- and mixed-species plantations across five sites in Brazil and Congo. *For. Ecol. Manag.* **2013**, *301*, 89–101. [CrossRef]
49. Epron, D.; Nouvellon, Y.; Mareschal, L.; Moreira, R.M.e.; Koutika, L.-S.; Geneste, B.; Delgado-Rojas, J.S.; Laclau, J.-P.; Sola, G.; Gonçalves, J.L.d.M.; et al. Partitioning of net primary production in Eucalyptus and Acacia stands and in mixed-species plantations: Two case-studies in contrasting tropical environments. *For. Ecol. Manag.* **2013**, *301*, 102–111. [CrossRef]
50. Ryan, M.; Binkley, D.; Fownes, J.H. Age-related decline in forest productivity: Pattern and process. In *Advances in Ecological Research*; Elsevier: Amsterdam, The Netherlands, 1997; Volume 27, pp. 213–262.
51. Aerts, R.; Honnay, O. Forest restoration, biodiversity and ecosystem functioning. *BMC Ecol.* **2011**, *11*, 29. [CrossRef] [PubMed]
52. Villela, D.M.; Nascimento, M.T.; De Aragao, L.E.O.; Da Gama, D.M. Effect of selective logging on forest structure and nutrient cycling in a seasonally dry Brazilian Atlantic forest. *J. Biogeogr.* **2006**, *33*, 506–516. [CrossRef]
53. Tilman, D.; Reich, P.; Knops, J.; Wedin, D.; Mielke, T.; Lehman, C. Diversity and productivity in a long-term grassland experiment. *Science* **2001**, *294*, 843–845. [CrossRef]
54. Cardinale, B.J.; Wright, J.P.; Cadotte, M.W.; Carroll, I.T.; Hector, A.; Srivastava, D.S.; Loreau, M.; Weis, J.J. Impacts of plant diversity on biomass production increase through time because of species complementarity. *Proc. Natl. Acad. Sci. USA* **2007**, *104*, 18123–18128. [CrossRef]
55. Lowman, M.D. Litterfall and leaf decay in three Australian rainforest formations. *J. Ecol.* **1988**, *76*, 451–465. [CrossRef]
56. Clark, D.A.; Brown, S.; Kicklighter, D.W.; Chambers, J.Q.; Thomlinson, J.R.; Ni, J.; Holland, E.A. Net primary production in tropical forests: An evaluation and synthesis of existing field data. *Ecol. Appl.* **2001**, *11*, 371–384. [CrossRef]
57. Bini, D.; Dos Santos, C.A.; Bouillet, J.-P.; De Morais Goncalves, J.L.; Cardoso, E.J.B.N. Eucalyptus grandis and Acacia mangium in monoculture and intercropped plantations: Evolution of soil and litter microbial and chemical attributes during early stages of plant development. *Appl. Soil Ecol.* **2013**, *63*, 57–66. [CrossRef]
58. Forrester, D.; Pares, A.; O'hara, C.; Khanna, P.; Bauhus, J. Soil organic carbon is increased in mixed-species plantations of Eucalyptus and nitrogen-fixing Acacia. *Ecosystems* **2013**, *16*, 123–132. [CrossRef]

15

Soil Elements Influencing Community Structure in an Old-Growth Forest in Northeastern China

Wei Xu [1,†], Minhui Hao [1,†], Juan Wang [1,*], Chunyu Zhang [1], Xiuhai Zhao [1] and Klaus von Gadow [2,3]

1 The key Laboratory for Forest Resources & Ecosystem Processes of Beijing, Beijing Forestry University, Beijing 100083, China; xuweistudy@163.com (W.X.); 960954652@qq.com (M.H.); zcy_0520@163.com (C.Z.); zhaoxh@bjfu.edu.cn (X.Z.)
2 Faculty of Forestry and Forest Ecology, Georg-August-University Göttingen, Büsgenweg 5, Göttingen D-37077, Germany; kgadow@gwdg.de
3 Department of Forest and Wood Science, University of Stellenbosch, Stellenbosch 7600, South Africa
* Correspondence: claire_wangj@163.com
† These authors contributed equally to this work.

Academic Editors: Scott X. Chang and Xiangyang Sun

Abstract: This study uses detailed soil and vegetation data collected in a 30-ha old-growth broad-leaved Korean pine forest to study the effect of soil properties on tree community structures. Spatial distribution patterns are simulated using a homogeneous Poisson process (HomP) and a homogeneous Thomas process (HomT). The simulated distributions are compared with the observed ones to explore correlations between certain tree species and several soil elements. The HomP model shows that all tested tree species are significantly correlated with at least one principal component in the upper-layer soil elements. The HomT model shows that only 36.4% of tree species are significantly correlated with the principal component of at least one upper-layer soil element. This result shows that the impact of dispersal limitation is greater than impact of environmental heterogeneity on species spatial distributions. The spatial autocorrelation of species induced by the dispersal limitation will largely conceal the plant-soil relationships caused by the heterogeneity of soil elements. An additional analysis shows that the elements in the upper soil layer which have the greatest impact on community niche structure are Pb, total phosphorus (TP), total nitrogen (TN), Cu, Cr, Zn and available nitrogen (AN). The corresponding elements in the lower soil layers are Pb, TP, Cu, organic carbon (OC), Mn, total potassium (TK) and AN. Different species seem to be complementary regarding the demands on the available soil resources. The results of this study show that the tree species in the different growth groups have different habitat preferences. Compared with subcanopy and shrub species, the canopy species have more significant correlations with the soil elements.

Keywords: plant-soil relationships; dispersal limitation; habitat filtering; soil elements

1. Introduction

The impact of species dispersal and habitat filtering on the species distribution at medium (1–100 km^2) and large landscape scales (100–10,000 km^2) can be easily quantified if the soil resources of a particular forest community show a mosaic pattern and exhibit obvious heterogeneity characteristics [1], in which case certain habitat factors may have a significant impact on the species spatial distribution [2–4]. At the local scale (<1 km^2), the pattern induced by a species dispersal limitation is very similar to that induced by the habitat heterogeneity. It is therefore relatively difficult to distinguish between the impact of species dispersal and habitat filtering on a species' spatial distribution. A considerable amount of evidence shows that the species distribution is often closely correlated with certain soil elements, as well as with the soil texture, terrain features, parent material and other factors at a local scale [5]. Such significant plant-soil relationships occur widely in temperate forests, subtropical forests and tropical forests [5–8].

Seed dispersal is an important ecological process affecting community structuring. Dispersal limitation will occur when the seeds cannot arrive at a new suitable location, and available evidence shows that species dispersal limitation is often found in the forest communities [9]. However, whether dispersal limitation will have a significant impact on community spatial patterns is still in dispute [10]. Habitat filtering is another important ecological process affecting community assemblage and species coexistence [11,12]. The habitat filtering effect may significantly impact the growth and survival of seedlings. A species adapted to a particular habitat can survive and coexist with other species, while a species which is not adapted to that habitat may be eliminated [13,14]. In tropical and subtropical forests, certain habitat factors such as topography and soil conditions may cause many plant species to show clustered distributions. The spatial distribution of plant species is often highly correlated with certain habitat factors [15,16]. Habitat heterogeneity isolates those species which are unable to coexist into different habitat patches. For example, different species have different competitive capacities in different habitat patches and respectively occupy the patches where they have the strongest competitive capacity [17]. In undisturbed forests, the resource heterogeneity shows a very low contribution to species diversity, while the resource quantity in heterogeneous habitats is the main driving force impacting on the species diversity [18].

Studies involving the relationship between species distributions and environmental factors often consider the topographic variables or habitat types based on specific topographic variables [8,19]. However, the topographical factors are not suitable to replace the soil property for disclosing the impact of environmental factors on the species distribution [20]. Therefore, it is necessary to adopt direct environmental factors such as soil elements in the study of plant-soil relationships [6]. In addition, different tree species contribute differently to the soil nutrient composition through their mycorrhiza and litter. Bonifacio et al. [21] found that the content of organic matter in the soil of a broad-leaved spruce mixed forest was lower than that in a pure spruce forest. Moreover, the organic matter in the local soil layer of the pure spruce forest was obviously enriched, but all of the soil organic matter in the broad-leaved spruce mixed forest was distributed uniformly.

The importance of the soil resource heterogeneity to the maintenance of community diversity has been widely recognized, but the impact of the soil resource quantity on the species distribution has not yet been evaluated systematically. Temperate forests account for a high proportion of the world's forest resources, and research regarding the relationship between soil resources and species distribution is helpful in revealing the structuring mechanisms of temperate forest communities. This study evaluates the relationship between species spatial distributions and soil resources in an old-growth broad-leaved Korean pine forest community and attempts to answer the following questions: (1) Do the soil elements significantly affect the species spatial distribution in a broad-leaved Korean pine forest? (2) If yes, which soil elements are closely related to the forest community niche structure? In order to determine the actual relationship between the plant and soil elements, we eliminate the disturbance resulting from the species dispersal using specific spatial simulation techniques.

2. Materials and Methods

2.1. Field Data Acquisition

The research area is located within the Jiaohe Forest Experimental Area Administration Bureau of Jilin province, where a temperate continental mountain climate affected by monsoons and an average temperature of 3.7 °C prevails. July is the hottest month, with an average temperature of 21.7 °C. January is the coldest month, with an average temperature of −18.6 °C. The average annual precipitation is 695.9 mm. The soil type is a dark brown forest soil according to the Chinese soil taxonomy [22], with an average depth of 45 cm. In winter, the soil surface freezing time is about

150 days, with a frozen layer depth of 1.0–2.5 m. Because of the strong biological accumulation of the humus layer effect, the soil has high organic carbon content.

In 2010, a 30-ha (500 m × 600 m) research plot was established in a typical old-growth broad-leaved Korean pine forest which is located at 43°57.928′–43°58.214′ N and 127°45.287′–127°45.790′ E, far from urban areas and is virtually unaffected by human disturbance. The altitude of the research plot is 576.03–784.18 m, and the altitudinal difference between the highest and the lowest point is 208 m. All woody plants with a breast height diameter (dbh) exceeding 1 cm were tagged and mapped, their species were identified, and their dbhs, tree heights, crown widths, and heights to live crown were assessed. In this 30-ha observational study, a total of 49,678 woody plants (belonging to 48 species) were recorded. The dominant tree species are *Pinus koraiensis*, *Tilia amurensis*, *Fraxinus mandshurica*, *Ulmus laciniata* and *Acer mono*.

A combination of systematic and random sampling was used to assess soil properties. We randomly selected one direction from the eight directions around each 40 m × 40 m grid intersection and then randomly selected two distances from 2 m, 8 m and 15 m distances to form three combinations (2 m and 8 m, 2 m and 15 m or 8 m and 15 m) in the selected direction (see [23] for a detailed description of this approach). Finally, a total of 540 sample points were obtained in the 30-ha plot. The plant root system in the plots was mainly concentrated in the 0–20 cm soil layer, especially in the upper 0–10 cm layer, where the root system was dominant. Therefore, soil samples were taken from the upper (0–10 cm) and lower soil layer (10–20 cm) at each sample point and 18 soil variables were assessed: available nitrogen (AN), total nitrogen (TN), available potassium (AK), total potassium (TK), available phosphorus (AP), total phosphorus (TP), organic carbon (OC), pH, Cu, Ni, Cd, As, Pb, Zn, Mo, Cr, Mn and Mg. We used ammonium nitrogen-indophenol blue colorimetry to measure the content of available nitrogen; the hydrochloric acid-ammonium fluoride method to assess available phosphorus; the ammonium acetate extraction method to assess available potassium; the Kelvin heating digestion method to assess total potassium, and the high-temperature external-heat potassium dichromate oxidation method to assess organic carbon. Heavy metal elements were extracted using the Mehlich3 general extracting agent. Concentrations were assessed using inductively coupled plasma atomic emission spectrometry (ICP-AES). The above experiments were carried out according to the recommendations of the China Soil Council [24].

2.2. Relationship between Species Distribution and Soil Elements

Principal component analysis (PCA) was used to identify several composite factors representing various original variables by means of the dimensionality reduction method, and to allow these composite factors to retain the greatest possible amount of information for a large number of original variables. These composite factors had to be mutually unrelated. The 18 soil variables used in this paper have a very high correlation, so there is also a serious multicollinearity problem. In order to compress and extract the main information of the soil elements, principal component analysis was used for these soil elements and the pH values.

The Monte Carlo method was used to simulate species spatial distributions. We assume that the heterogeneous spatial distribution of the soil elements causes specific species distribution patterns. To evaluate this assumption, we applied two spatial point process models, the homogeneous Poisson process (HomP) and the homogeneous Thomas process (HomT) for simulating specific spatial distributions. The homogeneous Poisson process (HomP) simulates a completely random spatial species distribution and is used to test the significance of the plant-soil correlation. The homogeneous HomP thus assumes that all individuals of a certain plant species are randomly distributed in the plots. However, most of the tree species are not randomly distributed, thus it is necessary to consider a non-random process [24]. Therefore, we also used a homogeneous Thomas process (HomT) to simulate the autocorrelation of the species spatial distribution, which may be the result of dispersal processes [25–27]. The Poisson process model is usually expressed as Equation (1):

$$\rho(\mu) = \alpha \cdot \exp\left(z_{1:k}(\mu)\, \beta_{1:k}^{T}\right) \tag{1}$$

When $\rho(\mu)$ is a constant, the point pattern is represented by a homogeneous Poisson process (completely random); $z_{1:k}(\mu)$ is the function of environment variables where μ is the location of a point; α is the average density of the species distribution and $\beta_{1:k}^{T}$ are the coefficients.

The homogeneous Thomas process is usually expressed by Equation (2):

$$\rho_c(\mu) = \alpha k(\mu - c; \delta) \tag{2}$$

where the parameter c defines the location of the parent tree, and $k(\mu - c; \delta)$ defines the dispersal model of the offspring around the parent trees.

The spatial point process model was used to simulate the distribution pattern of each species in the observational study area [28]. For each species, there are 1000 distribution maps, including 999 simulated maps and one observed species map. The simulated species distribution maps were then matched with the principle components (PCs) distribution maps of the soil elements. The Pearson coefficient between each simulated species distribution and the PC distribution map among the 40 m × 40 m square cells were calculated. Thus, the 95% confidence interval can be calculated based on 999 Pearson coefficients. When the Pearson coefficient between the observed species distribution and the PCs is greater than the upper limit of the 95% confidence interval, then the relationship between the species distribution and the PCs of the soil elements represents a significant positive correlation. The correlation will be negative if the Pearson coefficient between the observed species distribution and the PCs is smaller than the lower limit of the 95% confidence interval. Alternatively, the correlation will be non-significant.

To ensure reasonable sample sizes, the analysis was limited to the 33 most abundant species (Table S1) which are represented by at least 30 individuals in the study area. In accordance with the potential maximum height (hmax), the tree species were assigned to one of three groups: canopy species (hmax $\geqslant$ 15 m), subcanopy species (5 m $\leqslant$ hmax < 15 m) and shrub species (hmax < 5 m). The 33 species include 23 canopy species, 5 subcanopy species and 5 shrub species.

The Canonical Correspondence Analysis (CCA) was used to study the relationship between the species composition and the soil environment. The species composition was expressed by a species-site matrix, and the soil environment was expressed by 18 soil elements.

The ecological niche breadth, which reflects the degree of adaptation to different environments, is reciprocal to the degree of ecological specialization [29]. In this study, we evaluate Levins' ecological niche breadth of tree species represented by at least 30 individuals [30], and test the degree of ecological specialization of each species with regard to the soil elements. The Levins' ecological niche breadth indicator formula is specified in Equation (3):

$$B_i = \left(\sum N_{ij}\right)^2 / \left(\sum N_{ij}^{\,2}\right) \tag{3}$$

where B_i is the ecological niche breadth of species i, and N_{ij} is the resource value in the j'th resource grade for species i. In order to evaluate the overall impact of each soil variable on the community niche structure, a boxplot for the ecological niche breadth of all studied species was drawn for each soil variable.

All calculations were performed using the R statistical software [31]. The principal component analysis was implemented using the "vegan", "ade4", "gclus" and "ape" packages.

3. Results

The difference in the species compositions among the 40 m × 40 m cells was significantly correlated to the difference in the average concentrations of soil elements in those cells for both the upper and the lower soil layer. The CCA ordination analysis regarding soil nutrient and community composition shows that most soil variables are significantly correlated with the CCA ordination axes (Figure 1, Table 1). However, some soil elements show similar effects according to the lengths and directions of the soil nutrient vectors. Surprisingly, major soil elements (such as N, P and K) do not show a stronger influence than the metal elements (such as Ni, As and Mg) in both soil layers. The same nutrient shows a consistent direction in both soil layers, but the nutrient vectors in the upper soil layer are longer than those in the lower layer (Figure 1).

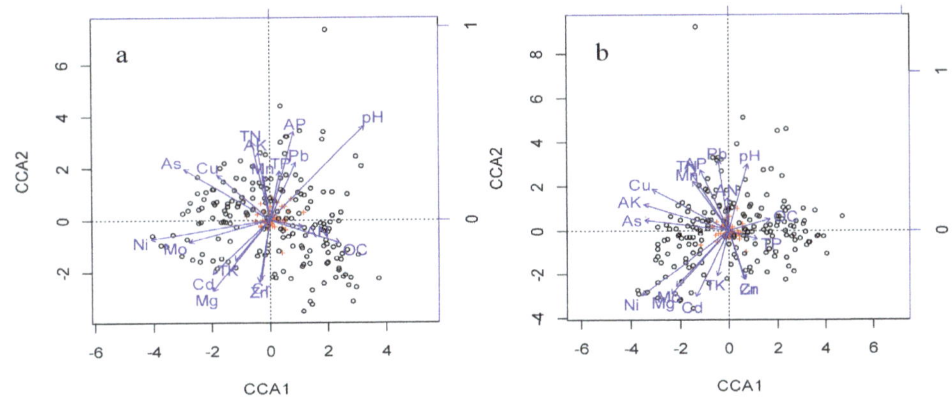

Figure 1. Canonical correspondence analysis (CCA) in two soil layers involving the upper (**a**) and lower layer (**b**). The species abundance and average values of soil nutrients in the 40 m × 40 m plots were used to test the relationship between species composition and soil elements. The species scores and site scores of the first two ordination axes are drawn in the figure. The arrow direction indicates the positive and negative correlation between the soil nutrient and ordination axes, while the length of the arrow represents the degree of correlation between certain soil nutrient variables and the community composition.

Table 1. Pearson's correlations between soil variables and the first two axes of canonical correspondence analysis (CCA).

Soil Variables	Upper Soil Layer		Lower Soil Layer	
	CCA1	CCA2	CCA1	CCA2
available nitrogen	−0.12	−0.13	−0.11	−0.14
total nitrogen	−0.91 ***	0.07	−0.91 ***	0.27 ***
available potassium	0.13	0.05	0.03	−0.05
total potassium	−0.02	−0.41 ***	−0.05	−0.36 ***
available phosphorus	−0.05	−0.36 ***	−0.22 **	−0.44 ***
total phosphorus	−0.05	0.12	0.00	−0.01
organic carbon	0.08	−0.51 ***	−0.07	−0.39 ***
Cu	0.07	−0.26 ***	0.08	0.10
Ni	0.15 *	0.19 *	0.05	0.18 *
Cd	−0.16 *	−0.23 **	−0.19 *	−0.41 ***
As	−0.34 ***	0.02	−0.30 ***	0.03
Pb	−0.10	0.17 *	−0.07	0.15 *
Zn	−0.21 **	−0.31 ***	−0.21 **	−0.29 ***
Mo	0.05	−0.26 ***	0.03	−0.35 ***
Cr	0.03	0.33 ***	0.05	0.28 ***
Mn	−0.16	−0.01	−0.09	−0.03
Mg	0.02	0.31 ***	0.04	0.29 ***
pH	−0.04	−0.23 **	−0.04	−0.36 ***

* indicates $p < 0.05$; ** indicates $p < 0.01$; *** indicates $p < 0.001$.

As soil nutrient variables are significantly correlated with each other (Table S2), we obtained a group of optimum orthogonal unit vectors based on principal component analysis. The first four PC_S (PC1, PC2, PC3 and PC4) are not related to one another and jointly explain 53%–54% of the variation in soil elements in the two soil layers. In the two layers, the first principal component (PC1) accounts for 20.3%–21.9% of the variation in soil nutrient concentrations, while the second (PC2) accounts for 11.3%–14.1%. The third principal component (PC3) accounts for 10.7%–10.9% and the fourth (PC4) for 8.9%–9.1% (Table 2).

Table 2. Soil variable loadings on the four principal components (PCs) in the two soil layers.

Soil Variable	Upper Soil Layer				Lower Soil Layer			
	PC1	PC2	PC3	PC4	PC1	PC2	PC3	PC4
available nitrogen			−0.387	−0.318		−0.445	−0.122	0.111
total nitrogen		−0.130	−0.370	−0.219		−0.419		0.164
available potassium	0.191		−0.306			−0.369	0.140	0.195
total potassium	−0.229		0.296	−0.156				0.260
available phosphorus		−0.379		−0.119		−0.190	0.145	−0.160
total phosphorus							0.142	−0.152
organic carbon	0.318	0.355	−0.203		0.202	−0.267	−0.308	−0.238
Cu		−0.252	−0.345	0.422	−0.103	−0.222	0.442	−0.356
Ni	−0.184	0.248		0.315	−0.259	0.153		−0.264
Cd	−0.389	−0.134	−0.150		−0.415		0.161	0.210
As	−0.270	−0.378	−0.154		−0.293	−0.165	0.318	0.159
Pb	0.113		−0.232	0.493		−0.103	0.393	−0.494
Zn	−0.318	0.401	−0.234	−0.141	−0.348		−0.379	−0.25
Mo	−0.421	−0.216			−0.430		0.127	0.317
Cr	−0.311	0.401	−0.238	−0.157	−0.334	−0.102	−0.380	−0.255
Mn	−0.158		−0.224	0.317	−0.196	−0.296	−0.174	
Mg	−0.292	0.176		0.306	−0.372			
pH	0.196		−0.318	−0.177	0.149	−0.384		
Variance (%)	21.9	11.3	10.9	8.9	20.3	14.1	10.7	9.1
Species ($n \geqslant 5$)	25	25	16	23	24	20	19	17
Species ($n \geqslant 30$)	24	23	14	20	23	20	17	16

PC1 is the most important principal component, reflecting the contribution of the metal elements. In the upper soil layer, PC1 shows a positive correlation with Pb, and a negative one with other elements. In the lower soil layer, PC1 has negative correlations with all metal elements except Pb. The loading values of each soil nutrient variable in PC3 show a large difference. PC3 exhibits a negative correlation with most soil variables in the upper soil layer, but a positive one with most soil variables in the lower soil layer. PC2 and PC4 together represent all the information on various soil elements.

When the Monte Carlo method was used to test the relationship between different principal components and species distributions, most species show a significant correlation with the principal components. The difference in species showing significant correlation with the PCs before and after eliminating rare species ($n < 30$) is minor. Only very few (1–3) rare species show significant correlations with the soil elements (Table 2).

When testing the tree species-soil relationships, four PC axes are used, which replace the individual soil elements to reduce the testing frequency. The HomP model was used to establish the confidence interval and to test whether the species distribution is closely related with the soil elements represented by the PCs. All studied tree species show a significant correlation with at least one principal component of soil elements in the upper soil layer, while 87.9% of the species are significantly related to at least one principal component of elements in the lower soil layer. The HomT model eliminates the effect of the species dispersal limitation, and the results show that only a few tree species exhibit a preference for particular soil elements. A total of 12 tree species are significantly related to at least one principal component of elements in the upper soil layer (accounting for 36.4% of all studied tree species), and 10 tree species are significantly related with at least one principal component of elements in the lower layer (accounting for 30.3% of all studied tree species). The results between the HomT and HomP models are very different. We assume that the spatial autocorrelation of species distributions induced by the dispersal limitation may confound the true plant-soil relationships induced by the spatial heterogeneity of soil elements (Tables 3 and 4).

Table 3. Relationships between species distribution and one of the first four PC axes of soil nutrients in the upper soil layer. HomP represents the homogeneous Poisson process; HomT represents the homogeneous Thomas process.

Tree Species	HomP				HomT			
	PC1	PC2	PC3	PC4	PC1	PC2	PC3	PC4
Betula platyphylla		−	−	−				
Acer mandshuricum	+	−	−				−	
Padus racemosa	−		−	−				
Abies nephrolepis	+	+		+				
Ulmus davidiana var. *japonica*	+		−	+				
Ulmus macrocarpa	+	−	−					
Betula costata	−				−			
Betula dahurica	+							
Pinus koraiensis	−	+	−					
Juglans mandshurica		+	−	−				
Sorbus pohuashanensis	+			+				
Phellodendron amurense	−	−		−				
Tilia mandschurica	+	+				+		
Ulmus laciniata	+	−		−	+			−
Quercus mongolica	+	−			+			
Carpinus cordata	+	+		−				
Acer tegmentosum	−	+	+	−	−		+	
Acer mono	+		−	−				
Abies holophylla		+		−				
Fraxinus mandshurica	+	−			+	−		
Sorbus alnifolia		−				−		
Populus koreana		−						
Tilia amurensis				+				+
Syringa reticulata var. *amurensis*	+	−	−	+		+		+
Acer barbinerve	−	+		+				
Tilia mandschurica		+		+				
Acer ukurunduense	−	+	+	+			+	
Lonicera maackii			+					
Euonymus macropterus		+						
Eleutherococcus senticosus	+		−	+			−	
Euonymus pauciflorus	−	+		+				
Corylus mandshurica	−	−	−	+				
Rhamnus davurica	−							
Significant negative correlation	10	11	11	9	2	2	2	1
Significant positive correlation	14	12	3	11	3	2	2	2
Non-significant correlation	9	10	19	13	28	29	29	30

Symbol "+" indicates significant positive correlation at the 0.05 level; Symbol "−" indicates significant negative correlation at the 0.05 level.

Table 4. Relationships between species distribution and one of the first four PC axes of soil elements in the lower soil layer. HomP represents the homogeneous Poisson process; HomT represents the homogeneous Thomas process.

Plant Species	HomP				HomT			
	PC1	PC2	PC3	PC4	PC1	PC2	PC3	PC4
Betula platyphylla	+	−	+					
Acer mandshuricum	+	−	+	−			+	
Padus racemosa	−	−						
Abies nephrolepis	+			−				
Ulmus davidiana var. *japonica*	+	−	+	−				
Ulmus macrocarpa	+	−						
Betula costata	−	+			−	+		
Betula dahurica	+							
Pinus koraiensis	−	+				+		
Juglans mandshurica				+				
Sorbus pohuashanensis								
Phellodendron amurense		−	+	+				
Tilia mandschurica	+	+	−	−				
Ulmus laciniata	+			+	+			
Quercus mongolica	+							
Carpinus cordata	+	+	−	−			−	
Acer tegmentosum	−	+	−	+		+		+
Acer mono	+							
Abies holophylla	−	+	−			+		
Fraxinus mandshurica	+	−	+	+				
Sorbus alnifolia			+	+				
Populus koreana		+	+				+	
Tilia amurensis								
Syringa reticulata var. *amurensis*	+	−	+	−		−		
Acer barbinerve	−	+	+					
Tilia mandschurica			−	−				
Acer ukurunduense	−	+	−					
Lonicera maackii			+	−				
Euonymus macropterus								
Eleutherococcus senticosus								
Euonymus pauciflorus	−	+		−		+		−
Corylus mandshurica	−	−	+	+				
Rhamnus davurica	−	−						
Significant negative correlation	10	10	6	9	1	1	1	1
Significant positive correlation	13	10	11	7	1	5	2	1
Non-significant correlation	10	13	16	17	31	27	30	31

Symbol "+" indicates significant positive correlation at the 0.05 level; Symbol "−" indicates significant negative correlation at the 0.05 level.

Significant species-soil relationships show up mainly for the canopy species. A total of 39.1% of canopy species (14 of 23 species) are significantly related to the PCs of soil elements in the upper soil layer, while 34.8% of canopy species (14 of 23 species) are significantly related to the PCs of soil elements in the lower soil layer. Two out of five subcanopy species show significant correlations with the PCs of soil elements in the upper soil layer, and only one out of five subcanopy species shows significant correlations with the PCs in the lower soil layer. Only one out of five shrub species shows significant correlations with PC3 of soil elements in the upper soil layer and with PC2 and PC4 of elements in the lower soil layer (Table 5).

Table 5. Species distribution-PC associations examined by the homogeneous Thomas process in different growth groups.

Soil Layer	Canopy Species	Subcanopy Species	Shrub Species
	23 Species	5 Species	5 Species
Upper			
Number of species related to PC1	5 (3+, 2−)	0 (0+, 0−)	0 (0+, 0−)
Number of species related to PC2	3 (1+, 2−)	1 (1+, 0−)	0 (0+, 0−)
Number of species related to PC3	2 (0+, 2−)	1 (1+, 0−)	1 (1+, 0−)
Number of species related to PC4	3 (3+, 0−)	1 (1+, 0−)	0 (0+, 0−)
Number of species unrelated to PCs	14	3	4
Lower			
Number of species related to PC1	2 (1+, 1−)	0 (0+, 0−)	0 (0+, 0−)
Number of species related to PC2	4 (4+, 0−)	1 (0+, 1−)	1 (1+, 0−)
Number of species related to PC3	3 (2+, 1−)	0 (0+, 0−)	0 (0+, 0−)
Number of species related to PC4	1 (1+, 0−)	0 (0+, 0−)	1 (0+, 1−)
Number of species unrelated to PCs	15	4	4

The first number in parentheses refers to the number of species that showed positive plant-soil relationships; the second number in parentheses is the number of species that showed negative plant-soil relationships.

The soil nutrient variables are highly correlated with each other (Table S2). It is thus difficult to evaluate the relative importance of individual soil elements in a community niche structure. We calculated the ecological niche breadth of all species for individual soil nutrient gradients and conclude that the smaller ecological niche breadth indicates a higher specialization degree of particular focal species. When the ecological niche breadth is relatively small, plant species show a weak adaptability to the variations in soil properties. Thus, the smaller the niche breadth for a given soil variable, the greater is the effect of that soil variable on the niche structure of the community. The soil elements which have the largest impacts on the community niche structure are Pb, TP, TN, Cu, Cr, Zn and AN in the upper soil layer and Pb, TP, Cu, OC, Mn, TK, AN, TN, Zn and Cr in the lower layer (Figure 2). The ecological niche breadths of AK, TK, AP, OC, AN, TN, Ni and pH show significant differences between the two soil layers. These nutritional elements, which are represented differently in the two soil layers (Table S4), are important for plant growth.

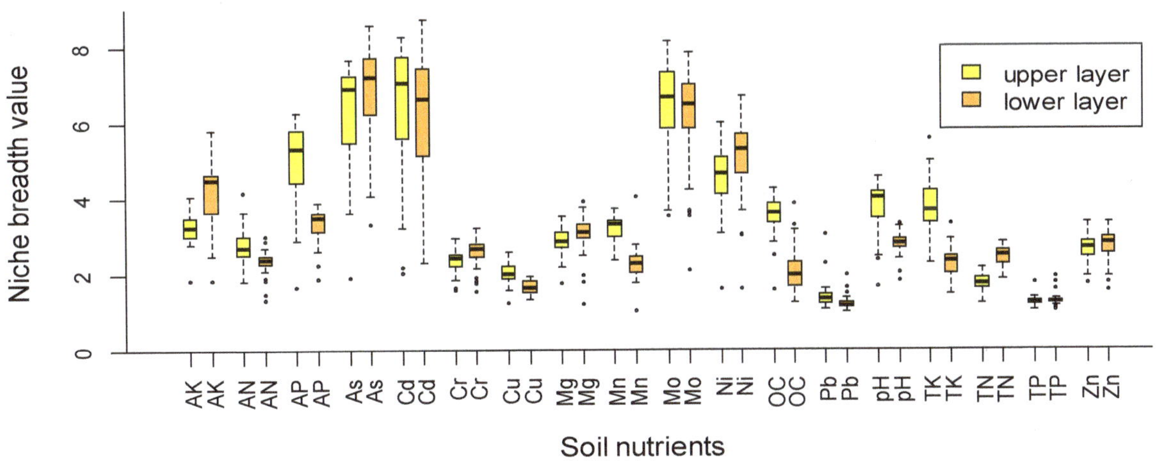

Figure 2. Boxplot of niche breadth values for 18 soil nutrients in two soil layers. The smaller the niche breadth for a given soil variable, the greater is the effect of that soil variable on the niche structure of the community.

4. Discussion

A particular community pattern involves specific ecological processes, and a particular process in turn represents the cause and driving force of many different patterns. Previous studies have shown that the old-growth broad-leaved Korean pine forest exhibits a strong dispersal limitation [32]. The joint effects of environmental heterogeneity and dispersal limitation can more effectively explain the species-area relationships and β diversity in the old-growth broad-leaved Korean pine forest [33]. Our study extracted four principal components to express the spatial variation of soil variables. Each principal component reflects different combinations of the soil characteristics. PC1 indicates the spatial variation of metal elements in soils such as Cu, Ni, Cd, As, Pb, Zn, Mo, Cr, Mn, Mg, while PC2, PC3 and PC4 indicate the spatial variation of the nutrients (nitrogen, phosphorus, potassium, organic carbon) and soil pH.

A species distribution simulation using a spatial point process model can directly evaluate the relationships with the nutrient PCs. The plant-soil relationships tested by the homogeneous Thomas model are less indicative than those tested by the homogeneous Poisson model: they are reduced by 63.6% in the upper soil layer, and by 57.6% in the lower layer. True effects of environmental filtering on the species distribution are reduced after eliminating spatial autocorrelation induced from a dispersal process. Therefore, we can conclude that the dispersal-limitation effect on a particular species distribution is greater than the effect of the environmental heterogeneity.

Previous studies have shown that soil elements strongly affect vegetation patterns in tropical and temperate forest communities [34–37]. Paoli et al. found that the spatial distributions of 18 out of 22 tree species in a tropical forest community were significantly associated with soil elements, especially with P, Mg and Ca [38]. They argued that dispersal and niche processes jointly determine mesoscale beta diversity in a *Bornean Dipterocarp* forest. John et al. [7] detected an effect of soil elements on the spatial distribution of particular tree species in three diverse neotropical forest plots. They found that the spatial distributions of 36%–51% of all the tree species in these plots show strong associations with soil nutrient distributions. B and K in their BCI plot, Ca and Mg in the Yasuni plot and K, P, Fe and N in the La Planada plot showed the strongest effects on community niche structure [7].

In a subtropical broad-leaved forest community, spatial heterogeneity of soil elements was associated with distributions of 88.2% (90 out of 102 species) of tree species after controlling the effects of dispersal limitation. The soil factors most strongly influencing species distributions were TC, TN, TP, K, Mg, Si, soil moisture and bulk density [6]. Our results showed that TP, TN, AN in the upper soil layer and TP, TK, TN, AN in the lower soil layer strongly affect species distributions. Nitrogen, phosphorus and potassium are the essential mineral elements for plant growth, as well as being rather limited soil elements, and their concentration directly affects plant growth and survival [39,40]. The variation coefficient of AN, AP, AK, TN, TP and TK in the broad-leaved Korean pine forest ranges from 0.16 to 0.44, and shows obvious spatial heterogeneity (Table S3). When the quantity of nitrogen, phosphorus and potassium in the habitat cannot meet the plants' demands, they will become important limiting factors for plant growth.

Metal elements show consistent effects on community structure in the two soil layers. Pb, Cu, Cr and Zn in both the upper and lower soil layers strongly affect the community niche structure of the old-growth forest. Previous studies have shown that metal elements have important effects on the growth, development and breeding of plants. Cu is a cofactor influencing the structure and catalytic compounds of proteins and enzymes and is thus essential for the normal growth and biochemical processes. However, excessive Cu will produce a large number of free radicals and malondialdehyde (MDA), causing metabolic disturbances and inhibiting plant growth [41].

Zn is an active cofactor for a large number of enzymes related to the metabolism of DNA transcription, protein, nucleic acid, carbohydrate and lipids [42,43]. Furthermore, Zn is the essential microelement for any living body, and Zn deficiency will cause a series of nutrition problems in plants. Pb is not an essential element and may cause enzyme function disorders, generating active oxygen in plant tissue cells, which will lead to membrane lipid peroxidation, and change the related enzymatic activity of active oxygen metabolism [44,45], which finally affects biochemical processes and the morphological structure of plants [46]. Cr is not an essential element, but may also affect plant growth and development. A high concentration of Cr may inhibit the root cell differentiation of plants, block water absorption, and have a toxic effect, while a low concentration of Cr can promote the growth of roots and root hairs, increase the cortical tissue layers in the roots, and thus facilitate plant growth [47,48].

A low concentration of Cr will also promote the formation of chlorophyll, while a high concentration will inhibit the formation of chlorophyll [49,50]. Mn in the lower soil layer has an especially severe impact on our community niche structure. Mn is an essential element in the PSII oxygen evolution complex [51–53], and its role is to realize water splitting in the reaction center of PSII and to provide electrons to the electron transfer chain as part of thylakoid coupling [54]. Therefore, an Mn deficiency will decrease the photosynthetic capacity of the leaves [55].

Different species groups were formed in accordance with their visible properties. Plant species usually belong to the same growth group under similar environmental conditions, which is the result of adaptive convergence [56]. Plant species in different growth groups have shown different soil habitat preferences in a subtropical forest plot. Only five shrub species were significantly related to the four soil property principal component axes, while most of the canopy species and subcanopy species were significantly related to 1–2 principal component axes [6].

Our results show that the tree species in the different growth groups have different habitat preferences. Compared with subcanopy and shrub species, the canopy species have more significant correlations with the PCs of the soil elements. In addition, different canopy species also show different soil preferences. Canopy species play a dominant role in structuring the forest community. Intense competition will occur when different tree species have similar demands for specific soil resources and some members of a species with weak competitive strength may die. On the other hand, the demands of different tree species on available soil resources may be complementary, which facilitates the full utilization of existing resources and the coexistence of species.

5. Conclusions

This work evaluates plant-soil correlations by comparing simulated the spatial distributions of tree species with observed ones in a large forest observational study. The impact of dispersal limitation is greater than impact of environmental heterogeneity on species spatial distributions. The spatial autocorrelation of species induced by the dispersal limitation will largely conceal the plant-soil relationships caused by the heterogeneity of soil elements. Certain elements in the upper and lower soil layers were identified to influence community niche structure. Tree species in different growth groups have different habitat preferences. Compared with subcanopy and shrub species, the canopy species have more significant correlations with the soil elements.

Acknowledgments: This research is supported by the Fundamental Research Funds for the Central Universities (2015ZCQ-LX-03), the State Key Program of National Natural Science Foundation of China (41330530) and the 12th five-year National Science and Technology plan of China (2012BAC01B03).

Author Contributions: W.X. and J.W. analyzed the data and wrote the manuscript. C.Z. and K.G. modeled and interpreted the results. X.Z. and M.H. performed the laboratory experiments.

References

1. Tuomisto, H.; Ruokolainen, K.; Aguilar, M.; Sarmiento, A. Floristic patterns along a 43-km long transect in an Amazonian rain forest. *J. Ecol.* **2003**, *91*, 743–756. [CrossRef]
2. Thessler, S.; Ruokolainen, K.; Tuomisto, H.; Tomppo, E. Mapping gradual landscape-scale floristic changes in Amazonian primary rain forests by combining ordination and remote sensing. *Glob. Ecol. Biogeogr.* **2005**, *14*, 315–325. [CrossRef]
3. Slik, J.W.; Raes, N.; Aiba, S.I.; Brearley, F.Q.; Cannon, C.H.; Meijaard, E.; Nagamasu, H.; Nilus, R.; Paoli, G.; Poulsen, A.D.; et al. Environmental correlates for tropical tree diversity and distribution patterns in Borneo. *Divers. Distrib.* **2009**, *15*, 523–532. [CrossRef]
4. Clark, D.B.; Olivas, P.C.; Oberbauer, S.F.; Clark, D.A.; Ryan, M.G. First direct landscape-scale measurement of tropical rain forest leaf area index, a key driver of global primary productivity. *Ecol. Lett.* **2008**, *11*, 163–172. [CrossRef] [PubMed]
5. Gleason, S.M.; Read, J.; Ares, A.; Metcalfe, D.J. Species-soil associations, disturbance, and nutrient cycling in an Australian tropical rainforest. *Oecologia* **2010**, *162*, 1047–1058. [CrossRef] [PubMed]
6. Zhang, L.; Mi, X.; Shao, H.; Ma, K. Strong plant-soil associations in a heterogeneous subtropical broad-leaved forest. *Plant Soil* **2011**, *347*, 211–220. [CrossRef]
7. John, R.; Dalling, J.W.; Harms, K.E.; Yavitt, J.B.; Stallard, R.F.; Mirabello, M.; Hubbell, S.P.; Valencia, R.; Navarrete, H.; Vallejo, M.; et al. Soil nutrients influence spatial distributions of tropical tree species. *Proc. Natl. Acad. Sci. USA* **2007**, *104*, 864–869. [CrossRef] [PubMed]
8. Zhang, C.; Zhao, Y.; Zhao, X.; Gadow, K.V. Species-Habitat Associations in a Northern Temperate Forest in China. *Silva Fenn.* **2012**, *46*, 501–519. [CrossRef]
9. Hubbell, S.P.; Foster, R.B.; O'Brien, S.T.; Harms, K.E.; Condit, R.; Wechsler, B.; Wright, S.J.; De Lao, S.L. Light-gap disturbances, recruitment limitation, and tree diversity in a Neotropical forest. *Science* **1999**, *283*, 554–557. [CrossRef] [PubMed]
10. Levine, J.M.; Murrell, D.J. The community-level consequences of seed dispersal patterns. *Annu. Rev. Ecol. Evol. Syst.* **2003**, *34*, 549–574. [CrossRef]
11. Beckage, B.; Clark, J.S. Seedling survival and growth of three forest tree species: The role of spatial heterogeneity. *Ecology* **2003**, *84*, 1849–1861. [CrossRef]
12. Lebrija-Trejos, E.; Pérez-García, E.A.; Meave, J.A.; Bongers, F.; Poorter, L. Functional traits and environmental filtering drive community assembly in a species-rich tropical system. *Ecology* **2010**, *91*, 386–398. [CrossRef] [PubMed]
13. Harper, J.L. *Population Biology of Plants*; Academic Press: London, UK, 1977.
14. Tilman, D.; Pacala, S. The maintenance of species richness in plant communities. In *Species Diversity in Ecological Communities*; Ricklefs, R.E., Ed.; University of Chicago Press: Chicago, IL, USA, 1993; pp. 13–25.
15. Webb, C.O.; Peart, D.R. Habitat associations of trees and seedlings in a Bornean rain forest. *J. Ecol.* **2000**, *88*, 464–478. [CrossRef]
16. Webb, C.O. Exploring the phylogenetic structure of ecological communities: An example for rain forest trees. *J. Am. Nat.* **2000**, *156*, 145–155. [CrossRef] [PubMed]
17. Williams, C.B. Patterns in the balance of nature and related problems of quantitative ecology. Academic Press: New York, NY, USA, 1964.
18. Bartels, S.F.; Chen, H.Y. Is understory plant species diversity driven by resource quantity or resource heterogeneity? *Ecology* **2010**, *91*, 1931–1938. [CrossRef] [PubMed]
19. Baldeck, C.A.; Harms, K.E.; Yavitt, J.B.; John, R.; Turner, B.L.; Valencia, R.; Navarrete, H.; Davies, S.J.; Chuyong, G.B.; Kenfack, D.; et al. Soil resources and topography shape local tree community structure in tropical forests. *Proc. R. Soc. Lond. B Biol.* **2013**, *280*, 20122532. [CrossRef] [PubMed]
20. Hall, J.S.; McKenna, J.J.; Ashton, P.M.S.; Gregoire, T.G. Habitat characterizations underestimate the role of edaphic factors controlling the distribution of Entandrophragma. *Ecology* **2004**, *85*, 2171–2183. [CrossRef]
21. Bonifacio, E.; Caimi, A.; Falsone, G.; Trofimov, S.Y.; Zanini, E.; Godbold, D.L. Soil properties under Norway spruce differ in spruce dominated and mixed broadleaf forests of the Southern Taiga. *Plant Soil* **2008**, *308*, 149–159. [CrossRef]
22. Institute of Soil Science, CAS. *Chinese Soil Taxonomy*; Science Press: Beijing, China, 2001.

23. Zhang, C.; Wei, Y.; Zhao, X.; von Gadow, K. Spatial characteristics of tree diameter distributions in a temperate old-growth forest. *PLoS ONE* **2013**, *8*, e58983. [CrossRef] [PubMed]
24. China Soil Council. *Soil Agricultural Chemical Analysis Procedure*; Chinese Agricultural Science Press: Beijing, China, 1999.
25. Harms, K.E.; Condit, R.; Hubbell, S.P.; Foster, R.B. Habitat associations of trees and shrubs in a 50-ha Neotropical forest plot. *J. Ecol.* **2001**, *89*, 947–959. [CrossRef]
26. Waagepetersen, R.P. An estimating function approach to inference for inhomogeneous Neyman-Scott processes. *Biometrics* **2007**, *63*, 252–258. [CrossRef] [PubMed]
27. Shen, G.; Yu, M.; Hu, X.S.; Mi, X.; Ren, H.; Sun, I.F.; Ma, K. Species-area relationships explained by the joint effects of dispersal limitation and habitat heterogeneity. *Ecology* **2009**, *90*, 3033–3041. [CrossRef] [PubMed]
28. Lin, Y.C.; Chang, L.W.; Yang, K.C.; Wang, H.H.; Sun, I.F. Point patterns of tree distribution determined by habitat heterogeneity and dispersal limitation. *Oecologia* **2011**, *165*, 175–184. [CrossRef] [PubMed]
29. Kohn, A.J. Microhabitats, abundance and food of Conus on atoll reefs in the Maldive and Chagos Islands. *Ecology* **1968**, *49*, 1046–1062. [CrossRef]
30. Levins, R. *Evolution in Changing Environments: Some Theoretical Explorations (No. 2)*; Princeton University Press: Princeton, NJ, USA, 1968.
31. Team, R.C. *R: A Language and Environment for Statistical Computing*; R Foundation for Statistical Computing: Vienna, Austria, 2012.
32. Li, B.; Hao, Z.; Bin, Y.; Zhang, J.; Wang, M. Seed rain dynamics reveals strong dispersal limitation, different reproductive strategies and responses to climate in a temperate forest in northeast China. *J. Veg. Sci.* **2012**, *23*, 271–279. [CrossRef]
33. Wang, X.; Wiegand, T.; Hao, Z.; Li, B.; Ye, J.; Lin, F. Species associations in an old-growth temperate forest in north-eastern China. *J. Ecol.* **2010**, *98*, 674–686. [CrossRef]
34. Read, J. Soil and rainforest composition in Tasmania: Correlations of soil characteristics with canopy composition and growth rates in *Nothofagus cunninghamii* associations. *Aust. J. Bot.* **2001**, *49*, 121–135. [CrossRef]
35. Lusk, C.H.; Matus, F. Juvenile tree growth rates and species sorting on fine-scale soil fertility gradients in a Chilean temperate rain forest. *J. Biogeogr.* **2000**, *27*, 1011–1020. [CrossRef]
36. Baltzer, J.L.; Thomas, S.C.; Nilus, R.; Burslem, D.F.R. Edaphic specialization in tropical trees: Physiological correlates and responses to reciprocal transplantation. *Ecology* **2005**, *86*, 3063–3077. [CrossRef]
37. Pyke, C.R.; Condit, R.; Aguilar, S.; Lao, S. Floristic composition across a climatic gradient in a Neotropical lowland forest. *J. Veg. Sci.* **2001**, *12*, 553–566. [CrossRef]
38. Paoli, G.D.; Curran, L.M.; Zak, D.R. Soil nutrients and beta diversity in the Bornean Dipterocarpaceae: Evidence for niche partitioning by tropical rain forest trees. *J. Ecol.* **2006**, *94*, 157–170. [CrossRef]
39. Reich, P.B.; Hobbie, S.E.; Lee, T.; Ellsworth, D.S.; West, J.B.; Tilman, D.; Knops, J.M.; Naeem, S.; Trost, J. Nitrogen limitation constrains sustainability of ecosystem response to CO_2. *Nature* **2006**, *440*, 922–925. [CrossRef] [PubMed]
40. Thuiller, W. On the importance of edaphic variables to predict plant species distributions-limits and prospects. *J. Veg. Sci.* **2013**, *24*, 591–592. [CrossRef] [PubMed]
41. Foyer, C.H.; Lelandais, M.; Kunert, K.J. Photooxidative stress in plant. *Physiol. Plantarum* **1994**, *92*, 708–719. [CrossRef]
42. Broadley, M.R.; White, P.J.; Hammond, J.P.; Zelko, I.; Lux, A. Zinc in plants. *New Phytol.* **2007**, *173*, 677–702. [CrossRef] [PubMed]
43. Marschner, H.; Rimmington, G. Mineral nutrition of higher plants. *Plant Cell Environ.* **1988**, *11*, 147–148.
44. Chaoui, A.; Mazhoudi, S.; Ghorbal, M.H.; El Ferjani, E. Cadmium and zinc induction of lipid peroxidation and effects on antioxidant enzyme activities in bean (*Phaseolus vulgaris* L.). *Plant Sci.* **1997**, *127*, 139–147. [CrossRef]
45. Gallego, S.M.; Benavides, M.P.; Tomaro, M.L. Effect of heavy metal ion excess on sunflower leaves: Evidence for involvement of oxidative stress. *Plant Sci.* **1996**, *121*, 125–159. [CrossRef]
46. Shah, K.; Dubey, R.S. Cadmium elevates level of protein, amino acids and alters activity of proteolytic enzymes in germinating rice seeds. *Acta Physiol. Plant.* **1998**, *20*, 189–196. [CrossRef]
47. Gupta, S.; Srivastava, S.; Saradhi, P.P. Chromium increases photosystem 2 activity in *Brassica juncea*. *J. Biol. Plantarum* **2009**, *53*, 100–104. [CrossRef]

48. Suseela, M.R.; Sinha, S.; Singh, S.; Saxena, R. Accumulation of chromium and scanning electron microscopic studies in *Scirpus lacustris* L. treated with metal and tannery effluent. *Bull. Environ. Contam. Toxicol.* **2002**, *68*, 540–548. [CrossRef]
49. Pandey, V.; Dixit, V.; Shyam, R. Antioxidative responses in relation to growth of mustard (*Brassica juincea* cv. Pusa Jaikisan) plants exposed to hexavalent chromium. *Chemosphere* **2005**, *61*, 40–47. [CrossRef] [PubMed]
50. Shanker, A.K.; Djanaguiraman, M.; Sudhagar, R.; Chandrashekar, C.N.; Pathmanabhan, G. Differential antioxidative response of ascorbate glutathione pathway enzymes and metabolites to chromium speciation stress in green gram (*Vigna radiata* (L.) R. Wilczek. cv. CO 4) roots. *Plant Sci.* **2004**, *166*, 1035–1043. [CrossRef]
51. Spencer, D.; Possingham, J.V. The effect of Mn deficiency on photophosphorylation and the oxygen—Evolving system in spinach chloroplasts. *Biochim. Biophys.* **1961**, *52*, 379–381. [CrossRef]
52. Gerretsen, F.C. Manganese in relation to photosynthesis. II. Redox potentials of illuminated crude chloroplast suspensions. *Plant Soil* **1950**, *11*, 159–193. [CrossRef]
53. Homann, P.H. Studies on the manganese of chloroplast suspensions. *Plant Physiol.* **1967**, *42*, 997–1007. [CrossRef] [PubMed]
54. Yachandra, V.K.; Sauer, K.; Klein, M.P. Manganese cluster in photosynthesis: Where plants oxidize water to dioxygen. *Chem. Rev.* **1966**, *96*, 2927–2950. [CrossRef]
55. Bottrill, D.E.; Possingham, J.V.; Kriedemann, P.E. The effect of nutrient deficiencies on phosynthesis and respiration in spinach. *Plant Soil* **1970**, *32*, 424–438. [CrossRef]
56. Rowe, N.; Speck, T. Plant growth forms: An ecological and evolutionary perspective. *New Phytol.* **2005**, *166*, 61–72. [CrossRef] [PubMed]

Permissions

All chapters in this book were first published in MDPI; hereby published with permission under the Creative Commons Attribution License or equivalent. Every chapter published in this book has been scrutinized by our experts. Their significance has been extensively debated. The topics covered herein carry significant findings which will fuel the growth of the discipline. They may even be implemented as practical applications or may be referred to as a beginning point for another development.

The contributors of this book come from diverse backgrounds, making this book a truly international effort. This book will bring forth new frontiers with its revolutionizing research information and detailed analysis of the nascent developments around the world.

We would like to thank all the contributing authors for lending their expertise to make the book truly unique. They have played a crucial role in the development of this book. Without their invaluable contributions this book wouldn't have been possible. They have made vital efforts to compile up to date information on the varied aspects of this subject to make this book a valuable addition to the collection of many professionals and students.

This book was conceptualized with the vision of imparting up-to-date information and advanced data in this field. To ensure the same, a matchless editorial board was set up. Every individual on the board went through rigorous rounds of assessment to prove their worth. After which they invested a large part of their time researching and compiling the most relevant data for our readers.

The editorial board has been involved in producing this book since its inception. They have spent rigorous hours researching and exploring the diverse topics which have resulted in the successful publishing of this book. They have passed on their knowledge of decades through this book. To expedite this challenging task, the publisher supported the team at every step. A small team of assistant editors was also appointed to further simplify the editing procedure and attain best results for the readers.

Apart from the editorial board, the designing team has also invested a significant amount of their time in understanding the subject and creating the most relevant covers. They scrutinized every image to scout for the most suitable representation of the subject and create an appropriate cover for the book.

The publishing team has been an ardent support to the editorial, designing and production team. Their endless efforts to recruit the best for this project, has resulted in the accomplishment of this book. They are a veteran in the field of academics and their pool of knowledge is as vast as their experience in printing. Their expertise and guidance has proved useful at every step. Their uncompromising quality standards have made this book an exceptional effort. Their encouragement from time to time has been an inspiration for everyone.

The publisher and the editorial board hope that this book will prove to be a valuable piece of knowledge for researchers, students, practitioners and scholars across the globe.

List of Contributors

Daniel M. Alongi
Tropical Coastal & Mangrove Consultants, 6 Cobb Court, Annandale, QLD 4814, Australia

Paul F. Hendrix
Odum School of Ecology, University of Georgia, Athens, GA 30602, USA

Ching-Yu Huang
Odum School of Ecology, University of Georgia, Athens, GA 30602, USA
Department of Biology, University of North Georgia, Dahlonega, GA 30597, USA

Grizelle González
United States Department of Agriculture, Forest Service, International Institute of Tropical Forestry, Jardín Botánico Sur, 1201 Ceiba St.-Río Piedras 00926, Puerto Rico

Junyi Xie, Haifu Fang, Qiang Zhang, Mengyun Chen, Xintong Xu, Jun Pan, Yu Gao, Xiangmin Fang, Xiaomin Guo and Ling Zhang
Jiangxi Provincial Key Laboratory of Silviculture, Jiangxi Agricultural University, Nanchang 330045, China

Christine Knust and Karoline Schua
TU Dresden, Institute of Soil Science and Site Ecology, 01737 Tharandt, Germany

Hongwei Xu
State Key Laboratory of Soil Erosion and Dryland Farming on Loess Plateau, Institute of Soil and Water Conservation, Chinese Academy of Sciences and Ministry of Water Resources, Yangling 712100, China
Institute of College of Natural Resources and Environment, University of Chinese Academy of Sciences, Beijing 100049, China

Qing Qu, Peng Li, Ziqi Guo and Entemake Wulan
Institute of Soil and Water Conservation, Northwest A&F University, Yangling 712100, China

Sha Xue
State Key Laboratory of Soil Erosion and Dryland Farming on Loess Plateau, Institute of Soil and Water Conservation, Chinese Academy of Sciences and Ministry of Water Resources, Yangling 712100, China
Institute of Soil and Water Conservation, Northwest A&F University, Yangling 712100, China
Institute of College of Natural Resources and Environment, University of Chinese Academy of Sciences, Beijing 100049, China

Dorit Julich, Stefan Julich and Karl-Heinz Feger
Institute of Soil Science and Site Ecology, TU Dresden, Pienner Str. 19, 01737 Tharandt, Germany

Jack M. Bridges
Geography and Earth Sciences, University of Aberystwyth, Wales SY23 3DB, UK

George P. Petropoulos
Department of Soil & Water Resources, Hellenic Agricultural Organization (HAO) "Demeter" (former NAGREF), 1 Theofrastou St., 41335 Larissa, Greece

Nicola Clerici
Biology Program, Faculty of Natural Sciences and Mathematics, Universidad del Rosario, Carrera 26 # 63B-48, Bogotá 111221, Colombia

Xiaolu Tang, Mingpeng Xia, Fengying Guan and Shaohui Fan
Key laboratory of Bamboo and Rattan, International Centre for Bamboo and Rattan, Beijing 100102, China

Jian Deng and Xianfang Li
Shaanxi Key Laboratory of Chinese Jujube, Yan'an University, Yan'an 716000, China
College of Life Science, Yan'an University, Yan'an 716000, China

Sha Wang and Dan Zhang
College of Life Science, Yan'an University, Yan'an 716000, China

Chengjie Ren, Wei Zhang, Xinhui Han and Gaihe Yang
College of Agronomy, Northwest A&F University, Yangling 712100, China

Fazhu Zhao
College of Urban and Environmental Science, Northwest University, Xi'an 710069, China

Jason James and Rob Harrison
School of Environmental and Forest Sciences, University of Washington, Seattle, WA 98195, USA

Dongxing Yang, Hongrui Mao and Guangze Jin
Center for Ecological Research, Northeast Forestry University, Harbin 150040, China

D. Jean Lodge
United States Department of Agriculture, Forest Service, Northern Research Station, Luquillo 00773-1377, Puerto Rico

Dirk Winter
Department of Psychiatry, Columbia University, 180 Ft. Washington Ave., New York, NY 10032, USA

Naomi Clum
Carmel Valley Middle School, 3800 Mykonos Lane, San Diego, CA 92007, USA

Benbo Xu
College of Life Science, Yangtze University, Jingzhou 434025, China

Haichao Zhou and Yiming Lin
Key Laboratory of the Ministry of Education for Coastal and Wetland Ecosystems, Xiamen University, Xiamen 361005, China

Hui Chen and Shudong Wei
College of Life Science, Yangtze University, Jingzhou 434025, China
Key Laboratory of the Ministry of Education for Coastal and Wetland Ecosystems, Xiamen University, Xiamen 361005, China

Lihua Zhang
Key Laboratory of Coastal Environmental Processes and Ecological Remediation, Yantai Institute of Coastal Zone Research (YIC), Chinese Academy of Sciences (CAS), Yantai 264003, China
Shandong Provincial Key Laboratory of Coastal Zone Environmental Processes, YICCAS, Yantai 264003, China

Jessie C. Buettel, Elise M. Ringwaldt, Mark J. Hovenden and Barry W. Brook
Discipline of Biological Sciences, School of Natural Sciences, University of Tasmania, Hobart 7000, Australia

Wei Xu, Minhui Hao, Juan Wang, Chunyu Zhang and Xiuhai Zhao
The key Laboratory for Forest Resources & Ecosystem Processes of Beijing, Beijing Forestry University, Beijing 100083, China

Klaus von Gadow
Faculty of Forestry and Forest Ecology, Georg-August-University Göttingen, Büsgenweg 5, Göttingen D-37077, Germany
Department of Forest and Wood Science, University of Stellenbosch, Stellenbosch 7600, South Africa

Index

www.ingramcontent.com/pod-product-compliance
Lightning Source LLC
LaVergne TN
LVHW081037210926
848485LV00004B/1040